CD-R/DVD: Disc Recording Demystified

CD-R/DVD:
Disc Recording Demystified

Lee Purcell

McGraw-Hill

New York San Francisco Washington, D.C. Auckland Bogotá
Caracas Lisbon London Madrid Mexico City Milan
Montreal New Delhi San Juan Singapore
Sydney Tokyo Toronto

McGraw-Hill
A Division of The McGraw Hill Companies

Copyright © 2000 by The McGraw-Hill Companies, Inc. All rights reserved. Printed in the United States of America. Except as permitted under the United States Copyright Act of 1976, no part of this publication may be reproduced or distributed in any form or by any means, or stored in a data base or retrieval system, without the prior written permission of the publisher.

1 2 3 4 5 6 7 8 9 0 AGM/AGM 0 6 5 4 3 2 1 0

P/N 135716-5
PART OF
ISBN 0-07-135715-7

The sponsoring editor for this book was Stephen S. Chapman and the production supervisor was Sherri Souffrance.

Printed and bound by Quebecor/Martinsburg.

 This book was printed on recycled, acid-free paper containing a minimum of 50% recycled, de-inked fiber.

McGraw-Hill books are available at special quantity discounts to use as premiums and sales promotions, or for use in corporate training programs. For more information, please write to the Director of Special Sales, McGraw-Hill, Inc. 11 West 19th Street, New York, NY 10011. Or contact your local bookstore.

Information contained in this work has been obtained by The McGraw-Hill Companies, Inc. ("McGraw-Hill") from sources believed to be reliable. However, neither McGraw-Hill nor its authors guarantee the accuracy or completeness of any information published herein, and neither McGraw-Hill nor its authors shall be responsible for any errors, omissions, or damages arising out of use of this information. This work is published with the understanding that McGraw-Hill and its authors are supplying information but are not attempting to render engineering or other professional services. If such services are required, the assistance of an appropriate professional should be sought.

To Mom and Dad for teaching me my first words.

Table of Contents

Preface xix
   CD-ROM Contents xx
Acknowledgments xxi
About the Author xxii

Chapter 1
A Concise History of Optical Recording 1
   Audio Roots 1
   The First Compact Discs 2
      Early Research 4
   Other Devices, Other Formats 6
   Recording Goes Mainstream 7
   Related Recording Technologies 8
   The Laserdisc 9
   DVD Arises 10
   Growth of Publishing and Title Development 12
      Case Study: Surviving in the Electronic Publishing Industry 12
      New Era, New Tools 16
   Ongoing Development 17

Chapter 2
Overview of CD-R and Writable DVD Technology 19
   Understanding the Technology 20
   Black Box Engineering 21
      Mastering the Levels of Complexity 21
      Developer Considerations 23
      Dealing with Performance Issues 23
   By the Light of the Shimmering Laser 24
      A Closer Look at the Disc 26
      Turning Pits into Data 28

Table of Contents

    Physical Differences in Recordable CDs 29
  Optical Storage Advantages 31
    Access-Time Considerations 33
    Data Transfer Rates 33
  Permanence of Data 34
  Data Storage on CD-ROMs 35
  Logical and Physical Components 37
  Audio Origins 37
  Computer CD-ROM Standards 38
  DVD Technology 39
    Storing Content on DVD 44
    Reading All Types of Discs 45
    Writable Forms of DVD 46
    DVD-R for Write-Once Applications 47
  The Zen of Data Flow 49
  Summary 50

Chapter 3
CD Standards **51**
  Why More Than One Standard? 52
  The Course of Standards Development 53
    Maintaining Compatibility 54
  Applying the Standards 55
  Using the Standards 57
  Red Book 60
    Uses 60
    Data Types Supported 60
    Implementation Issues 60
    Examining the Red Book Standard 61
    Red Book Error-Correction Techniques 62
  The Audio Origins of CD-ROM 62
  Yellow Book 63
    Uses 63
    Data Types 64
    Implementation Issues 64
    Examining the Yellow Book Standard 64
    Examining the Yellow Book Extended Architecture 65
    Yellow Book, ISO 9660, and the High Sierra File Format ... 67
  Green Book 67
    Uses 68

Data Types.	68
Implementation Issues.	69
Examining the Green Book Standard.	69
Another Offshoot: CD-I Ready.	69

White Book 70
    Uses 70
    Data Types.. 70
    Examining the White Book Standard 71
    PhotoCD 71
    VideoCD 73

Orange Book 73
    Uses 73
    Data Types.. 73
    Implementation Issues 74
    Examining the Orange Book Standard. 74
    Multi-Volume Discs. 76
    Modes and Multi-Session Problems. 77

The Frankfurt Group 78
ISO 9660 79
    Uses 79
    Implementation Issues... 79
    Examining the ISO 9660 Standard 80
    One File System/Many Platforms 80
    File Organization on Cross-Platform Discs........................... 82
    Operating System Extensions 83
    Resolutions of Earlier Problems 83

New and Evolving Standards 84
    CD-PROM 84
    Picture CD.. 85

Summary 86

Chapter 4
DVD Standards 87
An Evolving Set of Standards 87
DVD-ROM 88
    Data Storage Techniques... 89
    UDF.. 90
DVD-Video.. 92
    File Formats under DVD-Video 93
    Navigating DVD-V Content 94

Table of Contents

  Authoring for DVD-V 96
 DVD-Audio 96
 DVD-R 98
  Playback Compatibility 99
  File System for DVD-ROM 99
  Writing to DVD-R Media 100
  Uses for DVD-R 101
 DVD-RAM 103
  Uses for DVD-RAM 104
 Summary 104

Chapter 5
Optical Recording Equipment 105
 Selecting a Computer Interface 107
  Performance Issues 107
  Interface Options 109
  SCSI Considerations 110
  Rules of Thumb for SCSI Daisy Chains 112
  Portability of SCSI Drives 115
  SCSI Connectors 115
  Platform-Specific Issues 115
 ATAPI IDE Considerations 116
 Optical Disc Recording Issues 117
  Non-Erasable Media 118
 Rewritable CDs and DVDs 118
  Disc Formatting Considerations 119
  Multi-Session Recording and Packet Writing 120
  PacketCD 121
  DirectCD 121
  CD-R FS Packet Writing 122
 Selecting a Host Computer 123
 Minimum System Requirements for CD Recording 123
 Distributing Files to Replicators 124
 Other Hardware Considerations 125
  Uninterruptible Power Supply 125
  Hard Disk Considerations 126
  Virtual versus Physical Images 127
 Selecting a CD Recorder 127
  Pricing of CD Recorders 128
  Recording Speed 128

Table of Contents

    Onboard Buffers. 129
    Software Support 129
    Read Speed. 130
    Easily Upgradable Firmware 130
    Direct Overwrite Feature 131
    SCSI Version Supported 131
    Laser Power Calibration. 131
    Running OPC. 132
    Track-at-Once or Disc-at-Once 132
    Additional High-End Features. 133
Current Examples of CD Recorders 134
    Sony Spressa Professional (SCSI) CRX140S/C 134
    HP SureStore CD-Writer Plus M820e 135
    APS DVD-RAM External SCSI Drive 137
    APS CD-RW 8x4x32 FireWire 137
    Yamaha CRW6416sxz External CD-RW Drive. 138
    Young Minds, Inc. CD Studio and DVD Studio 140
Guidelines for Hardware Installation 141
From Hardware to Software. 143

Chapter 6
Disc Recording Software **145**
Evolution of Recorder Software. 146
    Early Travails. 146
    Development Platforms 147
    Terminology of Recording Software. 148
Types of Disc Recorder Applications. 149
    Backup and Archiving 150
    Basic Disc Recording Tools 150
    Professional-Caliber Applications 151
    Disc Recording Components. 152
Bundled CD Recorder Applications 152
Disc Recording Software Features 153
    Getting the Most Benefit from Recorder Software 154
    Interface Considerations 155
    File Format Support. 156
    ISO 9660 Issues. 157
    Recorder and Tape Support 158
    Simulation. 159
    Recording Simulations. 159

Table of Contents

    Performance Simulations 160
    Multisession Support. 161
    Kodak Multisession 162
    Hardware Multisession 163
    ECMA 168 163
    Autonomous Multisession 163
Online Assistance 164
Using Wizards. 164
    Hewlett-Packard CD-Writer Plus Wizard 164
Examples of Disc Recording Applications 171
    HP Disaster Recovery Wizard 171
    Automated Backup with HP Simple Trax 174
    Adaptec Easy CD Creator 177
    HyCD Publisher 183
    GEAR PRO DVD 185
    ISOMEDIA Buzzsaw CD-R Recording Software. 190
Deciding on an Application 193

Chapter 7
Recordable Media **195**
CD-R Media. 196
    Dye-Polymer Variations 196
    Licensed Dye Formulas 197
    Recording Process 198
    Extended Capacity Discs 198
CD-RW. 199
    Formatting CD-RW 201
    Limited Rewriting 202
    MultiRead Compatibility. 203
Hybrid CD-R 203
High-Speed Recording on Low Quality Discs 204
DVD-R Media 205
    How Long Do DVD-R Discs Last? 206
DVD-RAM Media 207
Magneto-Optical Media 207
    MO Media Lifespans 208
Printable Media Surfaces 209
    Poor Man's Labeling 209

Chapter 8
CD Duplicators 211
  Duplication Terminology 212
  Some Basic Concepts 213
    Recordable Media 213
    Production Efficiency 213
    System Configurations 214
  Uses for Disc Duplication 214
    Short-Run Disc Production 214
    On-Demand Disc Publishing 214
    Workgroup Disc Recording Applications 215
    High Security Publishing 215
    Time-sensitive Disc Production 215
  Workflow Issues to Consider 216
    Data Image Mastering Time 216
    Recording Throughput Time Requirements 217
    Disc Labeling Time 218
    Packaging 218
  Choosing Recordable Media for Duplication 218
    Labeling Issues: To Print or Not to Print 218
    Pre-screened Labels 219
    Disc Printer Options 219
    Wax Transfer 219
    Inkjet 219
    Paper Labels 220
  Media Packaging 220
    Jewel Boxes 220
    Spindles 220
    Bee-hives 220
    Shrink-wrapped 221
  Duplicator Classifications 221
  Disc Copiers 222
    Examples of the Disc Copier Class 222
  Tower Duplicators 223
    Examples of Tower Duplicators 223
  Automated Duplicators 225
    Examples of the Autoloader Class 226
    MediaFORM 226
    Microtech ImageAutomator 227
    Cedar 228

Table of Contents

    Rimage Protogé 229
    Trace Digital. 230
  Disc Printers 230
    Inkjet Disc Printers 230
    Wax Transfer (Ribbon-Based) Printers 231
  Guidelines for Media Handling with Duplicators 232
  Issues Involved in Printing on CDs 233
    Pre-screened CD-Rs 234
    Using Pre-labeled Media 234
    Print-alignment Devices 235
    Rimage AutoPrinter. 235
    Paper Labels. 236
  Packaging Duplicated Discs. 236
    Jewel Boxes. 237
    Disc Sleeves and Envelopes. 237
  DVD-R or DVD-RAM for Duplication? 237
    When Will It Happen?. 238
  Summary. 238

Chapter 9
Practical Applications **239**
  Email Archiving 240
  Exchanging Musical Project Data 243
  Distributing Cost-Effective Digital Press Kits 245
  Lightweight Digital Video Exchange 246
  Case Study: Optical Storage for Digital Video Applications 247
  Other Uses 250
  Summary. 251

Chapter 10
Simple Authoring Techniques **253**
  Publishing to Disc with Adobe Acrobat 253
    PDF Writer or Distiller 254
    Indexing a Document Set 255
    Post-Production Processing in Acrobat 256
    Automated Acrobat Production 257
  Inmedia Slides and Sound. 259
  Blue Sky Software RoboHelp HTML 2000 260
    Tapping into the Java Virtual Machine 261
    RoboHelp Development Environment 261

    Developing Content with an HTML Editor 263
    Case Study: Blackhawk Down. 265
    Summary 268

Chapter 11
Audio Recording for Music Enthusiasts 269
    Vinyl Restoration. 270
        Performing the Restoration. 272
        Producing an Audio CD. 274
    Case Study: Creating an Enhanced CD 279
    Case Study: Making a Web-Enabled CD 287
    Mechanics of Building an Enhanced CD 294
    Expert's View: Working with Surround Sound Audio 296
    Summary 304

Chapter 12
Interactive Music Design 305
    Expert's View: George Sanger Talks about Interactive Music 305

Chapter 13
Business Uses for Optical Recording 321
    Networking Optical Disc Drives. 322
        Advantages of Networking CDs and DVDs 323
        Cost Factors 324
    Computer Output to Laser Disc. 324
    Network Storage Solutions 325
        SciNet CD-Manager 5. 326
        MediaPath MA32+ 326
    Examples of Network Ready Optical Disc Units 327
        Plasmon AutoTower. 327
        Axis StorPoint CD E100 328
        Quantum CD Net Universal XP Cache Server 329
    Increasing Storage Requirements 330
        Particle Beams for Data Reading. 331
        Fluorescent Multi-layer Discs. 331
        Optical Super Density Format 331

Chapter 14
Using DVD with Video in a
Corporate Environment 333
    Applications for Corporate Video 333

Table of Contents

Delivery Options for Corporate Video 335
   Delivery via Analog Video 335
Digital Video 337
   Why Digital? 337
   How Digital Video Works 338
   Production Process for Digital Video 340
   Capturing and Assembling Digital Footage 341
   Getting the Best Compression 346
   Delivery via DVD video 347
   Playback Set-up for Corporate DVD Video 355
Producing a DVD video 357
   Preparing the Video and Audio Tracks 358
   Prepare Subtitles 359
   Prepare Graphics 359
   Create the DVD Video File Structure 359
   Create a Master DVD-Video Disc 363
   Delivery via DVD-ROM and DVD-RAM 363
Summary of Recommendations 366

Chapter 15
Interactive Multimedia on Disc **369**
  Case Study: Macromedia Add Life to the Web CD-ROM ... 369
  Case Study: The Creation of DroidWorks 380
  Summary 391

Chapter 16
Disc Replication,
Printing, and Packaging **393**
  Replication Overview 394
    Preparing Files for the Replicator 394
    Preparing the Artwork 395
    Delivering the Files to the Replicator 398
    Manufacturing the Discs 399
    Packaging the Discs 399
  Expert's View: Preparing Artwork for CDs 399
    Summary of Disc Artwork Tips 407
  Packaging Options 409
    Steps in the Packaging Process 409
  A Gallery of Package Types 412
  Summary 419

Chapter 17
DVD Creation 421
   Expert's View: Developing Titles for DVD 421

Chapter 18
Independent Marketing and Distribution 437
   Targeting Niche Markets 438
   Applying Permission Marketing 440
      The Flaws in Interruption Marketing 440
      Permission Marketing for Independent Developers 441
   Using the Internet as a Leveler 441
   Constructing a Web Storefront 443
      Online Auctions and Other Sales Outlets 444
   Targeted Press Releases 445
   Bidding on Search Terms 446
   Summary of Techniques 448

Chapter 19
Responsible Media Use 449
   The Revolution That Never Came 450
   A Model for Sustainability: The Natural Step 451
   Recycling Discs and Packaging 453
      Recycling Locations 454
   Summary 454

Chapter 20
Entrepreneurial Possibilities 455
   Case Study: The Entrepreneurial Possibilities of DVD 457

Appendix A
Resources 465
   Trade and Standards Organizations 465
      Optical Storage Technology Association 465
      SIGCAT 466
      DVD Forum 466
      Digital Video Professional's Association 467
      PCFriendly 467
   Stock Photos, Fonts, Media Assets 468
      EyeWire Studios 468
      The Stock Market 468
   Authoring Tools 468

Table of Contents

CD and DVD Recorders	469
CD and DVD Recorder Applications	471
Adaptec, Inc.	471
CeQuadrat	472
GEAR Software, Inc.	472
Computer Output to Laser Disc	473
Thin Server Technologies	474
Library Systems, Jukeboxes, and Towers	476
DVD-Video Tools	477
Media	478
Packaging Materials	479
Replication and Production Services	480
Glossary	**483**
Index	**505**

Preface

Optical recording has become the sturdy workhorse of the digital information revolution. The World Wide Web may be the standard bearer, sexier and clearly more high profile, but when it comes to distributing extremely large volumes of information, the information conduit of the Web begins to look more like the trickle of a small creek after a Summer drought. Streaming media makes it possible to deliver a small digital video image with sound over the Web, but if you want to watch a full-screen movie with CD-quality sound, you turn to DVD. Audio distribution of music on the Web represents an interesting novelty, but if you want to hear the latest work of James Taylor or Sheryl Crow in stunning clarity, an audio CD or audio DVD becomes the medium of choice. Someday broadband content distribution will change this picture, but right now we're waiting for the tools and technology to come of age. We'll probably still be waiting several years from now. In the meantime, optical recording is already here.

In a technology that has soared from complete obscurity to universal mainstream acceptance in a little over two decades, the evolutionary line has never been a clear, straightforward path from lizard to bird, but a dazzling array of subspecies have emerged, some flourishing, some falling from the sky and sinking back down into the mud. The evolutionary line still isn't settled. While the CD-ROM world has finally achieved relatively stability, some manufacturers are still pushing the edge of the envelope, embracing higher recording speeds for CD-RW formats and CD-R alike. The quest for the unified field theory of the optical world continues in earnest, with many companies diligently working towards designing a single piece of equipment that will both read and record the vast majority of CD and DVD formats. Like small independent-minded evolutionary creatures, these formats have arisen to meet the expanding needs of the digital storage community, moving beyond simple audio to include synchronized multimedia content, digital video, photographic images, compressed audio, multilingual text, Web-enabled content, multiplatform file formats, locked software content, and other interesting and sometimes short-lived variations. Much of the energy of the industry is now being poured into making DVD the inheritor of the optical recording mantle, the standard vehicle for distributing software applications, databases, libraries of content, and other data currently being distributed on CD-ROM. Within the next two years or so, most estimates see the DVD-ROM as eclipsing the CD-ROM, but the incredibly large installed base of CD-ROM players will ensure the life of this medium for some years into the millennium.

This book isn't intended to be a comprehensive reference of the technical aspects of optical recording, but a practical guide to the uses and possibilities of this new technology. While I use optical recording nearly every day in my work as a writer and consultant, the scope of my uses doesn't begin to encompass the full range of possibilities in the industry. For this reason, I've included interviews with many of today's leading developers, audio engineers, entrepreneurs, software programmers, project leaders, and others to offer perspectives based on experience that I could not hope to obtain in a lifetime. To those who freely offered their time to patiently explain concepts or detail the technical aspects of projects they have completed, I extend my grateful thanks. This book is much richer for the insights and guidance that they have provided.

CD-ROM Contents

The CD-ROM bundled with this book includes trial versions and demos of many of the leading disc recording applications, multimedia development tools, sound processing applications, and utilities designed to make the work of a title producer easier.

As of press time, the contents of the CD-ROM were still being determined. For full details on the disc contents, refer to the README.HTM file on the CD-ROM.

Acknowledgments

While the author gets to bask in the glory of having his name imprinted on the front cover, in reality a book is the collaborative effort of many individuals. I extend my sincere gratitude to those who contributed their time and knowledge to shaping and directing this work. Thanks to Stephen Chapman of McGraw-Hill for his patience while I scrambled to incorporate the latest changes happening in the DVD world. Thank you to Katherine Cochrane of The CD-Info Company for contributing Chapter 8 and her extensive knowledge of the CD duplicator market. Thanks to David Martin, who co-authored an earlier book about recordable CD with me, for writing Chapter 14 and contributing both copy and photographs for Chapter 16 on replication and packaging. Thanks also to Scott McCormick for weeding out the worst excesses of my prose and skillfully tending to the copy editing.

Many developers, programmers, project leaders, and others took time from their deadline-driven schedules to share their insights into the processes involved in creating CD-ROMs, DVD-ROMs, and DVD titles. Thanks to Marc Randolph of Netflix, Matilda Butler of Knowledge Access Publishing, Chris Andrews of The Andrews Network, Jeff Southard and Chris Xiques of 415 Productions, Collette Michaud of Lucas Learning, Ltd., John McQuiggan of *philly.com*, George Sanger (the Fat Man), Kevin Deane of Genesis, fellow author Rudy Trubitt, Lance Svoboda of Disc Makers, Kelly Meeks of Right Angle, Inc., Blaine Graboyes of Zuma Digital, Inc., and everyone else who generously shared their knowledge and time.

About the Author

Lee Purcell has been writing about computers and emerging technologies since the distant glory days of the Osborne 1 computer, the first machine that honestly deserved the title *personal computer*. He is the director of Lightspeed Publishing LLC, a Vermont-based company that strives to wean companies away from paper-based publishing to the freedom of electronic publishing. Lee can be contacted by email at: *leep@well.com*

1

A Concise History of Optical Recording

The idea of recording information using the energy of concentrated light isn't a new one. A boy focusing the rays of the sun through a magnifying glass to burn a pattern in a block of wood has learned the basics. Substitute a laser to precisely align the beams of light, point it at a layer of dye coating a reflective disc, burn some microscopic marks in the surface, and you have the essence of optical recording.

Around this fairly simple technique, an industry has grown—refining the methods by which binary data can be stored and retrieved using various forms of optical media. In the single flash of a red laser, earlier methods of archiving information have been rendered largely obsolete. The microfiche, magnetic tape, and other forms of magnetic storage, though still in use, have been relegated to greatly reduced roles in the record-keeping pursuits of the human species. Light moves in a swift, silent stream to etch the binary language of the computer into a light-weight, malleable medium. Optical recording takes center stage in the digital media and electronic publishing extravaganza that is changing the way we work with words, sounds, and images.

Audio Roots

A technology born out of a desire to produce pristine audio material in digital formats has expanded far beyond the expectations of its makers. Over the last two decades, CD-ROMs and other means of optical recording have grown to become the leading method of data distribution in the computer world. The next-generation version of this medium, DVD, promises to open up even more avenues of inexpensive communication as it becomes the primary vehicle for distributing motion pictures, audio works, and interactive multimedia content to a whole new generation.

No one could have predicted the impact that the humble compact disc would have on twentieth century communications back in the early 1980's when engineers from Sony and Philips struggled with the problems of reducing music to a series of digital patterns and storing it compactly for playback. The key to solving this equation was the laser—an intensely focused stream of parallel light beams, energy that could be used both to create a microscopic pattern in dye and to be read as reflections off a shiny surface imprinted with digitally coded content. The laser has been another technology that has moved in directions that surpassed the dreams and expectations of its inventors.

This chapter offers a concise perspective on the origins of the CD-ROM and DVD, and a compact history of optical recording. Some thoughts on the future of digital communication on disc are also included.

The First Compact Discs

We traditionally credit Alexander Graham Bell for inventing the telephone and Thomas Edison for fashioning the first light bulb, but no single inventor can step forward and take credit for the compact disc. The collaborative efforts of many engineers and technicians—and hundreds of thousands of hours of research and experimentation—resulted in the first disc: a half-ounce piece of plastic and aluminum that could store 74 minutes of music. The engineering effort required the studied application of principles that had evolved and been refined over many years. To become useful and practical in the real world, the compact disc also demanded the cooperation of numerous manufacturers (who produced the playback equipment), the music industry (who had to be convinced of the viability of this method of recording sound), and the music-listening public (who had to be persuaded to give up their turntables and vinyl records).

The clear advantage that the compact disc offered music listeners was the almost total absence of background noise. Proponents of this new music distribution medium emphasized this advantage by demonstrating compact disc playback using music pieces with quiet passages, where the absence of the usual white noise intruding on the silence was startling to listeners, most of whom had become used to it as an inescapable aspect of recorded music. Although analog recording devices using noise-reduction circuitry had successfully reduced the amount of background hiss present in recordings, the compact disc improved the signal-to-noise ratio substantially. Levels jumped from 50 or 60 dB to over a 100 dB. This jump was significant enough to be immediately noticeable to anyone listening to a piece of music—not just audiophiles. Consumers flocked to the medium in record numbers and the compact disc industry became the most rapidly growing communication technology ever introduced. This achievement has only recently been eclipsed by the meteoric rise of DVD. By the end of 1999, DVD player manufacturers had sold about ten times as many units (close to 5,000,000) as had CD player manufacturers within the same introductory time frame.

The compact disc had another major advantage over vinyl records: a long life-span. Vinyl record playback relies on a stylus that physically travels over a series of bumps and indentations, converting this microscopic topography into an electronic signal, which is then converted into sound waves. In comparison, the laser read head of a CD player scans the disc surface optically. Since the laser beam reflects off the surface, but never touches it, no wear to the disc occurs. Audiophiles tired of listening to their music collections gradually picking up snaps and pops and clicks over the years could enjoy recordings that remained unchanged through thousands of playings. Though some critics commented on the cold, metallic quality of the sound in comparison to analog recordings, the public found the advantages of the medium persuasive and began scrapping their vinyl collections in favor of CDs. Arguments as to the relative merits of analog versus digital sound continue to this day. There is clearly a perceptual difference between the characteristics of these two forms of audio recording and playback; whether this difference is significant enough to seriously affect the enjoyment of recorded music is part of the running debate.

The same engineers who wrestled with the issues surrounding the digital storage of music began to adapt the storage techniques so that they could be applied to data. One of the most significant problems was the approach to error correction and data integrity. The correction codes required on a music CD are minimal. If a disc gets scratched and several hundred bits become unreadable, the playback units can extrapolate from the surrounding data and basically fill in the missing information so that there are no audible artifacts. Computer data is not so forgiving. A single byte missing from an application program could make the program unusable, so preventing any data loss becomes absolutely critical to the usefulness of the medium. The solution would require a new format with expanded codes that could protect data integrity more effectively.

Chapter 1

Early Research

In 1972, Philips Corporation of the Netherlands issued announcements that foreshadowed the optical storage of audio content, but their initial experiments involved the use of analog modulation techniques. This poorly conceived approach was soon abandoned in favor of more promising digital signal encoding methods and Philips began work in earnest to define a signal format to use for audio storage. During this same time period, Sony Corporation of Japan was engaged in research to perfect error-correction methods that could be applied to digitally encoded audio. Collaboration between Sony and Philips resulted in the merging of Philips' signal format with Sony's error-correction method and in June of 1980 the two companies introduced their proposal for the Compact Disc Digital Audio system. With support from 25 manufacturers, the Digital Audio Disc Committee adopted the proposed standard and the work effort shifted towards retooling the industry to support the manufacturing effort and designing the components that would make the standard possible, including Digital-to-Analog circuitry for converting the digital audio signals, laser read heads integrated into semiconductor designs, and other kinds of circuit designs that could be implemented inexpensively in a mass market product introduction.

During the three-year period leading to the launch of the first audio CD, other companies—notably Hitachi and JVC—made an effort to introduce a competitive form of digital audio disc. Their approach used capacitive storage techniques rather than optical storage. Capacitive storage has the advantage that discs can be manufactured very inexpensively, but the disadvantage is that the charged particles embedded on the disc lose their charge gradually over time. The format was abandoned before it ever became an actual product.

Towards the end of 1982 the first audio CDs began shipping in Japan and Europe. The new music storage format caught hold quickly and in March of 1983 the United States joined in the product introduction fervor, snapping up hundreds of thousands of players in the first year (even though the cost of the early players was around $1000 each). By the end of 1984, more than a million players had been sold worldwide and CD sales exceeded 20,000,000 units.

It didn't take long before engineers at Sony, Philips, and other companies began to develop techniques for storing data on disc. One of the most pressing needs was to devise a more robust error correction scheme than the one that was used for digital audio. Lost bits during audio playback might cause an indiscernible blip in the audio content, but lost bits on a data disc could cause an application to fail or incorrect information to be used in a key process.

The CD-ROM format, tagged Yellow Book, was introduced in 1985, but initially the $1000 drive costs and dismal performance discouraged many potential users. Nonetheless, developers and manufacturers persisted and a variety of new formats appeared as drive costs came down, performance steadily increased, and an interested variety of titles came onto the market demonstrating the viability of interactive multimedia. This period in the history of the medium is well documented in Chris Andrews' work, *The Education of a CD-ROM Publisher*, a book that charts the erratic and often amusing course of the early participants in title development process.

The term *shovelware* was born as many CD-ROM title companies dumped any kind of content they could find onto disc, making consumers wary of purchasing any titles, since it was so difficult to identify those that contained anything of genuine value. The shakedown in the industry that followed the gold rush mentality of the late 1980's and early 1990's made it clear that successful title development and sales required a balanced mix of realistic budgeting, innovative marketing and distribution, and worthwhile content development.

Early in the life cycle of the CD-ROM industry, the emergence of a number of proprietary disc formats threatened to set back the progress that had been made. In 1986 a number of industry representatives gathered in the High Sierra Hotel and Casino close to Lake Tahoe, California, and hammered out a common file system structure that became known as the High Sierra format. Refined over a number of months, this format became solidified by the International Standards Organization in the form of ISO 9660, the file interchange standard still in common use today. More information about the evolution of the various file systems and standards appears in Chapter 3, *CD Standards*.

Chapter 1

Figure 1 - 1   **Optical recording timeline**

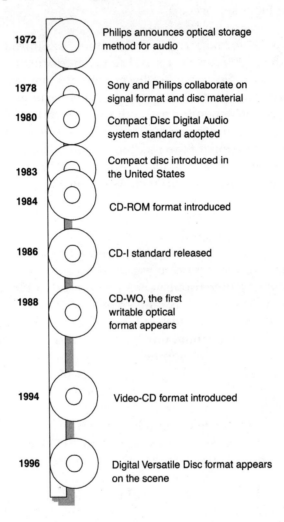

Other Devices, Other Formats

The remarkable success of the audio CD and eventual acceptance of the CD-ROM spurred manufacturers to introduce and promote numerous other types of digital storage products, some of which languished in the market, others suffered almost immediate market deaths, and some others that still exist in various forms.

Philips, for example, attempted to create a new market around a format that it designed: CD-i. The "i" stood for interactive and Philips created setup players that could play MPEG-1 video content, as well as interactive multimedia titles, such as games. Though the format still lingers in small numbers, it has been eclipsed by other approaches to delivering this same type of content that are much more effective.

Sony devised a recordable format and new media known as the Mini-Disc, banking on the desire for home recordists to embrace a technology that lets them create near-CD quality musical recordings. The original compressed music format used with the MiniDisc, tagged ATRAC by Sony, was distinguished by mediocre quality, but later versions were significantly better and the MiniDisc maintains an ongoing market share as an audio storage and distribution medium. Despite the lukewarm acceptance, MiniDisc has found some haven with home recording applications for semi-professional musicians and in the pre-recorded music market in Japan.

Much of the early research devoted to refining the methods for converting and storing analog audio data in a digital framework resulted in the development of the Digital Audio Tape (DAT), a cartridge-based recording and playback system initially introduced by Sony and Philips. DAT failed to make major inroads in the consumer audio marketplace, but has become a very common audio standard in the professional music industry. DAT cartridges are often used as masters when submitting music to a replication service for compact disc manufacturing.

Recording Goes Mainstream

If you know anything about the history of recordable CD, you've heard the stories on how expensive the early CD recorders were. Those in the industry reflect on the early pricing of recorders with a mixture of nostalgia and a hint of bragging rights—like the story about the good old days before the lake was overfished and you landed that 18-pound bass. Early recorders were sized like IBM mainframes and priced in the hundreds of thousands of dollar range. Premastering tools were largely non-existent. With the span of a decade, the prices would drop a hundred-fold. Today, you can get a CD-RW drive for under $100 and a DVD-RAM drive for under $500.

In his book, *The Education of a CD-ROM Publisher*, author Chris Andrews described his initial meeting with Allen Adkins, who designed the first premastering software for a CD recorder. His work was incorporated into the first complete systems produced by Optical Media International and shipped to numerous companies eager to create their own CDs inhouse.

Chapter 1

I arrived at his office late in the morning, and was greeted by Allen Adkins, an intense, distant man with large glasses. His office was located in an elementary school that had been converted into offices. The desks, tables, chairs, and general feel were still that of an elementary school. There were even some children's drawings still on the wall.

He sat me down next to a computer, which was on a table that looked like it had been the teacher's desk, and proceeded to type in several cryptic commands.

"There's more than a gigabyte right here," he said, with all the pride of an inventor showing off his creation for the first time. "The real problem is the premastering process," he continued. "There aren't any good tools, and that's why I'm building some. Here, look at this. I can put it all on a 3/4-inch Umatic tape and send it off to the factory to make a CD."

He showed me a chart that looked like a flowchart, but it had words I didn't understand—OCR, Premaster, Encode, Uninterrupted Pseudo-audio, Replication. These were strange words coming from the world of information where "update," "numeric," and "text" were common terms. He then took the chart back as though he regretted ever showing it to me.

As I looked around his desk, I noticed not only the quantity of equipment, but the diversity and size. He truly was part inventor, mixed with some of the absent-minded professor, and a little high-tech entrepreneur.

Refinements made in the original CD recorder and the premastering software resulted in the first shippable product, which was launched by Optical Media International in 1989. Despite the high cost and the significant technical knowledge required to successfully burn a blank disc, the early units were popular with companies that saw the advantage of being able to generate their own CD-ROMs inhouse.

Related Recording Technologies

A variety of spin-off optical recording technologies have emerged and become important, though more for archiving than for data distribution. Magneto-optical drives use phase-change techniques to store data, allowing the laser to both imprint data on the media surface and to erase it with a different intensity beam.

Rewritable techniques have also been employed in CD-RW drives, which rely on a crystalline alloy layer within the disc that can be returned to an unrecorded state by carefully controlling the power setting of the write laser. CD-RW discs can be rewritten around 1000 times. CD-RW disc readability has gotten a boost in popularity as the latest generation of CD-ROM drives—dubbed MultiRead capable—can interpret the disc contents, but earlier generation drives cannot handle the differences in the reflectivity of the media surface.

The key factor for data distribution, of course, is having a piece of equipment on the receiving end that can actually read the data. Nearly everyone today has a CD-ROM drive in their system and many people have DVD-ROM drives. But it's a different story when you try to send a magneto-optical cartridge or SyQuest cartridge or a floptical cartridge. The widespread installed base of CD-ROM drives makes this data distribution medium a near universal choice, whether you're attempting to deliver data domestically or internationally. DVD-ROM, by incorporating engineering features that make it possible to read all of the CD formats, will certainly inherit this mantle of acceptance and become a more compelling choice for universal data distribution as the installed base of equipment grows.

The Laserdisc

The Laserdisc is an odd mixture of analog recording and laser technology. It has become the pre-eminent storage medium for high quality motion picture storage, attracting a small, but consistent, audience of videophiles who value the clear images and excellent sound quality offered by the medium. The platter-sized discs, which require fairly expensive playback equipment, never quite caught on with the general public, but there are enough advantages to the medium that it has endured, even though growth has been less than breathtaking. Thousands of motion pictures have been transferred to this format and players can still be purchased in most electronic outlets.

The interactive training community also recognized the advantages of the Laserdisc for producing multimedia educational works, but the difficulty of authoring to Laserdisc and the expense of the tools involved has discouraged many potential trainers from using this medium. Nonetheless, for those who have invested in the technology, there is a small but consistent group of advocates who continue to make use of the tools. With the multiple advantages of DVD, however, it is unlikely that this format will persist very much longer.

Chapter 1

DVD Arises

When the first wave of personal computers were introduced in the early 1980's, many users relied on floppy diskettes for launching applications and storing data. Some of the first (highly expensive) hard drives for personal use offered storage capacities starting at around 5MB and sometimes reaching as high as 20MB, if you could afford the luxury. Given this framework, it's hard to believe that less than two decades later, the 650MB of data storage offered by the CD-ROM would be considered inadequate. The quest for even higher levels of storage lead to the design of the Digital Versatile Disc, or DVD in common parlance.

DVD formats range from 4.38GB to 15.90GB in the latest standards, but future variations may reach even higher capacities as new technologies are explored. Rather than expanding the disc surface by increasing the diameter, DVD uses multiple internal layers to encode additional data on two or more surfaces. The format has gone through considerable flux as computer industry leaders and motion picture industry officials sparred off over the nature of copy protection methods that would be incorporated into the discs and a multitude of other design issues.

In the early 1990's, several manufacturers were working on ways to increase the capacity of the standard CD-ROM. Nimbus Technology and Engineering devised a technique for producing a double-density CD and introduced it in 1993. This particular format was tailored to the delivery of MPEG-1 video content, providing a storage capacity of up to 2 hours of compressed video. It provided a valuable proof-of-concept that the compact disc could be a useful delivery medium for video content, particularly in the highly lucrative consumer marketplace where VHS reigned supreme. Digital video promised significantly improved moving pictures and sound, as well as potential cost savings in production and manufacturing of the discs on which content would be distributed. Hollywood representatives entered into discussions with computer makers and consumer electronics companies to forge a set of guidelines for this new medium.

The advisory committee that was assembled to define the goals of this project had a number of recommendations. First and foremost, the capacity of this proposed high-density disc had to accommodate at least a full 133-minute movie on a single side. The resolution of the video was targeted to the CCIR-601 broadcast standard, a high-resolution form of video in comparison with VHS. Copy protection had to be integrated into the medium to alleviate fears of easy-to-copy digital content being pirated and illegally distributed. The medium should also support very high quality audio—including six-channel surround sound. Additional languages were to be supported by providing three to five language streams and up to 30 individual subtitle streams. Other issues, such as parental protection mechanisms and screen aspect ratios, were also discussed.

Two very different digital video disc formats emerged in January of 1995. Led by Toshiba and several of its partners, one camp introduced the Super Density (SD) format. Sony and Philips devised their own approach—christened the Multi Media Compact Disc (MMCD) and enlisted their own band of supporters. For a time, it appeared that the battle lines that had been drawn would completely halt development efforts for everyone, but the charter for the DVD Consortium was drawn up and the dissension among the industry leaders diminished as the standard for the Digital Versatile Disc—DVD—was formalized in December of 1995.

As early as 1996, several companies, including Philips and Thomson, announced the release of DVD players for late in the year. It was at this point that the Consumer Electronics Manufacturers Association (CEMA) and the Motion Picture Association of America (MPAA) began pushing in earnest for restrictions on the DVD hardware and software that would discourage illegal duplication of copyrighted material. The industry fracas that resulted from their actions fragmented the development efforts of several companies and has required years of diplomacy and negotiation to reach a point where viable DVD products could be introduced into the marketplace.

The Optical Storage Technology Association (OSTA) played a key role in furthering the DVD design process by devising the Universal Disk Format (UDF) to serve as a fully cross-platform file system. By gaining the cooperation of the principal companies involved in the technology— including Sony, Philips, Toshiba, Fujitsu, Sun, Pioneer, Thompson, Mitsubishi, Hitachi, and Matsushita—a solid bridge was created to bring the computer industry and the television/movie industry one step closer to settling differences. While fractious differences continue even today, the stakes in this potential multi-billion dollar industry have pulled the opposing camps into some semblance of common purpose.

With the DVD-Video format now solidly entrenched, after years of standards battles, players and DVD titles have been selling strongly throughout the world. The DVD-ROM format has also gained a fairly stable footing and increasing percentages of new PCs are being sold with an installed DVD-ROM drive, which can, of course, also read the full range of CD-ROM formats.

Ongoing struggles, however, continue in the arena of the rewritable formats, with several competing formats vying for market share. As was the case with early recordable CD technologies, incompatibilities with the range of installed players—in this case, DVD players and DVD-ROM drives—have caused disenchantment with users of many of the first- and second-generation rewritable drives. Because of these incompatibilities, use of the rewritable DVD technologies has been primarily limited to testing during title development and large-scale archiving. More on this topic appears in Chapter 4, *DVD Standards*.

Growth of Publishing and Title Development

No history of optical recording technology would be complete without considering the impact of this medium on the publishing world and game development universe. From the early days of CD-ROM, optical discs have been used for a diverse and imaginative array of entertainment and business uses. Many of the companies that forged bold paths in the 1980's no longer exist, victims of the hyperbole that surrounded the industry like a stagnant air mass. From the survivors, valuable lessons can be learned.

Case Study: Surviving in the Electronic Publishing Industry

One survivor is Knowledge Access Publishing, originally founded as Knowledge Access, with corporate roots that go back to the birth of the industry. How did they survive when so many others failed? I talked to co-founder, Matilda Butler, to get her insights into how to stay on track in the CD-ROM publishing industry.

Could you offer a capsule history of Knowledge Access from it's founding to where you are now?

Matilda: Bill Paisley and I started Knowledge Access International in late 1984 (incorporating in 1985). In many ways, Knowledge Access was the outgrowth of a previous company, Edupro, that developed software for the education market. One of our contracts was to develop College Explorer for the College Board. Using College Board's database of information of US colleges, we designed a product for high school students to help them find schools that matched their interests in majors, sports, setting, size, etc. This was in the days of the Apple II. However, even at that point in history, we began to see the shape of technology. The IBM PC was released the second year of that contract. The cost of storage was dropping. (Remember our frame of reference was the Corvus 10MB hard disk that we bought for $4,500.) Floppies were storing more, hard drives were storing more and there was this new storage technical being discussed that could store hundreds of megabytes.

We wanted to combine our passion for getting information to people at a reasonable cost with the promising distribution technology CD-ROM. Bill had been the director of the ERIC Clearinghouse at Stanford University when Roger Summit said he wanted a Beta site for his idea of Dialog—an online information service for librarians. That means Bill had been involved with information delivery technology since the mid-1960s.

In the Spring of 1984, we went to Hawaii and left our programmers to finish the next release of College Explorer. The staff always hated it when we went on vacation because we came back with lots of new ideas. It was the only time we could get away from the daily deadlines to think about the future. This time, while sitting on the balcony of our room, we got talking about the idea of putting all kinds of information initially on floppy disks and eventually on CD-ROM. We talked about a search engine that could accept lots of different types of information—fielded, full-text, images—so that our cost to develop products could be kept reasonably low. We talked about changing the algorithm of the online information business that equated more use of information with more money. We thought the offline information business should create an opportunity that rewarded use of information—the fixed price CD-ROM meant that the more information was used, the less it cost.

We returned to Palo Alto and told our staff we had decided to start a new company: *Knowledge Access*. We would move away from custom education products to business and professional information products that would use a single search engine. In hindsight, we were blessed that we didn't have lot of venture capital although it made life much more difficult at the time. We had to do projects where we could make money because that was the only way we could meet the payroll and pay the rent. This meant we sought customers who would pay us to create their electronic information products. The well-financed competitors had fancy booths, expensive promotion materials, delegations of salespeople, and the ability to do products "for free" in order to seed the marketplace. Many of those companies folded or were taken over by others when the venture capitalists grew weary of trying to get their investment back. In the meanwhile, we kept right on going. Our customers were government agencies, commercial publishers, and corporations. Initially we created the titles for them. Eventually we were able to offer customers our OmniSearch Disk Publisher if they wanted to create their own titles or our OmniSearch Publishing Services if they wanted us to create their titles for them. The business settled into about half of each.

We tried creating some titles of our own, but that less-than-successful experience caused us to stay narrowly focused on our original idea of creating products for others who already had customers for their information.

In 1994, one of our customers came to us and said they were interested in purchasing our company. Believing Knowledge Access was a good match with our customer and recognizing that we would need additional resources to continue to stay up with the changing technologies, we sold the company. Bill and I agreed to stay on for two years and continue to manage Knowledge Access. Just about the time that our contracts were up, our parent company decided they would take our technology for in-house use and close the rest of Knowledge Access. Although Bill and I had other professional plans in mind, we didn't like the idea that our loyal publishing services customers would suddenly be without a solution. So, in agreement with our parent company, we formed a new company that would continue to provide publishing services.

Thus, Knowledge Access Publishing was begun. We publish about a dozen CD-ROMs per year. OmniSearch enhancements are now made on the basis of specific customer needs rather than for the general marketplace. What will the future bring? That's hard to say, but for now we find ourselves in the happy position of working with customers we really enjoy and creating products that their customers want and use.

Do you think that the DVD-ROM will have a significant impact on publishing to optical media in the near term, or will it have a long, slow start-up period, much as CD-ROM publishing did?

Matilda: We used to say never over-estimate the success of a technology in its first five years and never under-estimate its success in the second five years. That certainly was true of CD-ROM. However, there is an increasing rate of change and acceptance of change. There are some differences between CD-ROM and DVD-ROM. CD-ROM readers were expensive in the early days. The first reader Knowledge Access bought cost $2400. It operated at 1x. DVD readers, even in the early days, were not very expensive. CD-ROM had a hard time with distribution channels. DVD has been able to secure distribution much earlier—including drives built into PCs.

However, DVD drives are not as fast as CD-ROM drives. That presents a problem for the publisher who wants to put lots of information on DVD media since the read will be slower. The only reason for publishers to turn to DVD is if they need more than 670MB. The most likely candidates are publishers who have both text and video. The catch is that media products combining text and video are quite expensive to produce. Therefore, there has to be a large market for the product or the publisher can't get their money back. This seems a lot like the chicken-and-egg problem we worried about in the early days of CD-ROM.

Do you have any guidelines or advice that you would offer to independent developers who are planning on releasing their own title?

Matilda: Know your market. Figure out your distribution channel early. The best title in the world will languish without adequate marketing. Marketing costs can dwarf production costs. It is easy to get caught up in the product idea and figuring out the best way to create it. However, that is actually the easiest part. Skillful product promotion usually needs seasoned talent. I'd look for a good strategist who could deliver distribution contracts and I'd find that person early in the development cycle.

What do you see as the most interesting or useful trends in the industry?

Matilda: CD-ROM is still the preferred distribution technology for large amounts of information as well as for expensive information. I don't see a lot of examples of combining CD-ROM with Web updates, but I'm sure there will be more of that in the future. The Web is so page-oriented—so small information-nugget oriented—that it doesn't seem to directly compete with many CD-ROM titles. However, as Internet access becomes faster and as within-site search engines become more powerful, the Web may replace many CD-ROM titles. Why? Users are becoming so accustomed to thinking up a research question and then expecting to find a variety of answers immediately via the Internet that they may eventually not want to purchase and load a CD-ROM just to get a single answer. Over time, each of these technologies will be used for those publishing products that have the best fit. The installed base of CD-ROM drives means some titles will stay around for a long time.

Chapter 1

New Era, New Tools

Anyone with a good idea and some initiative can get involved in CD-ROM and DVD-ROM publishing today—the tools have improved to the point that a typical desktop computer can host an 8x CD recorder, 4x DVD-R recorder, or a CD duplicator that can produce 8 or more discs at a time. You don't have to look back very far in the short history of this industry to get a sense of how different it was in the beginning. In another short excerpt from *The Education of a CD-ROM Publisher*, Chris Andrews describes one of the first serious CD publishing tools:

> *If you worked for Meridian Data in the late '80's, you were cool. You were hip. Meridian was a very tiny company with an extremely big presence in the CD-ROM industry. It seemed as if every major announcement in CD-ROM involved Meridian Data. Part rebel, part industry leader, part corporate ass-kiss, everyone had an opinion of Meridian Data, whether they worked for the company or not.*
>
> *Meridian's main product was CD Publisher. It had been introduced in 1986, and about three-quarters of the world's CD-ROM producers used CD Publisher to build their CD-ROMs. As the salespeople at Meridian used to say, "If you want to be a player, you need to step up to a CD Publisher." Building on the fear that most people had about making their own CD-ROM, that sales tactic worked superbly.*
>
> *CD Publisher was deceptively simple: it was a large hard disk with some software for formatting a CD-ROM before sending it to the factory for pressing. It debuted in 1986, giving CD-ROM publishers the ability to become "in-house publishers." It included an assortment of utilities and hardware that would be helpful to a CD-ROM publisher, like a nine-track tape drive so that when your CD-ROM information was the way you wanted it on a hard disk, you could copy it on tape to send to the factory. It was big and bulky and looked more like a washing machine than anything high-tech. It was sold as a "desktop" CD-ROM publishing product. What was on your desk, though, was a computer. The computer was hooked up to the 300-pound CD Publisher that sat next to your desk. One of Meridian's ad campaigns made fun of this, and showed a box of Tide sitting on a CD Publisher. That campaign was symbolic of not only the size of the product, but the attitude of the company.*

Ongoing Development

Each time someone makes a bold pronouncement that we've reached the ultimate compression limits for a particular type of signal or that we've exhausted our options for improving performance or data storage capabilities of a format, some clever group of engineers finds a way to sidestep the obstacles and produce another evolutionary improvement in existing technologies. Such is bound to be the case with optical recording techniques, as research continues into new types of encoding, alternative varieties of lasers (such as the blue laser), advanced compression techniques, new motors and servo techniques, and other ingenious adaptations to existing methods.

Right now, the momentum is fully behind the passing of the baton from CD-ROM to DVD, the data storage medium that will eclipse the CD sometime early in the new millennium. This does not mean that the CD-ROM will then disappear from the face of the planet. The millions of installed CD-ROM drives and the extremely low cost of this data distribution method will keep the technology vibrant for years to come. The latest DVD equipment is designed to read all of the earlier CD formats and the recordable equipment is increasingly being designed to be able to write to earlier forms of media, such as CD-R and CD-RW, as well as the DVD formats. The most compelling reason for moving to DVD is still the significantly higher capacity, so many individuals and companies, whose data distribution requirements are below 650MB, will continue to use CDs. The two different media types are complementary and they should happily co-exist for some years to come.

2

Overview of CD-R and Writable DVD Technology

With the current generation of hardware and software, using a laser beam to imprint data on a compact disc can be as easy as copying to a diskette or transferring a set of files from one hard disk to another. Many of the performance-related issues of early generation CD-R drives have diminished or disappeared in the era of PowerPC G4, and Pentium III processors. Although some of the early DVD-RAM, DVD+RW, and DVD-R drives suffer from the kinds of problems that troubled users of the early CD-R equipment, you can still get a system set up for premastering or archiving with far less difficulty than in the not-so-distant past.

The basic division that arises when examining the approaches to optical recording techniques is whether or not the format is rewritable. Write-once media (such as CD-R and DVD-R) allow only one set of patterns to be burned to the disc surface. Rewritable forms of media (such as CD-RW and DVD-RAM) make it possible to effectively erase data that has been recorded, essentially returning the media to the equivalent of a blank unrecorded state. Data can then be rewritten hundreds or thousands of times.

To accommodate newer rewritable formats, some changes in the way that data is structured and accessed from disc have been instituted. Widespread acceptance of packet-writing as a means for performing incremental write operations to disc has made optical recording practical as a backup storage medium when a particular application requires that the media be reusable. Overcoming the write-once limitations of CD-R has helped gain even more acceptance for optical recording technology within organizations, and also helping lead to the demise of magnetic tape as the predominant backup and archiving medium. Optical recording continues to make inroads in the corporate environment, where networked DVD-RAM units are providing a replacement for earlier generations of CD-ROM and CD-R jukeboxes. The stability and flexibility of this method of data storage offers a suitable alternative in many instances for expensive RAID storage units and similar approaches to organizational storage requirements.

Understanding the Technology

To take full advantage of recordable CD and DVD technology, it helps to have some understanding of the foundations of optical recording. Many different elements combine to make optical recording possible—successful recording requires that each of the components in the process be tuned and operational. This includes the recordable media, the optical recorder, the I/O interface, the computer hosting the recorder, and the computer operating system. A weakness or deficiency in any one of these components can cause the process to fail.

The first choice you generally make when creating a disc is to choose the best data format for the data with which you are working. Two or three formats are commonly used, several others are used less frequently for specialized applications. Most of the premastering applications on the market require that you choose the appropriate disc format as step number 1, before the application will let you start selecting files and organizing directories to be written to the disc. Being able to intelligently select among the various possible formats for a CD or DVD project is the best way to get the recording process launched smoothly.

This chapter and the following two chapters provide a foundation for understanding the hardware, software, and data formats associated with recordable CD and DVD technology. Some of this material is fairly technical, so you may want to skip over it and return later if you need to sort out the practical differences between Red Book audio or Blue Book, for example, or to determine the best way to place photographic images on disc, or to get a better feeling for why one CD-ROM format may provide better performance for multimedia presentations than another. In any case, the information will be here when you need it, and, as with most works of this type, this book is designed so that you can skip around freely between chapters to access information as necessary.

Black Box Engineering

One of the techniques that has been used very effectively when engineering complex computer systems and internetworked applications is to reduce elements in the system to black boxes. You don't particularly care what goes on inside the box. From an engineering standpoint, you know that if you send a certain combination of signals into the box, you get a certain combination of signals back out. If you had be fully aware of every detail about what was happening inside the box, you'd never be able to complete any kind of complex project. By partitioning intricate systems into discrete functional elements, engineers and designers are able to focus on specific areas of a design, while being confident about the behavior of other parts of the system. This approach works very well and has made it possible to engineer astoundingly complex interactions, even when the working components are stretched across the breadth of the globe (as is the case with the World Wide Web).

In some cases, however, it helps to know what's going on inside the black box, particularly with optical recording technologies where you may be faced with making decisions as to what equipment to purchase, what media to purchase, or how to set up a networked disc array within your organization. For the benefit of those curious souls who would like to examine the inner workings of this box, the next few sections explore the inner workings of recorders and players, and also investigate the properties of the recording media.

Mastering the Levels of Complexity

Much of the design effort over the last few generations of optical recording hardware and software has been focused on ease of use. First generation recordable CD equipment was usable by engineers only. With recorder costs in the hundreds of thousands of dollar range and recordable media selling at $100 a pop, most of those in the business community and audio community did not have the stomach or cash resources to attempt the error-prone process.

Gradually, improvements in the design of the recording software simplified the process of burning discs, making many operations as simple as drag-and-drop file transfers. Performance improvements in the computers hosting the optical recording equipment eliminated the grueling process of fine tuning the system operation so uninterrupted recording could predictably take place. Intimate knowledge of the data formats became less important as software wizards guided new users through the premastering process, and new hardware and software drivers became capable of reading and writing all of the existing data formats.

The current generation of CD-R equipment fits this description in a number of ways. Software applications designed for use with CD-R equipment typically treat recorders as just one more drive in your system. You can transfer files to a recorder in the same way as you would copy a set of files to your hard disk drive. Multi-session write compatibility is necessary to accomplish this feat, but the software removes most of the intricacies of the transfer process and hides the differences inherent in the compact disc medium from your attention.

In contrast, DVD equipment is still maturing. Authoring tools, hardware interfaces, software drivers, and the standards themselves are being hammered on by a diverse army of forces intent on shaping the technology to suit their own individual ends. As with recordable CD, the recordable forms of DVD will eventually emerge from this evolutionary storm in more stable form. For the near term, the technology is primarily useful when you have control over the playback equipment (eliminating compatibility concerns), such as inhouse archiving and testing of DVD-ROM titles, DVD-Video projects, and DVD-Audio mastering.

Drag-and-drop file transfer may be suitable for storage and archiving, but if your goal is to create high-performance CD-ROMs or DVD-ROMs that integrate a variety of multimedia assets, or to ensure that users on different computer platforms will be able to gain the maximum benefit from a cross-platform presentation, the task becomes more demanding. Knowledge of file formats, understanding how to squeeze every spark of performance from the medium during playback, cross-platform challenges and solutions, all become paramount to the task of completing a polished and universally usable title on optical media. Business users of optical recording technologies face many of the same issues with which commercial developers must deal, although business users also encounter other issues and difficulties as well, as covered in later chapters.

Someday the recording and development process may be so completely seamless that anyone can create any kind of project, aided by software that makes it virtually impossible to create bad discs, but—until that day—those who understand the nuances of CD-R and DVD equipment and formats will have a definite advantage in the field.

Developer Considerations

While much of this book is targeted for business and corporate users, whose primary goal may be using recordable discs to store and distribute organizational data, many topics also address the unique challenges encountered by the developer—whether one individual or a team of artists, programmers, and technicians working to create the next best-selling interactive adventure game or a database of automotive parts. For the most demanding applications, knowing the underlying mechanics of optical recording becomes the solid foundation on which you can confidently construct a digital work, fully understanding the shifting and unpredictable tides to which your completed disc will be subjected. The variables encountered can be anything from irregular platform playback on different equipment, including quirks caused by different operating systems, to the emerging standards that offer developers exciting new capabilities, but strain compatibility.

To be successful developing and producing a title on disc, you should have a sense of the performance constraints of the medium, the techniques used for data storage, the file structures commonly used, the physical mechanisms of data access and retrieval, and those bus transfer limitations imposed by the host computer. Knowledge of the various standards (and the evolution of these standards), as described in Chapter 3 and Chapter 4, can also help eliminate many potential development problems.

Can you successfully create CDs or DVDs without becoming an expert in all these areas? Absolutely. In fact, Chapter 10, *Simple Authoring Techniques*, provides a number of tips on how you can create CD-ROMs and DVD-ROMs using simple tools and simple processes. These tips and guidelines should be helpful to everyone from the businessperson who needs to put a PowerPoint presentation on a CD for the next corporate meeting to the experienced developer who needs to best way to encode video to produce a training session on DVD-Video.

Dealing with Performance Issues

For all the promise of optical storage, and despite the fact that 20x, 24x, and even 32x CD-ROM drives have appeared on the market, you are still dealing with a storage device that must work vigorously to deliver data fast enough so that your system can play it back smoothly. Playback issues on DVD-ROM are comparable; similarities in the hardware of these devices force developers to consider many of the same issues that affect CD-ROM development.

Chapter 2

The performance issues escalate when you are intermingling video and sound and data and then trying to play back the entire mix in fluid fashion. The fundamental goal is to keep the laser read head from traversing large areas while trying to access files and to funnel data through the interface bus as soon as it is needed (or before it is needed, in the case of some caching utilities).

One of the goals of this chapter is to illuminate those underlying details that will affect your decisions as you are developing and mastering a CD-ROM or DVD-ROM. In many cases the level of detail presented may exceed your needs. If your goal is to create a CD-ROM that contains all of the word-processing correspondence files you've generated over the past 12 years, you may have no interest in what goes on inside the hardware and software, unless, perhaps, curiosity motivates you. For any uses other than just pure archiving (and even for some instances where you want to archive certain kinds of data), you will inevitably be faced with making some decisions based on the nature of CD-R, DVD-R, and DVD-RAM technology. Whether you're motivated by curiosity or by necessity, I hope that this technical discussion helps answer your questions.

The next two chapters, which discuss the standards in some detail, may also seem a trifle dry to those who are eager to start toasting blank discs. The same admonition, however, holds true as for the hardware discussion. Skip over this material if you must; you can always return to it when you have a pressing need to address or a strategic issue to resolve.

With all this in mind, prepare to enter the aluminum and polycarbonate recesses within the interior of an optical disc. The journey promises to be an enlightening one.

By the Light of the Shimmering Laser

Without the invention of the laser, there would be no CDs, CD-ROMs, or DVD-ROMs. Through its intense and narrowly focused beam, the laser provides a means for precisely detecting and registering the passage of millions of tiny impressions upon a rapidly spinning disc surface. The process generates no friction, since the detection is based on the measurement of phase shifts in reflected light. The technique allows considerable data compaction, since the carefully focused laser beam is capable of responding at the speed of light to extremely small variations in the disc surface. The process works not only because the laser mechanism is capable of reading back signals in this manner, but because a laser can be used to accurately record the disc impressions in the first place as a part of the mastering procedure. If you own a desktop CD recorder or DVD recorder, the equipment itself confirms the fact that these principles work quite well.

Light derived from conventional natural or artificial sources consists of photons that move in random wave patterns, even when they originate from light beams of the same frequency. Light beams of this sort of are considered incoherent—the uncoordinated waves travel freely in all directions. In contrast, light that originates from a laser is coherent. Coherent light moves in neat, orderly waves, which makes it useful for a host of scientific applications. Coherent light is created through stimulated emission, a fact which is noted in the acronym *laser: light amplification by stimulated emission of radiation.*

A laser beam is created when a source of energy is introduced into what is called an *active medium*. A pair of mirrors positioned on each side of the active medium is used to channel a portion of the radiation that strikes it. The active medium can consist of a gaseous mix (such as helium and neon) or ions within a crystalline matrix (such as is found in the gallium-arsenide lasers typically used in CD-ROM drives and recorders). The materials and the energy source used to stimulate the light determine the strength and intensity of the resulting beam. The lasers used within CD-ROM and DVD-ROM equipment are of extremely low power. If you boost the power sufficiently, more energy-intensive forms of the laser can used to burn holes in things like '59 Cadillacs and to melt roofing shingles.

The CD-ROM drive laser is directed at the spinning disc and the reflected light passes through a lens and strikes a photodiode. Data on the disc surface is encoded in the form of *pits* (indentations in the disc) and *lands* (the level surface of the disc). Logic timing circuits coupled to the photodiode can register the difference between the distance the light has traveled when it strikes the disc surface and the distance it has traveled when it strikes an indentation in the disc surface. This difference is detected as a *phase shift* in the light beam. This basic technique and some closely related variations make it possible to store and retrieve data from the surface of an optical disc.

As with all things digital, the pattern composed of pits and lands—relayed as a electronic string of 1's and 0's by the photodiode—can represent much more complex analog equivalents, such as the envelope of a soundwave or the passage of shapes and colors in a video sequence. Figure 2 - 1 illustrates the fundamental components of a CD-ROM drive (which are not drawn to scale).

Chapter 2

Figure 2 - 1 **Fundamental components of a CD-ROM drive**

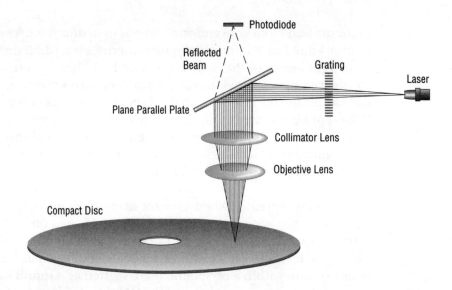

A Closer Look at the Disc

So far we've observed the principles that let us bounce focused light off a disc and detect any phase shift of the returning light waves. This simple principle makes the whole optical recording process possible. The nature of the recording medium, the optical disc itself, also bears further examination.

A compact disc measures 4.72 inches (120 millimeters) in diameter and weighs approximately 0.6 ounces. The standard supports variations, such as smaller format discs known as Card Discs that have a non-circular form factor. On a Card Disc two of the disc edges are removed; the disc rides in the central receptacle in a CD-ROM player. Smaller, 80-millimeter discs are also used in some applications.

The reflective disc surface, visible on the side opposite the label, is very striking (aesthetically speaking) for the prismatic effect demonstrated if you shift it under a light. You can see rainbow patterns by shifting the disc slightly under strong light.

One remarkable aspect of a compact disc is the density of the microscopic array of data patterns embedded within the disc materials. The ability to compress data to such a fine degree and read it back accurately gives the CD-ROM one of its defining characteristics: the capability to store 650MB of data (or 680MB of digital audio data). Translate this storage capacity to more commonly understood terms and the utility of this storage medium becomes very clear. You can store 600,000 pages of text or 74 minutes of music or the entire contents of a 20-volume multimedia encyclopedia. DVD exceeds this storage ability with a variety of disc types storing from 4.38 Gigabytes to 17.9 Gigabytes. The significantly greater storage capacity for DVD make it possible to store digital video versions of entire films or multimedia titles with contents designed for playback in several different languages.

The extra capacity available on a DVD-ROM comes primarily from the addition of extra layers within the disc subsurface. The number of layers varies depending on the DVD variation being discussed. For more details, refer to Chapter 4, *DVD Standards*.

Compact discs are primarily made of a form of plastic known as polycarbonate, a petrochemical marvel usually delivered to disc replication houses in the form of small beads, which are melted down to the consistency of honey and then injected into molds. The molds already contain the surface irregularities—the pits and lands—representing the data that was recorded by the laser on a circular glass master. When the fluid polycarbonate conforms to the mold, it acquires the same pattern of indentations. The resulting plastic disc, consisting of the bulk of the material from which a CD-ROM is composed, is called the *plastic substrate*.

As previously mentioned, indentations in the plastic substrate are called pits; they extend in a continuous spiral from the innermost area of the disc to the outer boundaries of the recording area. The level areas between the pits are called lands. A reflective coating of aluminum is sputtered over this surface, capturing the surface geography of pits and lands in precise fashion.

The reflective aluminum layer is extremely shallow, measured in units called angstroms. A single angstrom is one ten-billionth of a meter; the aluminum layer is a few hundred angstroms thick. To prevent this very thin layer of aluminum from being scarred and scratched, which would essentially obliterate the data patterns residing on it, a clear protective layer of lacquer is laid down over its surface. The laser beam can read through this layer without difficulty to detect the microscopic impressions in the disc surface.

Chapter 2

The final step in the disc creation process is to apply artwork to the surface, either through a silk-screen process or lithographic printing, as discussed in Chapter 16, *Disc Replication, Printing, and Packaging*. The disc can then be packaged in any one of a number of different packaging types, as discussed in same chapter.

Turning Pits into Data

To successfully communicate by means of nothing more than a series of pits in a disc requires a good deal of computer processing and some high-technology wizardry. At no point in the data exchange process does the laser's read mechanism ever touch the disc surface—all information is conveyed by reflections of the laser. The data detection process is based on this principle: the light beamed from the laser takes a certain amount of time to return when it is reflected off the lands, but it takes longer to travel when it is swallowed up and reflected by the pits. The depth of the pit is engineered to be one-quarter the wavelength of the laser light. If the reflected beam from the pit cancels out the beam from the land, this indicates a signal transition. Signal transitions (marked by either the beginning or the end of a pit) represent binary 1's. If there is no signal transition, this indicates a binary 0. Figure 2 - 2 shows the relationship between transitions and the resulting binary values.

Figure 2 - 2   **Signal transitions within a track**

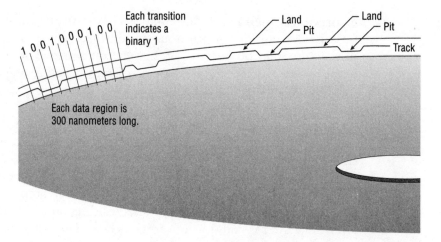

This may lead you to wonder how the microcontroller embedded in the CD-ROM drive can determine how many individual 0's are strung out in a long sequence with that has no transitions. Clearly, there must be some type of clock signal to distinguish between 0's.

To maintain synchronization while reading the data patterns, the CD-ROM drive uses a self-clocking mechanism that is commonly found in hard disk drives; it's called *Run Length Limited*. Because data exists within finite divisions on the spiral track—each data division extends approximately 300 nanometers—the CD-ROM microcontroller can produce regular clock signals by synchronizing to the speed of the disc rotation and the occurrences of transitions. Although many forms of data storage use an 8-bit sequence for storing data bytes, the CD-ROM requires a 14-bit pattern to avoid creating combinations of 1's and 0's that would prevent decoding of the stored data. This modified form of storage is called EFM (Eight-to-Fourteen Modulation). An additional 3 bits, called *merging bits*, act as separators between the 14-bit patterns, resulting in a 17-bit pattern to represent a single 8-bit byte of data.

Another significant division of data at the bit level is the *frame*, which consists of 588 bits. The frame encompasses a collection of bits: some of them signify data, others allow the laser to be synchronized with the spinning of the disc, and still others contribute to the error-correcting capabilities within the CD-ROM equipment. Of this collection of bits, only twenty-four 17-bit units (408 bits altogether) can be translated into 8-bit bytes. As you can see, many additional bits are needed to convey the information contained in a mere two-dozen data bytes.

Physical Differences in Recordable CDs

Recordable CDs differ in some important ways from standard, mass-replicated CDs. When a blank, recordable CD is positioned in a recorder for the first time, a spiral groove is already embedded in the polycarbonate substrate, indicating the path that the CD recorder laser head will track while burning the data pattern. Instead of the aluminum layer used in standard CDs, a recordable CD has a fine gold layer. Gold not only offers a highly reflective surface, but it resists corrosion much more effectively than aluminum.

Another layer, composed of an organic dye sandwiched between the polycarbonate substrate and the gold layer, serves as the recording medium. As with standard CD-ROMs, a protective lacquer surface provides a degree of scratch resistance and durability to the outer layers of the disc. Figure 2 - 3 compare the materials used in the layers for CD-ROM discs and CD-R discs.

Chapter 2

Figure 2 - 3 **Compact Disc layers**

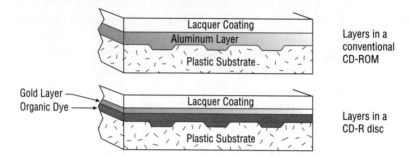

Note: The CD-ROM drive and recordable CD drive share fundamental principles of operation, but recordable CD equipment uses a more highly focused, more intense laser beam capable of searing impressions in the dye layer of a blank compact disc.

When recording begins, the CD recorder's laser beam penetrates the plastic substrate and heats the dye as it pulses a string of data patterns to the disc. The dye rises where it is heated, forming mounds that protrude into the gold layer. The resulting mounds correspond with the pits in a conventionally recorded CD-ROM. The laser beam is deflected when it strikes them and they are read by a CD-ROM drive in the same manner as pits. Rather than confusing an entire industry, by calling them "mounds" or "bumps," the mounds are still typically referred to as "pits." Figure 2 - 4 illustrates this concept.

Figure 2 - 4 **Writing data with a laser**

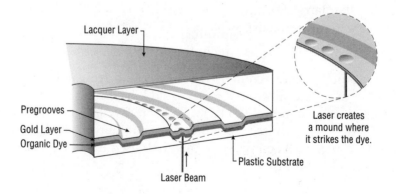

Compact discs are exceptionally durable when compared to hard disk drives or to diskettes, but there is a limit to their durability. Although they are immune to magnetic fields (since the data is not stored in a magnetic format), discs can be damaged by rough handling or extreme conditions. Most data formats include coded information that can fill in data bytes lost to scratches or debris, but discs can still be rendered unreadable by abusive treatment. The long-term archival properties of the medium are subject to the perils of oxidation, although this is more of a factor in commercially produced discs with their aluminum data layer, rather than CD-R one-offs with their gold reflective layer.

Optical storage also offers a significant environmental savings. Short-term paper-based materials (telephone books, catalogs, corporate telephone records, and so on) can be reduced to an extremely compact format. A single disc can contain the equivalent to a copy of all of the residential and business phone directories for the entire country. Choosing discs over paper for storage can clearly save a considerable number of trees from destruction in the pulp mill. By any method of accounting, it is easier to produce and dispose of a half-ounce piece of plastic and aluminum than the equivalent several hundred pounds of telephone directories.

Optical Storage Advantages

The CD-ROM (Compact Disc-Read Only Memory) belongs to a family of optical data storage devices. Unlike electromagnetic methods of storage, which depend on reading and writing polarized signals on a magnetic surface, optical storage uses a laser beam to detect impressions in the surface of a reflective disc. Another member of this family, the WORM drive (Write Once, Read Many), initially made some inroads in the archiving and mass storage game, but incompatibilities and dissimilar standards handicapped its utility as a common medium for exchanging data; it remains an active archival option, however, using 12-inch discs that can hold up to 30MB. The magneto-optical (MO) drive also belongs to this family; it offers the advantage of being able to re-record data on the same media using optical techniques. Magneto-Optical drives still thrive in specialized environments—such as large art departments and photographic agencies—where a reusable medium comes in handy for storing very large quantities of data. Media costs, however, are significantly higher than the recordable forms of CD and DVD.

The defining characteristics of the CD-ROM—its light weight, durability, long media life, and considerable capacity—have led to its increasing popularity as an archiving medium for government and business data. Tape backup has become a less popular choice for archival storage, largely due to the inaccessibility of data in the linear storage format and the lifespan issues presented by magnetic signals that diminish over time. Looking at the full range of data-storage options, CD-ROMs consistently rank among the best data-storage bargains.

Note: Another benefit of the CD-ROM as a storage medium is that it shares its lineage with the audio CD. See *The Audio Origins of CD-ROM* on page 62 for more background information. This means that CD-ROM replication can take advantage of the cost reductions that have evolved over more than 20 years of producing audio CDs. Often, the same equipment used to master and replicate the hundreds of millions of audio CDs being produced today can serve double duty as a production platform for CD-ROMs. The evolution of the processes and technology for accomplishing high-volume replication has helped drive the costs of manufacturing CD-ROMs down to levels where they now represent one of the best values for cost-per-megabyte of storage in the industry.

Beyond the twin advantages of durability and compactness, the CD-ROM has another significant advantage over other forms of data storage: CD-ROMs provide *random* access to any information stored on them. In comparison, linear methods of storage, such as streaming tape devices, cannot easily access a block of data unless it is in close proximity to the current position of the read-write head over the tape. If you're attempting to read a block a data near the end of the tape and the read-write head is located at the start of tape, the drive motor is going to be spinning for a few moments to get there. If the next piece of data is back at the beginning of tape, even more time is consumed getting back.

To achieve random data access, CD-ROM drives use a servo mechanism to position the laser's read head, which is sometimes mounted on an apparatus that looks like a miniature sled riding on a pair of rails. This assembly ranges over the surface of the disc, on command, to locate blocks of data. By referring to a directory area that lists the file contents of the CD-ROM and indexes the paths, any given piece of information can be quickly accessed and retrieved. This form of random access has certain similarities to the technique used to locate and retrieve data on a hard disk drive, but also some notable differences. For more details, refer to *Data Storage on CD-ROMs* on page 35.

Access-Time Considerations

The time required to pinpoint a required bit of information, prior to retrieving it, is referred to as *access time*. Access time is such an important part of the overall performance and responsiveness of a CD-ROM drive that it is usually featured prominently in the drive specifications and product advertising. Access times for CD-ROM drives are noticeably inferior to hard disk drives; while it is not uncommon for a hard disk drive to have an access time below 8 milliseconds, the fastest 32x CD-ROM drives can typically do no better than 100 milliseconds.

If you're presenting material on CD-ROM, particularly material with media-rich content, you'll be working continually to overcome the inherent sluggishness of the CD-ROM for delivering data to your system (and your customers' or clients' systems). The problem is becoming less severe with the universal proliferation of 24x and 32x CD-ROM drives, but since the nature of multimedia content is also pushing the bandwidth envelope, performance considerations continue to be an important issue for CD-ROM developers. A number of tips to address this problem appear in later chapters.

Data Transfer Rates

Another key consideration in the ability of the CD-ROM drive to deliver information is the *data transfer rate*. In many ways, the data transfer rate is more important than the access time in assessing performance, since it measures the ability of the CD-ROM drive to move data from the disc surface to the host computer. Mainstream 24x CD-ROM drives theoretically deliver data at an impressive 3600K bytes per second. However, if you develop a CD-ROM in anticipation of this kind of performance, you may suffer seriously degraded performance on the multitude of low-speed drives still installed in the marketplace. In many cases you will want to deliver content to the largest installed base of equipment, a factor that imposes rather stern limitations on the scope and execution of your project. While the bulk of the installed base is rapidly shifting from 12x and 16x drives to 24x and 32x drives (largely because of price reductions in high-performance equipment), you can never reliably count on the majority of your audience having high-performance CD-ROM drives.

At this point in time, your best bet is to optimize your CD-ROM production to perform at its best on a 12x drive. This may handicap you in some ways, but it will also challenge you to find the best techniques to keep data barrelling through the pipe. For certain types of productions, perhaps those where the mass market is less important and you have control over the presentation tools, you may want to optimize for 24x speed or even 32x speed performance. If you do, make sure that your playback equipment can match the anticipated performance level.

Permanence of Data

Optical storage offers a certain degree of precision and permanence that is difficult to obtain in magnetic media. Because the optical data is represented by indentations in the surface of the recording medium, the only dangers to its permanence are corrosion or physical distortion of the surface (such as might be caused by scratching or exposure to intense heat).

Data stored on magnetic media, in comparison, undergoes a slow, inevitable diminishing of the signal from the first day it is recorded. Adjacent polarized fields—such as the encoded signals on the surface of a hard disk drive—tend to influence each other over time and are subject to drastic change if passed through a strong magnetic field (like the one generated by your handheld vacuum cleaner, if you get overzealous about cleaning the interior of your computer cabinet). The lifespan of data stored on magnetic tape (depending on who you talk to) ranges from about 6 to 12 years.

Estimates for recordable compact disc lifespans generally suggest a century of stable data storage. CD-ROMs created through mass replication techniques don't fare quite as well for long-term archiving, primarily because of differences in the substrate materials. Estimates range from about 10 to around 25 years for mass-replicated CD-ROMs. Since many of the discs being used for computer data have barely been around that long, it may be some time before we start getting real-world reports on CD-ROM longevity. The enemy, in this case, is aluminum corrosion. When the microscopic pits in the disc surface start competing with the microscopic pits of aluminum corrosion, your data may be in serious jeopardy. The gold layer used in the blank discs manufactured for CD recorders offers much more stability—this is the primary reason for the improved lifespan of CD-R media.

While it may be reassuring to archivists that CD-ROMs can effectively store data for more than 100 years, industry pundits have pointed out that it is unlikely that the same data storage playback equipment will be available that many years down the road, so the archival properties are essentially moot. It's more likely that valuable digital data—the cultural records of our age—will be shuttled from format to format as new storage methods evolve that can encompass greater and greater amounts of data.

Data Storage on CD-ROMs

The data stored on CD-ROMs does not exist in the familiar concentric circles (referred to as tracks) of the hard disk drive world, but instead in a continuous spiral like the phonograph records of days past (as shown in Figure 2 - 5). CD-ROM data is partitioned into tracks. This similarity with hard disk drive formatting standards allows the CD-ROM to be easily integrated into a framework containing other "track-based" devices. To accomplish this kind of integration, some logical leaps are required to overstep the boundaries of the physical reality of the data storage method.

Figure 2 - 5 **Tracks and spirals**

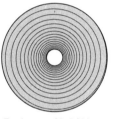 

Tracks on a CD-ROM are a continuous spiral.

Tracks on a hard disc are concentric circles.

Before getting into the architecture, let's compare the characteristics of CD-ROM and hard disk data storage. The first major difference between hard disk drives and CD-ROM drives involves the *rotation* of the media. The magnetic disks that contain the data on a hard disk drive spin up to a predetermined speed and they maintain that same speed through disk operation. Tracks are circular and they extend from an area of densely packed data, near the center of the disk, to less densely packed data, near the outer edge of the disk. Each track contains an identical number of sectors that act as addressable units for locating files and portions of files. The regularity of the arrangement of tracks and sectors and the constancy of the rotation speed allows the controller logic that positions the hard disk read-write head to predict when and where data can be found on a given track within a particular sector. Once it is given the track and sector address of a particular piece of data, the head can be swung into position and can begin reading that data very quickly. The term *Constant Angular Velocity* is applied to this technique and you may encounter the acronym CAV as the shorthand reference to it. CAV is largely responsible for the superior data-access rates associated with hard disk drives, which have improved to the point that modern drives can locate and prepare to read a specific piece of data anywhere on the disk within 2 to 10 milliseconds.

While this technique for storing data on hard disks allows fast access, it sacrifices some storage capacity. On the other hand, the storage techniques for CD-ROM sacrifice some access speed while delivering greater storage capacity. Tracks are artificial constructs on the CD-ROM, since you're working with a continuous spiral of data. Nevertheless, that the spiral is partitioned into sectors that are equally spaced and, like the hard disk drive, contain address information to help locate data. Files are never fragmented as they often are using hard disk storage techniques; they are laid out in contiguous sectors at the time the compact disc is recorded. Several of the innermost sectors contain detailed information about the location of each CD-ROM file in a directory table and a path table.

The drives varies the speed at which the compact disc is rotated to ensure that data can be read from the disc at a constant rate. To capture data at the outer edges of the disc requires a different rotational speed than to capture it at the inner edges, so the disc is continuously being sped up and slowed down to accommodate data retrieval at an unvarying rate. This technique is called *Constant Linear Velocity* (CLV).

The tracks containing CD-ROM data are packed much more closely than on a hard disk drive—there are 16,000 of them within the span of an inch as compared to only a few hundred within an inch of hard disk surface. The manner in which files are organized on the CD-ROM and the access constraints imposed by the inherent sluggishness of the CD-ROM drive form the basis of two of the cardinal rules for the CD-ROM developer:

- Always place the most frequently needed files near the center of the disc. That is where the files can be most quickly accessed.
- Always place files in close proximity on the disc when an application will need to access them consecutively. Keeping related files in close proximity will greatly reduce the time required to retrieve them while the application is running.

File placement becomes especially important for mixed-mode applications as well as applications where many resources are being pulled off the disc in rapid sequence (such as full-screen video or complex interactive multimedia). Many software applications for producing recordable CDs provide some technique for performing optimization, organizing corresponding and related files for the best possible access and smoothest playback. Even though improvements in equipment performance have made this consideration somewhat less important, most of the developers I talked with mentioned file placement as a useful technique for CD-ROM development.

Some premastering software applications relegate the placement of files to a less-important role or hide access to this feature from you. These kinds of software applications may not have the flexibility necessary for professional-level CD-ROM recording. In later chapters, techniques for optimizing discs are presented and the premastering packages that best deal with optimization issues are discussed.

Logical and Physical Components

The standards that specify the storage and organization of files on CD-ROM have both logical and physical components:

- The logical formatting dictates two things: the measurement units to be used and the system of organizing the data. Together, these make the data accessible to the computer to which the CD recorder is connected.

- The physical formatting determines how the data is structured on the physical media—in other words, the mechanics of how data is written and retrieved from the surface of the compact disc.

These components are at the heart of the various standards that have arisen for controlling the production of CDs, from audio CDs to computer CD-ROMs.

Audio Origins

The simplest of the compact disc formats (defined by the *Red Book* standard) is for audio CDs. Audio CD architecture, as opposed to CD-ROM architecture intended for computer use, includes only two layers: 0 and 1. The lowest layer—Layer 0—is the logical layer; it defines the bit structure, specifying the manner in which bits are combined to form bytes. Layer 1, the physical block structure, defines the basic addressable unit of the CD—the *block*, which is also sometimes called a *sector*. A block is subdivided into *frames*, and each frame consists of 24 bytes. With 98 frames to the block, you can see that each block contains 2,352 bytes; this value remains consistent from standard to standard, although the manner in which these 2,352 bytes are used differs in some important ways, as you'll see when we discuss the specifics of each standard in Chapter 3, .

Given that there are so many bytes of information in each block, how does the drive determine how to find individual pieces of data within the block? Here is where the audio roots of the compact disc become particularly evident. Data on an audio disc is located by means of a block address with a time offset. In other words, an address is specified by providing the format *minute:second:sector*. This format is referred to as the *absolute time scheme*.

The beginning of the very first block on the compact disc has the address 0:0:0. Ten seconds into that block, the address would be 0:10:0. Similarly, the address for data that is ten seconds into the second block would be 0:10:1. At standard transfer rates for audio CD, 150 kilobytes per second, 75 blocks can be read in the span of a second. Using the absolute time format of *minute:second:sector*, any block located on the compact disc can be located and the musical data within the block pinpointed.

So far we've described an addressing scheme that lets us start playing music located anywhere on the compact disc, but what happens when computer data is involved?

Computer CD-ROM Standards

Computer CD-ROM architecture consists of four layers: two layers that specify the physical ordering of data on a disc and another two layers that standardize the logical layout of files and other information required to locate files. Layer 0 is left unchanged from the audio CD standard, but the specs for Layer 1 have undergone a few improvements to make data storage less error-prone (more details on this topic appear in Chapter 3, *CD Standards*). The next two layers, Layer 2 and Layer 3, specify the organization of *files* on a CD-ROM: Layer 2 defines the logical organization of sectors on the disc, and Layer 3 creates a logical file structure. Taken together, Layers 0 through 3 form the basis for storing and accessing digital sound and computer data on compact disc.

These additional specifications were tacked on in the *Yellow Book* standard (but, fortunately, tacked on thoughtfully) to expand the capabilities of the compact disc from merely a digital phonograph record to a serious data-storage medium. The additions that the Yellow Book standard added to the error detection and correction scheme place the CD-ROM at the top of the list for data reliability in storage devices.

Since the advent of the Yellow Book standard, the specifications have needed to be expanded and redefined in order to support previously unimagined applications for CD-ROM. This adaptation of the original standard has resulted in some incompatibilities, but compared to other standards battles in the computer industry, the evolution of CD-ROM standards has progressed in reasonably polite and orderly fashion. Consequently, you can pop most off-the-shelf CD-ROMs into your drive with confidence that they will perform as expected. However, as a developer, you need to be aware of the changing standards in order to choose the appropriate one for each project to ensure that you reach your intended audience. You must also take into account both your own and your customers' playback devices and computer platforms. These considerations and other related issues are discussed in the chapters that follow.

DVD Technology

More is better, at least when it comes to data storage, and the goal of DVD from the beginning was to expand the storage capabilities of a disc that remained the size of a standard compact disc. The two obvious ways to do this were to compress more data within the span of an individual track and to add additional layers within the disc itself, all without expanding the actual size of the media. As a result, enterprising manufacturers have produced a wide array of DVD implementations, ranging from single-sided, single-pressed discs to multiple layered DVD-18 discs that store a remarkle 17.9 Gigabytes of data.

Commercial digital video discs, such as the kind you buy DVD movies titles on, store as much as 17 billion bytes of data on the disc surface. This equates to a storage capability sufficient to handle nine hours of audio and video content, or 26 times more data than can be stored on an audio CD. The quality of the content is significantly better than other forms of video storage released to the consumer marketplace, surpassing the quality of the previous standard bearer—the Laserdisc. The ability to deliver multichannel audio has led to the creation of *theater* entertainment systems that position speakers around the viewer and produce a sensation of being in the midst of the onscreen action. Up to six channel digital sound can be delivered, generally subdivided for the following uses:

- A center channel for dialog
- Left channel for music
- Right channel for music

- Left rear channel for effects
- Right rear channel for effects
- Bass channel

Most of the information stored on a DVD-Video disc is in compressed format, with MPEG-2 being used for the video content and Dolby Digital (AC3) for the audio content. MPEG-2 digital audio is also sometimes used for audio material stored on disc. These compression technologies make it possible to provide the prodigious storage capabilities of the media, while delivering superior audio and video, since the loss due to the compression is minimal.

DVD technology was designed to be backwards compatible with CDs, which presented an immediate technical challenge since the pits embedded in a CD occur at a different level on the disc surface. There are also four different possible surfaces within a DVD disc, requiring that the read laser in a DVD player be able to adjust its focus to retrieve data from the various layers.

This technological difficulty was solved through the use of different lenses, including holographic lenses that can simultaneously focus on more than one distance at a time. The built-in backwards compatibility also embraces VideoCD 2.0, an earlier format for distributing video material in compressed MPEG format. VideoCDs achieved some success in Japan and Europe, but never made significant inroads in the United States.

The first step in increasing capacity over what is available on the audio CD is to tighten the spiral and reduce the size of the pits used to form the data impressions. This is illustrated in Figure 2 - 6.

Figure 2 - 6   **Comparison of pit spacing in CDs and DVDs**

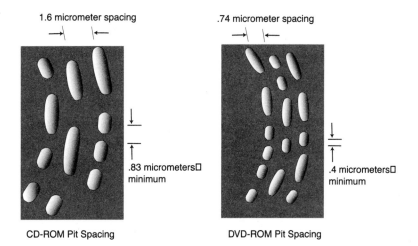

The second technique for increasing the storage capacity is to add additional layers—up to 4 layers total, two on each side of the disc. The laser beam focus is adjusted as necessary to access data on individual levels. Improved optics are required to achieve accurate data reading, as well as a reduction in the wavelength of the laser beam reflected off the disc surface.

The single-sided, single-layer DVD consists of two 0.6 millimeter polycarbonate substrates that are bonded together to form a disc that is 1.2 millimeters thick. The stamped surface of a single-layer disc is coated with a layer of aluminum through a process called *sputtering*. This aluminum forms the reflective surface upon which the laser beam is used to detect the data pattern. This type of disc has a capacity of 4.7 Gigabytes and is referred to as a DVD-5 disc (as shown in Figure 2 - 7).

Two methods are typical used to bond the two substrates together:

- Hot-melt method: A thin coat of melted adhesive is spread over each substrate and then the two surfaces are bonded by means of a hydraulic ram. This is the least expensive method of bonding.

- UV method: A thin layer of lacquer is distributed over the disc surface to be bonded, either by rotating the disc or through silk-screening. Ultraviolet light is then applied to harden the lacquer.

While this method is more expensive, it forms a bond that is resistant to temperature extremes.

The label surface of this disc can be printed using conventional printing techniques or silk-screening. Impressions molded into the blank substrate can also be used as a substitute for printing a label.

Figure 2 - 7 **Layers within a DVD-5 disc**

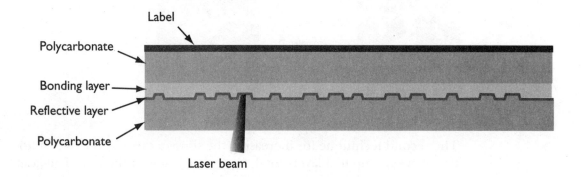

A DVD-9 utilizes two separate layers of data. The data layer closest to the laser is composed of a semi-reflective coating that enables the laser to focus on it for data reading or to focus through it to the next higher layer to read the data from the more reflective surface. The capacity of the DVD-9 disc, shown in Figure 2 - 8, is 8.5 Gigabytes.

The two substrates of a dual-layer disc are bonded together using one of two methods:

- By means of an optically transparent adhesive film that affixes the layers to each other

- Through a photopolymer material that combines the second layer on top of the first on a single substrate which is then bonded to the blank substrate

Conventional methods of lithographic printing or screened printing can be used to complete the label.

Figure 2 - 8 **Layers within a DVD-9 disc**

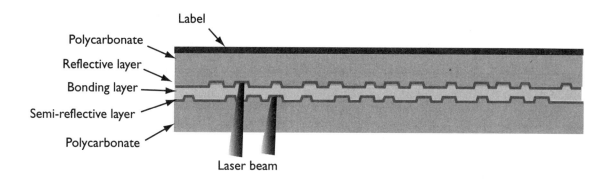

A DVD-10 disc has data on both substrates, using reflective layers on each so that the disc must be physically flipped over in the DVD player or DVD-ROM drive in order to read the data on the second side. This format has a capacity of 9.4 Gigabytes. Since the laser must be directed through both surfaces of the disc, no label is applied to either surface—this is a quick way to recognize this particular format, shown in Figure 2 - 9.

Figure 2 - 9 **Layers within a DVD-10 disc**

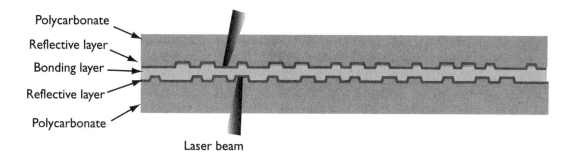

The most complex DVD format, and the most expensive to manufacture, is the DVD-18 disc, which includes both dual sides and dual layers, as shown in Figure 2 - 10. The two layers on each side must be manufactured on a single substrate. One layer is created on a substrate using a conventional stamper to produce the data pattern and then a second stamper creates a data image on a photopolymer material, which is then

Chapter 2

affixed to the substrate. This same process is followed for the second substrate, which also contains two layers of data, providing a total capacity of 17.9GB. As with a DVD-10 disc, this type of disc must be turned over in the player or the DVD-ROM drive in order to access the data on the second side. Similarly, neither surface can be printed with a label, since it must offer clear access to the laser for data reading.

Figure 2 - 10   **Layers within a DVD-18 disc**

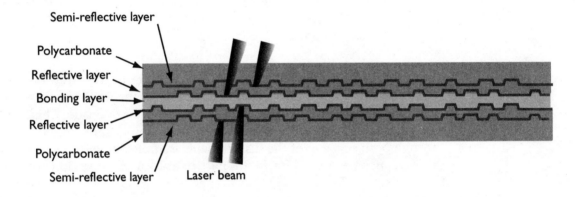

Storing Content on DVD

As you can see from the previous description, the addition of new layers to a disc, each with its own microscopic spiral of data, adds to the complexity of manufacturing. For this reason, and the additional costs associated with multilayer disc manufacturing, many commercially released DVD titles utilize only a single layer. This is sufficient to store a typical 135-minute film. For longer works, or for including multiple movie titles on a single disc, additional layers can be added to the disc to achieve the desired capacity. Approximately 2 Gigabytes of storage is required for each hour of video material compressed with MPEG-2. Title producers can determine the amount of storage needed for a project and then master the disc accordingly.

Unlike a compact disc, which employs a single substrate, a DVD is composed of two 0.6 millimeter substrates that are combined to increase the rigidity. The additional rigidity is also important for the overall disc balance and reducing the amount of wobble while the disc is spinning—both critical characteristics for ensuring accurate reading of data from the medium.

Reading All Types of Discs

In many ways, the DVD is an extension of the data storage techniques originally perfected for CD-ROMs and CDs, but with its own special characteristics. Just as the vast majority of CD-ROM drives are capable of playing audio CDs, DVD-ROM drives have been designed to be backward compatible with CDs and CD-ROMs. The following table summarizes the similarities and differences between the two types of discs.

Table 1: Comparison of CDs and DVDs

	CD	DVD
Diameter	120 millimeters	120 millimeters
Thickness	1.2 millimeters	1.2 millimeters
Data capacity	680 Megabytes	4700 Megabytes
Layers	1	1, 2, 4
Track pitch	1.6 nanometers	0.74 nanometers
Minimum pit length	0.834 nanometers	0.40 nanometers
Laser wavelength	780 nanometers	640 nanometers

The optical pickup designed for use in a DVD unit is mounted on an arm that positions the laser beneath the disc surface during playback. As you can see from the previous table, the required laser wavelength is different for CDs than it is for DVDs. One technique for handling this difference is to use a twin-laser pickup that features completely separate laser and lens fixtures. If the DVD player or DVD-ROM drive is attempting to read a CD or CD-R, it uses the fixture optimized with a laser wavelength for these media types. For DVDs, DVD-Rs, or DVD-ROMs, the unit employs the laser and lens with the wavelength optimized for DVD media.

A focusing control adjusts the depth of focus to be able to read the individual DVD layers. For a DVD, layer 0 is about 0.55 millimeters above the bottom surface of the disc. The second layer, if present, is another 55 micrometers higher. In comparison, data that is embedded in a CD or CD-ROM appears approximately 1.15 millimeters above the bottom surface of the disc. This difference, as well as the different laser wavelengths required for reading the data, is the reason that separate lens and laser fixtures must be used for DVD and CD media.

Writable Forms of DVD

Writable forms of DVD come in several varieties, a fact which has generated considerable confusion among those interested in obtaining a equipment to record to DVD discs. The two main categories are:

- DVD-R—as defined in Book D of the DVD Forum's specifications, this is the write-once form of the media with a standard capacity of 4.7 Gigabytes. The recording surface is a dye layer. Recorded discs are playable in standard DVD-ROM drives.

- DVD-RW—similar to CD-RW, this approach relies on phase-change technology to support the erasing and rewriting of data. The standard capacity is 4.7 Gigabytes.

- DVD-RAM—defined in Book E of the DVD Forum, this format also uses a phase-change recording layer, which may be single-sided or double-sided. Capacity is 2.6 Gigabytes per side. The RAM stands for Random Access Memory.

- DVD+RW—a latecomer to the rewritable market, this format achieves 3.0 Gigabyte capacities using a phase-change approach to storage. A 4.7 Gigabyte version of the media is also in the works.

Early in the design process for the writable form of CD-ROM (CD-RW), progress was hampered by the incompatibility of the rewritable discs with the majority of CD-ROM drives. This problem was overcome by the introduction of the Multiread specification for CD-ROM drives, an extension which ensured that drives certified as MultiRead ready could retrieve data from CD-RW discs.

A similar effort is underway to produce a MultiRead specification that encompasses the range of writable forms of DVD, so that the various formats will be readable in Super MultiRead-certified DVD-ROM drives. For the near term, however, if you want to produce a disc that will be readable in the majority of DVD players and DVD-ROM drives, use the DVD-R format for recording.

DVD-R for Write-Once Applications

DVD-R media and recorders produce discs that are suitable for premastering of DVD-ROMs or DVD-Videos, as well as discs intended for data distribution and exchange, document imaging, and archiving. The specification crafted by Working Group 6 (WG-6) of the DVD Forum includes provisions for single-layer, single-sided media or single-layer, dual-sided media. As with CD-R media, both 12-centimeter and 8-centimeter disc sizes are supported, although the large majority of applications will favor the more common 12-centimeter disc size.

Two polycarbonate substrates—one containing a dye layer and reflective coating and the other blank—are bonded together to produce a 1.2-millimeter thick disc for single-sided DVD-R applications. The first forms of DVD-R media used only cyanine dye, which appears violet on the recording side of the recordable disc.

A spiral pregroove extends from the center of the disc to the outer diameter to act as a guide for the laser during recording. A slight wobble in the pregroove in a pre-established pattern generates a frequency used as a carrier signal; the timing information helps regulate servo motors, tracking of the laser assembly, and focus of the beam. Land pre-pits molded into the substrate provide address information and pre-recorded data, used to initiate write operations.

Pulsed laser beams directed at the dye in the pregroove form impressions by searing variable length marks in the dye surface. These marks, consisting of deformation of the substrate material and bleaching of the dye, serve the same purpose as pits in a pressed DVD disc. Areas in the pregroove that are not exposed to the pulsed laser are interpreted as lands.

DVD-R recording requires a more complex write strategy to establish the appropriate lengths for the pits, which are approximately half the size of those on a CD-R disc. The spacing between the pits and lands within the spiral data pattern is also significantly less than on a CD-R disc. To compensate for the extra precision required during write operations, the laser pulses are very carefully controlled, both in terms of intensity and duration. During recording, the laser is rapidly modulated between the power setting required for writing and the setting used for reading to avoid overheating the media surface and to regulate the size of the mark seared in the dye. A technique known as Optimum Power Calibration (OPC) is used to perform test write operations to a specified calibration area on the recordable media surface and then to read back the test pattern and adjust the laser power settings to match the recorder to the media. Given the extra precision required for recordable operations using DVD-R, this feature becomes a highly desirable addition to any recorder and helps ensure the most consistent results when performing disc recording.

First generation DVD-R media offered 3.95 Gigabyte capacities and approximately 3.68 Gigabytes of usable space (considering the overhead required for lead-in and lead-out areas and other file system data). The recordable capacity of the second generation DVD-R discs is 4.7 Gigabytes, of which approximately 4.38 Gigabytes is available for data storage.

Data transfer rates for recording DVD discs are based on a nominal 1.32 Megabytes per second rate, which is considered 1x speed. At this data transfer rate, completing the recording of a 4.7GB DVD-R disc requires slightly less than an hour.

DVD-R serves a critical role in project prototyping for developers and title producers, since it is designed by definition to be playable in standard DVD players and DVD-ROM drives. Early recorder costs were in the $17,000 range, but as was the case with CD-R equipment, costs have been steadily declining. Second generation equipment, such as the Pioneer DVD-S201, are close to the $5000 mark. Blank media costs are approximately $40 for 4.7GB discs and $35 for 3.95GB discs.

Those who followed the development of CD-R technology witnessed the difficulty inherent in maintaining compatibility given the many variables in recordable media, playback equipment, recorders, premastering software, and so on. It took several years for all these varying characteristics to be tamed and controlled in such a way that recorded discs could be freely distributed among the vast majority of CD-ROM drives. A similar evolution is taking place with DVD-R as manufacturers, engineers, and developers refine the tools and techniques used to burn data in discs. Early adopters of this technology should be prepared for a variety of trials and tribulations as the compatibility problems are worked out and distribution becomes more universal.

The Zen of Data Flow

There is one aspect of CD and DVD recordable technology in its present state that you should be thoroughly aware of before you begin—the need for maintaining a consistent, unbroken flow of data from the time when you first start recording to the finalization of the disc. Successful recording demands delivering the data to the CD-R or DVD-R at a speed equivalent to the recording speed without interruptions. You can't go back and correct data that has been erroneously recorded—once the laser strikes the dye layer, the impression is formed and the result can't be erased. Rewritable forms of optical recording (CD-RW, DVD-RAM, and so on) don't suffer from this same limitation, but they also don't produce discs that are suitable for mastering or for widespread distribution to other platforms or systems.

The traditional allies of computer data-transfer techniques—high-speed bus interfaces, buffers, and caches—prove useful in this regard. Since in most cases the data you'll be recording will be initially stored on hard disk, your hard disk must be of reasonably modern vintage to support the necessary transfer rates. Many CD recorder software packages recommend data access rates of 16 milliseconds or better (in other words, less than 16 milliseconds) to support an uninterrupted data flow once recording starts. If you're pulling files from different locations on your hard disk, rather than sequentially from a dedicated partition serving as a virtual image of the CD-ROM, be sure to defragment your disk drive before recording. This helps ensure that the files will be transferred smoothly.

Most CD recorders and DVD recorders have some type of built-in cache buffer. This type of buffer can compensate for slight interruptions in the data flow from the hard disk. Since we're talking about half a Gigabyte or more being transferred over ten minutes to an hour, there are obviously a lot of data bits moving through the cable, resulting in a great potential for many things to go wrong. As recording speeds have gradually risen from 2x to 4x and now to 6x and 8x, many manufacturers have increased the size of the recorder buffer, with many units including a 2MB built-in buffer. Recorder software often sets aside an area on disc to use for buffering the data flow during transfers from the system to the recorder. This provides an additional level of insurance during the recording process to keep the steady flow of data from being interrupted. As discussed in later chapters, you have a number of options when you are recording to minimize the chance of failures, including turning off all necessary processes (screen savers, network monitors, and so on), disabling hardware on the same bus as the recorder (scanners, digital cameras), organizing your disc data by defragmenting your drive, and performing test recordings before actually burning a disc.

Some professionals recommend using two SCSI host adapters, one for the hard disk drive and another for the CD-recordable unit, but I've also seen contrary recommendations. Some premastering applications suggest you should only use a single host adapter. No single set of guidelines can cover every conceivable hardware configuration, and even the experts sometimes disagree on the right approach. Try to take each different recommendation with a grain of salt (including those presented in this book) and never discount your own experiences.

Summary

As a prerequisite to making interesting and useful CD-ROMs and DVD-ROMs, as opposed to merely copying data to disc, you need to know how to choose a data format and manage the files being recorded to the optical disc. The various formats evolved in response to requirements for storing and retrieving different kinds of data from disc—knowing the rationale behind the different modes of data storage can help you produce discs that are compatible with the widest range of equipment and that deliver data in the fastest and most effective manner.

3

CD Standards

The original compact disc standard—Red Book—has gradually evolved to encompass more and more standards under one umbrella, enlarging the range of compatible platforms and increasing the utility of data stored on CD-ROM. These burgeoning standards have not evolved without some species being killed off in the process, but the situation has stabilized to the great relief of developers wanting the reach the widest possible segment of the market.

The standards that apply to compact discs have been developed to encompass more efficient storage of different types of data, file structures that allow multi-platform access, directory indexing to allow recordings to be made in more than one session, and other modifications intended to keep CD-ROMs in the forefront of an increasingly multimedia-oriented world. Turf wars are still taking place over proposed new standards, such as double-sided video storage techniques (DVD), new data-compression schemes, and alternate disc sizes, but the existence of tens of millions of inexpensive CD-ROM drives does much to ensure that new standards will pay homage to (and ensure compatibility with) the existing standards.

If you plan to burn some compact discs, you'll need to give some thought to both the existing and proposed CD-ROM standards. You may have already heard terms sounding suspiciously like coloring books being tossed around in regard to CD-R use—Red Book, White Book, Blue Book, and so on—but if you don't know their significance to the type of recording you'd like to accomplish, you'll learn it in this chapter. This chapter guides you through a comprehensive description of all the significant formats associated with the CD-ROM standards and suggests the most effective ways you can use these standards when recording.

Why More Than One Standard?

The various CD-ROM standards have been designed to support a number of different data types on a storage medium that is formatted radically differently from familiar diskettes and hard drives. To a large degree, the standards have evolved to wring every ounce of performance out of a technology that initially had very slow data transfer and access rates.

Each different type of data has required a different approach. Through compression, careful data placement, interleaving of different data types, and other techniques, developers have been able to store and retrieve audio, video, high-resolution photographs, mixed media data, and all combinations of these data forms.

Since the early 1990s, CD-ROM drives have reached much higher performance levels, but the legacy of the earlier standards hangs on. One of the first decisions you must make before mastering a disc with virtually all of the CD recorder applications is which format you will use for the output. Figure 3 - 1 shows each of the primary standards and its typical applications. Each of these standards is presented in detail later in the chapter.

Figure 3 - 1   **Primary standards**

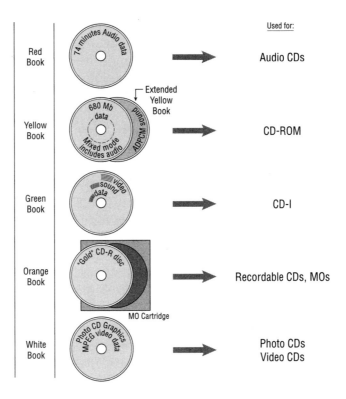

The Course of Standards Development

The manner in which standards are proposed and ratified in the electronics industry has all the political intrigue, power bartering, and deception of the fourteenth-century royal struggles over fiefdoms and land holdings. Every company has its own agenda and every organization would like to make lots of money from the acceptance of particular standards. Fortunes are won and lost on decisions made by standards committees, so it is not surprising that the process of reaching consensus can be so grueling. Alliances among industry giants trying to overcome their competitors often work at cross-purposes with the consumers, who just want to be able to buy products and software that work together effectively.

The standards affecting CD-ROMs have developed in more orderly fashion than some other electronic technologies, but they have also undergone a number of revisions and extensions to accommodate rapid changes, both in the medium's capabilities to accommodate rapid changes and in our expectations for the medium. CD-ROMs have evolved from simply being a storage medium for high-quality audio tracks to being the proposed delivery medium for motion pictures, complex interactive multimedia, and just about anything else that can be digitized. Consequently, the CD-ROM family tree looks like one of those biology charts showing the genetic mutations of fruit flies. Current standards committees are grappling with new ways to pack more data onto disc, including new compression standards, double-sided CD-ROM standards, and similar evolutionary trends. While these more or less polite debates go on, the installed base of CD-ROM drives (the number of drives actually hooked up to playback machines) is currently slipping past the two hundred million mark and still showing strong sales despite the encroachment of DVD onto the market. Such an impressive installed base of equipment provides assurance that standards committees will give some thought to maintaining compatibility with existing standards.

Maintaining Compatibility

As mentioned previously, CD-ROM standards have developed to a large degree through a series of enhancements and extensions, driven by market demands and the conversion of more and more types of information to digital form. As new needs arose, such as additional kinds of error correction or support for a new data type, the earlier standard was extended or modified rather than being completely rewritten. The end result of this foresight is that you can insert an audio CD into the CD-ROM drive installed in your computer and (with the correct drivers) play back music that was recorded using specificiations defined in the earliest Red Book standard. Similarly, most modern CD-ROM drives, under most operating systems and with the appropriate hardware drivers, can play back CD-ROMs containing data in any of the currently active standards.

A modern CD-ROM drive with current software drivers can typically play audio CDs, read ISO 9660 file structures (for cross-platform support), read Kodak's PhotoCD discs (for high-quality photographic display), and support CD-ROM XA (for faster synchronized playback of audio interleaved with computer data). While there are other capabilities that are high on everyone's wish list, this combination of features makes it possible to create and develop some exceptionally interesting material that can be enjoyed on the vast majority of existing CD-ROM drives.

The proliferation of enhancements, however, has created some incompatibilities. As you might expect, some of these come about from trying to play newer CD-ROMs on older drives. For example, the CD-ROM XA standard, which allows interleaving of of audio and data, is not universally supported by older equipment. The PhotoCD standard, institutued by Kodak as a method for storing photographic data, must be supported by drivers that read the PhotoCD format. More recently, the Blue Book standard, which lets music producers create discs that play music tracks in a standard CD player and multimedia tracks in a computer's CD-ROM drive, also relies on drivers installed on the user's computer for multimedia playback. Though these drivers are commonly supplied with modern operating systems, users on older systems may have a difficult time locating the appropriate driver.

You must also consider the suitability of standards on the equipment that you plan to use for recording. For example, some of the software packages for recording CDs support the option for incremental write operations, allow you to divide a data track into smaller packets and write these packets individually. Unless the CD recorder that you are using supports incremental write operations, you cannot use this feature, despite the fact that it may be available in the software.

Applying the Standards

As a developer, you have control over a number of factors that are directly influenced by the standards. Your decisions on these issues will not only affect the compatibility of your title with the installed base of playback devices but also the overall performance of your work.

Much of the material in the following chapters deals with the decisions you face when using recorder software to create a virtual "image" of the disc, on your hard drive or some other fast medium, before writing it to the recordable disc. The term *image* refers to an exact block-by-block copy of the data, including correction codes and addressing information, that precisely matches the form in which the data will appear on the CD-ROM. (The size of the image file will be the same size as the data contents of the disc—so if you have 600MB of data to be written to disc, the image file will require 600MB of your hard drive space.) The CD recorder application does the work of determining how to structure all this data and then stores it in a single image file on your hard drive. Recording the actual CD-ROM disc is then mainly a matter of copying that file.

Depending on the software application with which you are working, you may have control over the following factors:

- Geographic placement of the files upon the disc.

- Choice of file and directory names, and organization of the directory hierarchy.

- Interleaving of compressed audio content with computer data. (Depending on the degree of compression, interleaving of sectors should occur at predetermined intervals.)

- Selection of search engines, indexing applications, multimedia players, and similar applications that can link to multiplatform files stored on the CD-ROM.

- Creation of a Red Book audio region on disc and the combination of this region with Yellow Book CD-ROM data (as in Mixed Mode applications).

- Storage of MPEG-1 and MPEG-2 files on discs intended for cross-platform playback.

- Capacity and compatibility issues surrounding recording of Mode 1 and Mode 2 multi-session discs.

- Storage considerations for images in any one of the PhotoCD compatible formats.

- Compression levels to use for ADPCM data.

- Choices when considering use of multi-session versus multi-volume recording techniques.

- Use of native file formats or hybrid file structures under certain circumstances.

- When to pay attention to these low-level format and file issues and when to relinquish these kinds of decisions to an application that ensures ISO 9660 compatibility.

Many of the latest generation of software applications remove most of the burden from worrying about disc formats and file systems. Depending on your application, these may work admirably for your particular application project.

Programs such as Adaptec Easy CD Creator (for Windows users) and Toast (for Mac users) provide a simplified view of the recording process, which can guide even first-time users through the creation of flawless CDs. Other programs, such as Asimware HotBurn and GEAR Software PRO DVD contain both features for novices as well as additional and control features that may appeal to more professional users. GEAR PRO DVD software, in particular, adds an additional level of flexibility and control; this control gives you a degree of precision over how a disc is recorded, but less-experienced users can easily fall into a number of traps that may result in unusable compact discs.

Using the Standards

Perhaps the best way to become familiar with the ancestral history of the CD-ROM standards is to view the family tree. Each of the current offspring of the Sony and Philips union is shown in Figure 3 - 2 on page 58.

The physical sectors in each of the different formats contain a total of 2,352 bytes of information. Depending on the standard, these bytes are distributed between the data carrying the audio or program information and additonal bytes that handle error correction, synchronization, and identification of the information contained in the sector. Figure 3 - 3 on page 59 shows the layout that applies to the primary standards.

Note: The descriptions of the standards provided in this chapter are for conceptual understanding only. If you need detailed information on the standards for implementing a device driver or other low-level application use, consult the individual organizations that maintain the standards. Be forewarned, though: obtaining the full specifications for some of these standards can be an expensive proposition—some of the manufacturers holding copyrights on portions of this technology expect steep reimbursement. It can cost in the range of several hundred dollars to receive the specifications containing the full details of their intellectual holdings.

Chapter 3

Figure 3-2 **Compact Disc Evolution**

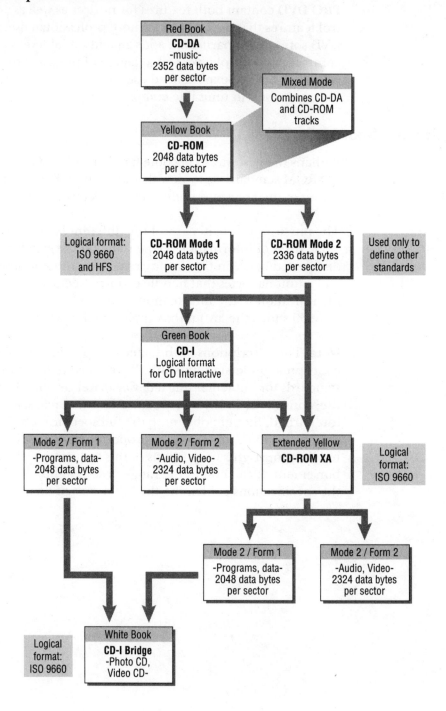

Figure 3 - 3   Physical sector layout

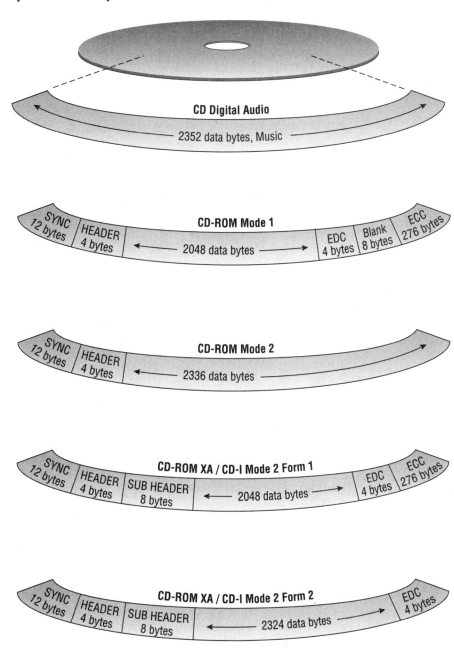

Chapter 3

Red Book

Red Book, sometimes called CD-DA for Compact Disc Digital Audio, predates all of the other standards. The compact disc masterminds at Sony and Philips drew up specifications for the way audio CDs would be recorded, and subsequently the music industry was reborn in digital form. Red Book support can be found universally on all recent CD-ROM drives, although some of the extensions for adding graphics to a Red Book disc (Compact Disc plus Graphics or CD+G) have not been readily used or supported.

Uses

The one prevalent use for CD-DA is for recording music, sound effects, and voice. This use has been extended from simply recording digital audio for music albums to including music and voice in mixed-mode applications for CD-ROM.

Data Types Supported

Red Book defines the recording of audio material that is digitized using Pulse Code Modulation (PCM) format. This is a non-compressed format produced by sampling sound waves at discrete intervals and producing a 16-bit value at each sample point.

Implementation Issues

Generally, if you are recording Red Book audio tracks on a compact disc, the entire disc must be recorded in a single session. Multiple sessions are not available and incremental writes are not supported under the Red Book standard. Because the data region is longer (2,352 bytes) than the other standards, data throughput is a greater concern when recording a compact disc under this standard.

In its extended form, *Mixed Mode*, Red Book offers an additional challenge to developers. Audio data formatted under Red Book guidelines (in uncompressed, PCM form) exists on a separate area of the disc from computer data, which consists of programs, databases, images, or other binary information. The CD-ROM drive cannot read computer data while it is accessing audio information, and vice versa. This results in a synchronization problem: It is difficult to properly synchronize audio segments in the Red Book region with material being played back from the computer data region. The laser read head must alternately jump back and forth between the two areas to locate and read the necessary tracks. This problem spurred the creation of an extension to the Mixed Mode standard, called CD-ROM XA. This extension, discussed later in this chapter, interleaves data and audio for faster access.

Examining the Red Book Standard

At the time of its introduction, no one saw the compact disc as anything more than a vehicle for delivering music in a digital format. While experimenting with digital audio tape, Sony had come up with a system for converting analog sound for digital storage; this new method had many advantages over analog methods of storage. Research on optical laser technology demonstrated the capability of being able to precisely record signals at a density that easily supported the recording and playback requirements of this new medium (44.1Kilobytes per second, abbreviated as Kbps). Intent on avoiding the product introduction problems that had plagued video laserdiscs, Sony and Philips mapped out a strategy for unifying and standardizing the data requirements of compact discs so that all players would consistently play back any compact disc. Their strategy worked.

Audio data as specified by Red Book is partitioned into tracks, up to a maximum of 99 tracks per disc. The contents of each track might represent a single song or ten minutes of a speech or a 30-second sound clip of a dolphin's chittering. These tracks originate from a continuous spiral on the disc surface, starting from the innermost area of the disc where a table of contents indicates the location of each item. Locations are expressed using a time code that specifies minutes, seconds, and a sector number to pinpoint the start of a track.

Sectors (sometimes called *blocks*) can be read at the rate of 75 sectors per second. In other words, a block contains audio data that occupies 1/75th of a second. Audio data is always played back by CD-ROM drives at single-speed rates (150 Kilobytes per second), even on 12x or 24x drives.

Red Book further divides each sector into a group of frames. Each sector contains 98 frames, and each frame contans 24 bytes. A technique known as *Eight-to-Fourteen Modulation* (*EFM*) is used to encode the digital samples of audio waves into a series of transitions, hard-coded into the disc surface as pits and lands. During playback, this data is decoded back into the digital representation of the audio signal, which is then put through a Digital-to-Analog (D/A) convertor to produce the sound waves that drive audio speakers.

Digital sampling forms the basis of any audio information stored on compact disc. To accurately represent the characteristics of all sound waves within the range of human hearing (about approximately 20Hz to 20,000Hz), the Nyquist theorem was applied to determine that the sampling rate must be double the highest recorded signal frequency. By this means, a sample rate of 44,100Hz was selected to capture the the intricacies of the waveform. At each sample point, a 16-bit value records the waveform at that instant in time, providing a maximum of 65,536 discrete values. This level of resolution proves fully capable of smoothly and accurately depicting sound waves. The combination of these two elements—a 44.1Kbps sample rate and 16-bit sample values—produces what is generally referred to as *CD-quality sound*.

When WAV files that have been sampled at rates less than 44.1Kbps are to be transferred to a recordable compact disc, they must be first converted to the 44.1Kbps format. Many of the recordable-CD software applications include a conversion utility to do this.

Red Book Error-Correction Techniques

While not providing extremely sophisticated error correction (since it is not really required for audio data), Red Book defines a correction scheme using CIRC, Cross-Interleaved Reed-Solomon code, that can detect and correct up to 220 bad frames in each sector. CIRC is considered a Level 1 error correction scheme, and it is sufficient to for compensateing for minor scratches and imperfections in the disc surface.

Red Book also deals with more substantial errors, by specifying techniques for approximating any erroneous data that cannot be fixed through CIRC. In effect, data of this sort is restored by either replicating the same value as the preceding data, filling in a value by averaging the values on either side of the incorrect data, or silencing the data completely. In any of these cases, since we're talking about time spans in the milliseconds, you probably won't hear the results of the correction.

The Audio Origins of CD-ROM

Without the audio CD, there would be no CD-ROM. The technology has grown directly from the R&D work of Sony and Philips and their industry shaking product introduction of the 1980's.

The pioneering work done by Sony and Philips was originally focused on the benefits of the digital recording of music, which is an easier task than the digital recording of computer data. Digital representations of the soundwaves composing an Eric Clapton guitar riff or a synthesizer melody played by Peter Gabriel don't need to be completely error-free. If a few bits get scrambled here and there, odds are that you won't even hear the difference during playback. Digital computer data, on the other hand, demands absolute precision. Computer programs consist of instructions that directly control the computer processor operation. Dropping even a single bit from an instruction may change the nature of that instruction to something totally different. Instead of performing a logical-OR operation on the contents of two internal registers, for example, the processor may suddenly perform a RESET and terminate all activities. The result is that the program you're running on your computer may crash.

During the first wave of enthusiasm over the success of the digital recording phenomenon on compact disc, researchers at Sony and Philips noticed that the error rates of recorded digital data were extremely low. Somewhere along the line, they realized that if these low error rates could be maintained during manufacturing and replication, compact discs could be easily adapted for data storage as well as digitized music. Out of this simple discovery, the CD-ROM was born, and standards were established for the Yellow Book enhancements to the Red Book audio standard.

Yellow Book

Yellow Book, also called the CD-ROM standard, opens up the compact disc medium for computer data storage (as opposed to simply audio information). The physical sector format was modified to included additional error-correction fields. Yellow Book also defines methods for storing and locating data. An offshoot of Red Book and Yellow Book, the *Mixed Mode* standard, combines audio (in Red Book format) with computer data. A further offshoot, *CD-ROM XA*, interleaves audio, video, and computer data for quicker access.

Uses

The Yellow Book standard is used to specify storage of most types of computer information on compact disc, including applications, database information, indexed text, and so on. Mixed-mode applications can also include a separate region on disc containing standard Red Book audio data. Yellow Book was also extended through CD-ROM XA (the "XA" stands for extended architecture) to improve multimedia interaction; the extension improves synchronized playback of audio, video, and data. CD-ROM XA is often used by game developers, educational developers producing interactive works that involve multiple video or sound clips, and anyone constructing a complex multimedia title. In the early days of introduction, CD-ROM XA could not be played on some systems that lacked the compatible hardware and software for playback, but on modern systems and current generation CD-ROM drives, support is universal.

Data Types

In its pure form, Yellow Book supports binary text and computer data. In mixed-mode form, it includes Red Book audio. A logical layer of the architecture deals with the file structures and it is here that ISO 9660 (discussed later in this chapter) becomes important. Yellow Book includes two fundamental modes, Mode 1 and Mode 2, that rely on different methods of data correction. Mode 1 specifies an Error Detection Code (EDC) and an Error Correction Code (ECC); Mode 2 uses the previously discussed CIRC for less rigorous error correction.

Implementation Issues

When recording in the Yellow Book standard, you need to ensure that your file organization permits cross-platform access to disc contents under ISO 9660 (for more details, refer to *ISO 9660* on page 79). When recording Mixed Mode discs, you also need to consider the trade-offs in storage space between Red Book audio and computer data (the use of Red Book audio reduces the available data region at the rate of approximately 10 Megabytes of storage per 1 minute of recorded PCM audio).

Examining the Yellow Book Standard

Sometime in 1983, Sony and Philips realized that their digital music delivery creation, the compact disc, could do a reasonable job of storing data as well. The discussion and committee meetings that took place led directly to the birth of the CD-ROM in essentially the same form you now use today. Error correction was made more robust, which was essential for a medium where a bad bit could crash a program (as opposed to perhaps causing a minor "pop" in an audio sound track). Data storage also necessitates data retrieval, which dictated some system of indexing and locating data rapidly. Yellow Book provides these as well.

Yellow Book specifies a four-layer architecture:

- Layer 0. Identical to the Red Book definition of bit structure
- Layer 1. Specifies the sector layout, including the error detection and correction code use

- Layer 2. Defines the logical sector organization. While fixed in terms of the physical sector layout, the logical organization provides flexibility to adjust sector sizes to accommodate the requirements of certain types of software.
- Layer 3. Represents the logical file organization to standardize access to files across various computer platforms. ISO 9660 offers the method of organization, which is widely supported by CD-ROM developers.

Examining the Yellow Book Extended Architecture

Large corporations and government agencies found CD-ROMs a convenient repository for the tons of paper documents that they used to store in file cabinets. Early uses of CD-ROM emphasized text storage and database archiving. The original Yellow Book standard works just fine for most of these types of applications.

Text storage and retrieval, however, doesn't particularly drive the medium in any kind of demanding way—the original Yellow Book standard works just fine for most of these types of applications. As with all things computer-related, some developers started pushing the media in some new ways that rubbed up against the its inherent performance limitations and challenged the current methods of storing audio and video. Developers who were moving information to CD-ROM became more enamored of multimedia bells and whistles—such as video—but found it was difficult to both play back a computer application and access audio data from the Red Book region at the same time. The laser read head had to cover too much territory to retrieve data. The sound track often fell out of sync with a displayed animation or video clip, so that lips moved but no words came out, or the narration lagged the on-screen event by a few seconds. Multimedia presentations played like badly dubbed foreign films. No one was very impressed with the potential of the medium at this point.

At this time Microsoft joined into the discussions with Sony and Philips and crafted the CD-ROM XA standard, introduced in 1989. The primary enhancement of CD-ROM XA involves a redefinition of the physical sectors to include a special form of compressed audio. Audio in this format could be wedged between the computer data running the application—residing on contiguous sectors—and the interleaved audio. Data could be accessed much more quickly. Words flowed out of on-screen lips in sync and the climactic crescendo of music at the end of the presentation came right on cue, right down to the final cymbal crash.

CD-ROMs produced using the extended standard, however, could not be immediately played back by existing CD-ROM drives. Many manufacturers saw no reason to produce XA-compatible drives because there was no software developed for them. CD-ROM title developers didn't want to invest the time and effort in producing XA-compatible titles because none of the drives could read them anyway.

This situation has been eliminated with the current generation of CD-ROM drives and XA-compatibility is standard throughout the industry. If you are developing an application that doesn't require extensive use of multimedia, you can create your CD-ROM master using standard Yellow Book guidelines. Most CD-ROMs, in fact, are produced within the constraints of standard Yellow Book. If you want to optimize performance for multimedia playback, you are probably going to want to use the XA conventions.

The compressed form of audio data storage (ADPCM) used under the CD-ROM XA standard permits very lengthy monaural audio segments to be recorded. While sacrificing some of the dynamics of full CD-quality sound (but still getting better quality than your typical AM radio broadcast), you can create discs with up to 18 hours of sound. One excellent application for this feature could be books on disc, a compact-disc replacement for the popular audio books that are distributed on cassettes by a number of companies. If you coupled this feature with good voice navigation software, you could create interactive references or audio books on disc for physically challenged individuals.

The extended version of Yellow Book recognizes two different modes of data storage.

Mode 1, reserved for bit-sensitive computer data, relies on the error-correction schemes introduced in Red Book, EDC, and ECC, but increases the ECC value at the end of the sector to 276 bytes for more exacting correction. At the beginning of the sector, a 12-byte synchronization field and 4-byte header allow the CD-ROM drive to identify and get locked into the information contained within the sector. Headers and error-correction fields reduce the available data-storage region within the sector to 2,048 bytes.

Mode 2, dedicated to less-critical forms of data (where a bit or two can be misplaced without bringing down the program) uses the CIRC correction scheme.

Yellow Book, ISO 9660, and the High Sierra File Format

The Yellow Book standard unified CD-ROMs and drives to the degree that most CD-ROMs produced could be successfully played back on most drives, but the file systems used on early releases were individually designed by developers and it was difficult to find two that were alike. The situation was not unlike the time during the late 1970s and early 1980s with the first wave of personal computers, each with its own operating system or CP/M variation and each with its own way of formatting floppy diskettes. A genuinely uniform development environment for personal computers was not achieved until the release of the IBM PC, and after many small companies finally conceded that their individual directions were leading nowhere.

With this history in mind, an assembly of companies engaged in different aspects of the CD-ROM industry gathered at Lake Tahoe, California in November of 1985 to define the logical file structure for CD-ROMs. Among the participants were DEC, Sony, TMS, VideoTools, Xebec, Microsoft, 3M, Philips, Hitachi, LaserData, Apple, and Reference Technology. They succeeded at their task, despite the close proximity of the gambling casinos and other temptations on the south shore of the lake.

Rather than following the practice that resulted in the first two CD-ROM standards being named after the color of the binders that they were stored in, the participants in the Lake Tahoe conference decided to call the agreed-upon standard the *High Sierra file format* (HSF), perhaps because Lake Tahoe is in the High Sierra range. We can be thankful that this group wasn't meeting in Mount Lassen National Park to the north or the standard might have been named the Bumpass Hell file format after an aptly named region of steaming, volcanic pools and craters in the park. The High Sierra file format—with a few tweaks and modifications—became the basis for ISO 9660, officially approved in September of 1987. It has since been extended and expanded several times to adapt to new twists and turns in the industry.

Green Book

Green Book expands the realm of the CD-ROM into the interactive world of set-top playback equipment, such as the Philips CD-Interactive (CD-I) devices. CD-ROM drives that include CD-ROM XA compatibility can generally play back titles created under this standard. Green Book not only describes the CD-ROM storage characteristics, but specifies the hardware and the operating system to be used for playback.

Chapter 3

Uses

Green Book optimizes disc contents for real-time playback of complex elements, including compressed graphics, video, and sound, such as the loud shrieks of dying monsters struck by your photon torpedo when playing games. CD-I also supports multi-lingual applications; as many as 16 separate audio language tracks can be stored side-by-side. Creative use of this feature could allow a single CD-I title to be produced for a truly international audience and marketed in many different countries. This feature can also be handily applied to language-learning discs; for example, a single disc could help you learn Swedish, Russian, French, or Zulu.

Data Types

Green Book accommodates a mix of data types—in fact, the sector header contains an identification string to indicate the type of data that is stored within the sector. Different data types, primarily sound (four types), video, and computer data can be interleaved upon the same track.

The sound types supported include Red Book digital audio and three levels of sound compressed using ADPCM techniques. Speech-quality sound has the greatest degree of compression (and the greatest loss of quality). A middle level, usually called Mid-Fi, exhibits less compression and has improved sonic characteristics. The best quality is called Hi-Fi, and this is basically a compressed version of the PCM digitization used for Red Book. Although there is a slight loss in quality because of the ADPCM sampling technique (which records differences between successive digital samples), the difference in quality is slight, and it comes very close to what we regard as true "CD-quality" sound.

The video data specification includes full 24-bit color (allowing 16-million color variations) and the use of discrete "planes" or (layers) of video. By subdividing the video data into layers in this manner, many of the throughput problems and memory-use issues can be resolved (moving 24-bit color frames at 30 frames per second takes lots of memory and a high transfer rate). Areas within a video frame that are relatively unchanging, like the background, can reside on one plane, and this data does not have to be updated as frequently as a moving foreground object, such as a Formula 1 racer streaking towards the camera. Effective use of planes helps compensate for the performance limitations of compact discs.

Implementation Issues

Green Book is a far-reaching standard that ventures beyond the compact disc surface to include the hardware and the operating system. Use of this standard provides advantages only if you are trying to reach an audience using set-top CD-I players or CD-I compatible CD-ROM drives (generally covered by CD-ROM XA compatibility)

Examining the Green Book Standard

Philips introduced the Green Book Standard in 1987 to accommodate an expected flood of CD-I hardware and applications. The floodgates have yet to open. Green Book shares some characteristics with CD-ROM XA, and allows the interleaving of audio and video on data tracks. The operating system associated with CD-I disc playback is called CD-RTOS.

Because the table of contents does not include the CD-I track, audio players can handle these discs without outputting a shriek when they encounter computer data instead of audio material.

A properly equipped CD-I player can also play back PhotoCDs as well as audio CDs (Red Book audio can be read within this format without problems). CD-I also supports VideoCD, providing full-screen, full-motion video with the correct hardware adapter. MPEG-1 was originally adopted as the compression standard; MPEG-2 is an extension of this standard with improved transmission rates and higher resolution. A well-tuned system with support for MPEG-2 can deliver audio and video frames at the rate of 9.8 megabits per second. MPEG-1 is significantly slower, with maximum rates of about 1.8 megabits per second.

Storage of video data on a CD-I disc must use the Form 2 sectors. Within these sectors you can store any of the three varieties of MPEG data:

- MPEG video data
- MPEG audio data
- MPEG still-picture data (also called MPEG SP)

Another Offshoot: CD-I Ready

It should be obvious by now that manufacturers freely adapt each of these standards whenever they see something that looks like an open market. CD-I Ready is an example of this; it is an audio disc that includes additional features that can be accessed when it is inserted in a CD-I player. Basically, you can embed graphics and text into gaps in the audio data and index these in such a manner that they can appear on a display while the associated audio segment is playing. This supplementary material might consist of the lyrics of an individual song, the history of the band or biographies of the individual artists, psychedelic backdrops for quiet contemplation, or advertisements for additional albums or products. CD-I Ready has never really made much of an impact and has been superseded by other standards that do the same kind of things more effectively. Its description here is primarily to document the standards evolution.

White Book

White Book, sometimes called CD-I Bridge, defines an environment in which CD-ROM XA, CD-I, and PhotoCD discs can be freely exchanged. The recordable nature of PhotoCD discs is considered an important part of this standard, as is the VideoCD standard, which describes the use of MPEG compression to compactly store feature films and video data.

Because MPEG compression only allows 74 minutes of video per disc, and feature films generally last 90 minutes or more, the release of films on CD-ROM discs never became successful in the United States, although the market fared much better in Japan and in parts of Europe. The use of DVD has now surpassed the utility of this feature for most uses, other than training videos and video content that must be directed to a large installed base of equipment.

Uses

White Book permits a number of different types of digitally stored data to be conveniently exchanged between players. It also supports MPEG playback with the potential for distribution of short independent films, training materials, or corporate informational videos. Longer video releases are more reasonably packaged on DVD.

Data Types

White Book accepts most of the data types discussed previously—in fact, its appeal is that it unifies many of the other standards. The one omission is Yellow Book Mode 1, and this creates a problem when attempting to record true multi-session-compatible discs (as described in *Orange Book* on page 73). White Book handles Red Book audio, CD-ROM XA, or and CD-I formats. It is commonly used for PhotoCD storage, since it includes support for partially recorded discs. This allows photo processors to continue adding scanning photographs to a CD over two or three sessions, until the capacity of the disc has been reached.

Examining the White Book Standard

White Book originated from a collaboration involving Sony, Philips, JVC, and Matsushita. The primary goal was to create an extension of the existing standards that would permit full screen (640x480) video playback of compressed MPEG-1 video files. However, White Book has also had an impact on the way that Kodak PhotoCDs are written and read.

PhotoCD

PhotoCD techniques, which allow photographic images to be added and appended to a recordable CD over several sessions, ranks among the most important uses for this the medium. Digitized versions of photographs are sampled at several different resolutions and recorded to the compact disc, where they can be retrieved for viewing or insertion into a multimedia epic.

The PhotoCD specification (with some help from Philips and Kodak) has evolved into its own standard, sometimes called the Hybrid Standard, that combines elements of Green Book and Orange Book. This hybrid provides support for a wide variety of players, including devices like Panasonic's 3DO unit and equipment that handles only PhotoCD. The most important platform (from the standpoint of installed base) is CD-ROM XA compatible drives.

Kodak dreamed up this technology back in 1990 and they've gradually received more and more acceptance both from consumers, who can view PhotoCD portraits of their dog Bingo on television, and from the graphics arts wing of the computer industry, where professionals can easily store and access high-quality photographic images.

Kodak has actively forging alliances with a number of companies to increase acceptance of this storage format. They have worked with Adobe to provide extensive support for PhotoCD formats within Photoshop and PageMaker. They have worked with Hewlett-Packard to improve color printing of photographs on ink-jet printers. They are have offered solutions to rapidly transporting images across networks. The new Picture CD format brings compact disc storage of photographs to a wider base of consumers. Kodak seems to be intent on simplifying the tools for working with digital photographic images and returning to their roots. As Kodak founder George Eastman back in the late 1800s was fond of saying, "You press the button, we do the rest."

Compact discs recorded for PhotoCD applications use only the Mode 2 Form 1 sector layout of the CD-ROM XA standard. This creates a problem for multiple sessions that rely on Mode 1 as defined by the Orange Book. Some drives and software certified for PhotoCD multi-session use are not able to read multi-session recording created using Mode 1. Ideally, CD-R equipment and mastering software should support both modes.

PhotoCD specifications include 5 categories of images that can be stored within this framework:

- *Medical images.* This category includes things like X-rays, digital CAT scan data, and data from magnetic resonance scanners. Storage on compact disc permits easy distribution and convenient archiving.

- *Pro.* The Pro format provides a storage medium for professional photographers. It includes most of the major film sizes: 4x5-inch, 8x10-inch, 35-millimeter, 120 film.

- *Catalog.* The Catalog storage format includes a framework that can handle as many as 6,000 images within a linked environment, so that you can jump from a menu to any of the images. As you might guess by the name, it is useful for creating interactive catalogs that require high-quality images of products.

- *Portfolio II.* The replacement to the earlier Portfolio standard, Portfolio II supports Kodak's Image Pac files and digital sound files. This newer format covers much of the ground initially introduced dealt with by earlier formats, with tools for creating presentations accompanied by sound and manipulating archived images.

- *Master.* The PhotoCD Master disc stores 100 images in several different resolutions. This is the format most commonly associated with the introduction of PhotoCD processing of film. This is the category that most people associate with PhotoCD when they place an order for processing their film.

The most recent development in this area, Kodak's Picture CD, is discussed in the section titled *Picture CD* on page 85.

VideoCD

While White Book defines the use of VideoCD, a method for storing and playing video files using the MPEG compression scheme, this the standard is already being superseded by new storage formats and possibly new forms of compact disc media. The problem is that even with MPEG you can only store 74 minutes of full-screen, full-motion video. This might be fine for short documentaries or animated shorts, but most mainstream motion pictures run over 100 minutes (thus requiring at least two discs). MPEG also requires a particular type of controller for playback to handle the rapid decompression necessary to keep thirty 640x480 frames of video accompanied by sound unfolding every second. There are a few titles that have been released, showing the feasibility of VideoCD, and this format has achieved a fair amount of success in Japan and parts of Europe, but the extra capacity of DVD has made the format largely obsolete for mainstream film releases.

Orange Book

Orange Book defines recordable media used in both CD-R units and Magneto-Optical drives. The goal is to ensure compatibility of the recordable optical media among different players. Essentially, Orange Book specifies how each of the other standards (Red Book, Yellow Book, Green Book, White Book) should be recorded onto CD-ROM.

Uses

Most of the specifications that appear within the Orange Book standard are designed to keep hardware and software developers on track as they produce equipment and applications that support recordable optical media. By the time you have their products in your hands (or onto your computer workstation), there is not too much more you can personally do to ensure compatible playback. You can, however, be aware of the manner in which multi-session recording is specified and the effects of this specification on playback compatibility. The section *Examining the Orange Book Standard* on page 74 discusses this issue.

Data Types

Orange Book is not concerned with individual data types, but defining a framework for recordable media that unifies each of the other standards and ensures readability on multiple platforms, including the readability of multi-session recordings.

Chapter 3

Implementation Issues

Although Orange Book specifies how information should be recorded during multiple sessions (essentially appending data to a disc that contains data from earlier sessions), playback issues on some equipment remain unresolved. In other words, you may create a multi-session disc following all the rules of Orange Book, but that disc may still be unreadable by drives that are supposed to be "multi-session compatible" or "PhotoCD compatible" or "CD-ROM XA compatible." This issue is discussed in more detail later in this section.

Examining the Orange Book Standard

Once all the other standards were neatly encased in their binders and lined up in a multi-colored array on the bookshelf, the technology surrounding the use of optical techniques to record data on writable media began to mature. Philips came out with the definitive standard, the Orange Book, in 1990. which deals not only with recordable CD (called *Compact Disc Write-Once* or *CD-WO* in the standard), but magneto-optical recording techniques (*CD-MO*), as well.

In theory, Orange Book sets out to ensure that media recorded within the specifications defined by this standard will be playable on the widest range of equipment. In practice, some difficulties still remain, which will be discussed in the following sections.

Part I of the Orange Book describes recording to CD-MO systems (Compact Disc Magneto-Optical). While it may be fascinating reading, we don't discuss the details in this section.

Part II applies to CD-WO (Compact Disc Write-Once), which is simply an alias for recordable CD. This part of the standard addresses both single-session and multi-session write operations.

The disc layout under Orange Book contains some additional regions to handle the increased requirements of recording. As shown in *Orange Book disc layout* on page 75, the writable data area of the disc includes a Program Calibration Area. This area allows the laser write operation to be calibrated before more extensive writing takes place. The next region is the Program Memory Area containing a list of the tracks as referenced by their starting points and end points.

Within the Program Memory Area, a Lead-In Area precedes the start of the data region. The Program Area represents the portion of the disc reserved for the actual data being written. At the conclusion of the write operation, a Lead-Out Area is created to indicate the end of the recording session. The table of contents, located on the innermost track, remains unrecorded until all required sessions have been written and the disc can be "fixed," which compiles the table of contents and provides full access to all sessions.

Figure 3 - 4 **Orange Book disc layout**

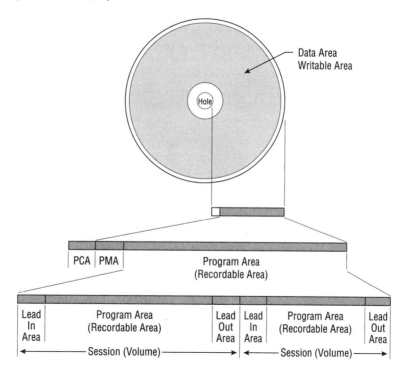

During single-session recording, the table of contents is created immediately after the data is recorded. This essentially "closes" the disc to any further write operations. Multiple-session recording leaves the table of contents unwritten from session to session, until the final session, when the disc is fixed.

For a CD-ROM drive to read a multi-session disc that hasn't been fixed (a disc that does not have a table of contents), the device driver for the drive must identify and access the last session that was recorded and examine the directory structure in that session. In other words, to access the information recorded over multiple sessions, the device driver must be able to

Chapter 3

reconstruct the disc contents from the individual session information. Many early CD-ROM drives (and their drivers) cannot do this, but most current generation drives can.

Those that can accomplish this operation, however, assemble the links and present the entire group of multiple sessions as if it was were a single session. With the right playback equipment, the user of a CD-ROM recorded in this fashion is never aware that the disc has been recorded over multiple sessions. Figure 3 - 5 illustrates this concept.

Figure 3 - 5 **Multi-session recording technique**

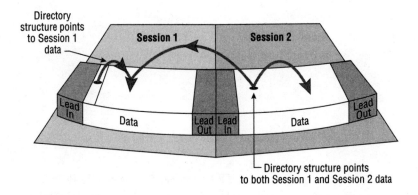

Multi-Volume Discs

Multi-volume recording is related to multi-session recording, but it has one key difference. Although volumes are created during separate recording sessions, they are not logically linked together—they are essentially independent groups of files. Recording and playing back multi-volume discs requires hardware and software that specifically include this built-in capability. This capability is largely universal at this time. The principle of multi-volume recording is illustrated in Figure 3 - 6.

Figure 3 - 6   **Multi-volume recording technique**

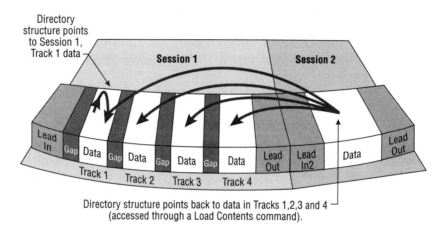

Modes and Multi-Session Problems

One of the sources of grief in ensuring multi-session compatibility among recordable discs and disc players has its roots in a slight omission in the Orange Book standard. From the earlier descriptions of Yellow Book physical sector layouts, you may recall that the standard specified two modes: Mode 1 and Mode 2, each handling the data within a track differently.

As Orange Book describes multi-session recording, it specifies that whatever track type—Mode 1 or Mode 2—is initially used for recording, that same track type should be used consistently throughout. The question is, how do you determine which one was used initially? Orange Book provides no clues to help resolve this question, which has led to varying interpretations by drive and software manufacturers.

Kodak's implementation of the PhotoCD standard, which uses an extension of the CD-ROM XA sector format (as described in the White Book standard), originated because of the fact that photographs being digitized and written to disc needed to be recorded over multiple sessions. One roll of film didn't produce enough images to fill a single CD-ROM. Rather than waste the remaining space on the CD-ROM, Kodak helped devise the method for performing more than one write operation to a single disc. When a customer brought in a second roll of film, those images could be appended to the same disc as contained the previous images.

Some drive manufacturers responded to the popularity of the PhotoCD standard by providing support for Mode 2 (XA) multi-session data, but failed to include a provision for responding to multiple sessions on a Mode 1 disc. These drives (and drivers) can only detect the first session of a Mode 1 multi-session disc. In some cases, however, you can update the device driver to make the drive recognize additional sessions.

The Lead In and Lead Out regions required for recording multiple sessions introduce a substantial amount of overhead to your disc. The initial session recorded requires nearly 24 Megabytes for the Lead In and Lead Out areas. Subsequent sessions add approximately 15 Megabytes of data overhead to your disc. If you plan to record numerous multiple sessions, this data overhead can clearly reduce the data capacity of the disc. The finite limitation, considering this overhead, is about 40 sessions; the Lead In and Lead Out regions in this case will consume more than half the available space on a 650MB CD in this case. If your data requirements are considerable, you will probably want to limit the maximum number of sessions to under 10.

The Frankfurt Group

Nine companies concerned with the future of recordable CD met in Frankfurt, Germany in 1991. Those companies were: Sun Microsystems, the Jet Propulsion Laboratory, DEC, Sony, Philips, Meridian Data, Hewlett-Packard, Ricoh, and Kodak. Their goal was to resolve some deficiencies in the original Orange Book specification to provide more details about the recording process and to extend some of the file system aspects of ISO 9660. This meeting resulted in The Frankfurt Group proposal.

The solution to the problem of performing multiple recordings on a disc that normally has a fixed file structure was handled as follows:

- Instead of having a single fixed set of disc contents on sector 16 of the first track, the Frankfurt Group solution allows sector 16 of any track to be used for writing a volume descriptor.

- Track 1 remains blank until the disc contents have all been recorded, after which it can be filled in with a master volume descriptor.

Under this proposal, a CD recorder must have the capability of reading existing data as well as writing, so that it can detect the contents of already recorded sessions. Earlier CD recorders and drives often do not have the ability to read anything but the first session and cannot be used with the modified approach to recording.

The Frankfurt Group also came up with a scheme for supporting UNIX files. This portion of the proposal specified the use of the extended attribute definition that is required for UNIX to handle directories and files. Another extension to ISO 9660, the *Rock Ridge Extensions*, further refines ISO 9660 for use with UNIX by employing a technique known as the *System Use Sharing Protocol*. This structure enables a variety of file-system extensions to all be resident on a single compact disc, thus permitting more flexible and more robust file structures.

ISO 9660

ISO 9660 provides the logical structure for cross-platform use of CD-ROMs. This standard has helped fuel the rapid growth of the CD-ROM industry by letting developers reach CD-ROM players on DOS, Windows, Macintosh, OS/2, UNIX, and VAX platforms. Most CD-R software mastering tools fully support ISO 9660, and often provide test utilities to ensure the that a disc image does not violate any of the constraints of this standard. A large proportion of the commercial CD-ROM titles are produced using ISO 9660 as their file structure. ISO 9660 covers the two logical layers of four-layer CD-ROM architecture, converting the data contained in individual sectors into a hierarchical arrangement of files, directories, and volumes.

Uses

ISO 9660 is used almost universally as the dominant file system for storing CD-ROM data in a platform-independent way. Unless you have a strong reason for adopting the file structure of a particular operating system, or you need to develop your own file system for a special-purpose application, we recommend that you master all CD-ROMs using ISO 9660. The newer UDF standard, in comparison, applies to DVD-ROM and DVD-Video releases. An interim standard, UDF Bridge, provides backwards compatibility to ISO 9660.

Implementation Issues

ISO 9660 puts a number of restrictions on file naming (with rules concerning allowable characters and filename lengths) and other file system conventions to ensure platform independence. Many of the software mastering applications provide screening of the CD-ROM image or virtual image to make sure that data being recorded to CD under ISO 9660 does not violate any of these conventions. Check your software package to see if screening is performed before a recording session; if not, check the restrictions that are described later in this section.

Examining the ISO 9660 Standard

ISO 9660 is a single file system that is designed for use under many different operating systems. Unlike a typical file system, which has to be updated and changed dynamically as the operating system adds and deletes files, ISO 9660 describes a read-only medium: the CD-ROM.

The original description of ISO 9660 had to be extended in Orange Book to account for additional recordable compact disc implementations. The extension allows magneto-optical media to accept changes to the file structure, and permits multi-session recording on compact disc (where the file structure is modified incrementally as each write operation takes place during a session).

The original specifications for ISO 9660 have gone through a series of permutations, primarily to improve the degree of native file support for UNIX and Macintosh systems and to make access of CD-ROMs under these systems more transparent.

As with most of the CD-ROM standards, ISO 9660 was born out of necessity. The Yellow Book standard unified CD-ROMs and drives to the degree that most CD-ROMs produced could be successfully played back on most drives, but the file systems used on early releases were individually designed by developers and it was difficult to find two that were alike. The situation was not unlike the time (during the late 1970's and early 1980's) when the first wave of personal computers were produced, each with its own unique operating system or CP/M variation and each with its own way of formatteding floppy diskettes. a little differently. A genuinely uniform development environment was not achieved until the release of the IBM PC (after many small companies finally conceded that their individual directions were leading nowhere).

One File System/Many Platforms

The problems faced by the originators of ISO 9660 were twofold:

- CD-ROM drives cannot access data very fast, and retrieving files takes longer than from a hard disk drive.

ISO 9660

- Each of the target platforms to be supported uses its own conventions for files, directories, and volumes.

As you might remember from the earlier discussions in Chapter 2, data stored on a compact disc is located using *A-time*—a combination of time (minutes and seconds) and sector numbers (ranging from 0 to 98). This system of pinpointing data has to be reconciled with a file system that can support the requirements of the major computer operating systems.

The Primary Volume Descriptor, located at Logical Sector 16 (in absolute time that is 00:02:16), contains the fundamental characteristics of the ISO-9660 disc, and it provides pointers to the start of the Root directory (where the hierarchical file system is anchored) and to the Path Table, (an indexed list of the directory contents on disc). The Path Table serves as an accelerated way to locate directory items. Figure 3 - 7 shows a simplified view of this file system organization.

Figure 3 - 7 **Simplified ISO-9660 file system**

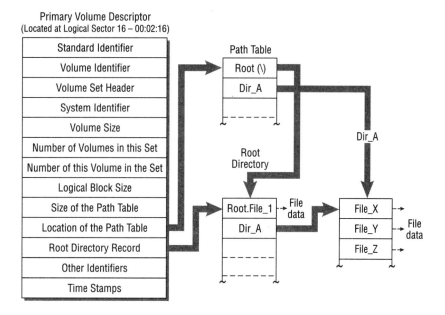

Within this structure, some basic rules and conventions apply.

- Files and directory names in ISO 9660 are limited to 8 characters for Level 1, and 30 characters for Levels 2 and 3.

- Files can have a 3-character extension, but extensions cannot be used in directory names (unlike DOS, which does allow this).

- Allowable characters are as follows: A through Z, 0 through 9, and the underline character (_). Alphabetic characters must be uppercase.

- The directory structure can include a maximum of 8 levels.

- Version numbers following a filename are optional. Version numbers can range from 1 to 32,767. (Version numbers that appear after filenames are separated by a semicolon. Since UNIX does not permit semicolons to be used as part of a filename, developers typically use some form of conversion to mask these characters from the UNIX operating system.)

- The total number of characters specifying the path to a file within a series of subdirectories is 256.

- Directories are sorted alphanumerically. Numerical entries (0 through 9) are sorted first. Underlines (_) are sorted last.

There are a couple of different ways that a file can be located within this structure. The CD-ROM drive controller can either trace a path through the directory hierarchy until it reaches the requested file. Or, a file can be located by referring or it can refer to an independently compiled path listing that serves as an index for all files on disc.

File Organization on Cross-Platform Discs

When creating a cross-platform disc under ISO 9660, you generally need to create a group of subdirectories to contain the executable files for each operating system that is supported. The type of applications stored in the OS-specific subdirectories will be search-and-retrieval engines, players for multimedia programs, and any files that are specific to a particular platform and processor.

The data files that can be accessed by each of the operating systems can be stored in a common subdirectory or a nested series of subdirectories. Using this system, a Macintosh user will be able to double-click and open a folder containing the access tools to view the data area of the disc. A UNIX user will access another subdirectory. A DOS user will find the necessary access tools in still another directory. In other words, the first level of subdirectories off of the root can partition the second level of the file system to provide entry points for users of each operating system. This form of organization is shown in Figure 3 - 8.

Figure 3 - 8   Organizing a disc for multiple operating systems

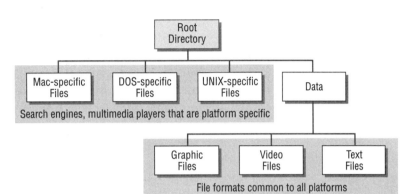

Operating System Extensions

To complete the logical framework allowing CD-ROMs to be read regardless of the platform, one more element is needed: an extension to each operating system that can mesh the requirements of the system software with the low-level hardware operation of the CD-ROM drive and its unique storage environment. Within the DOS environment (which, surprisingly, is still alive in some businesses that shun change), the software component is a device driver known as MSCDEX (Microsoft CD Extensions). MSCDEX serves as a bridge between the two environments, allows DOS to see the CD-ROM contents as a DOS volume, and relieves developers of the need to understand the individual device operations for each supported CD-ROM drive. A similar function is performed on the Macintosh platform by an extension called Foreign File Access. With the Foreign File Access extension installed in the Extensions folder of the Mac System folder, interaction with the CD-ROM drive becomes seamless. Similar software components exist for each of the operating systems that can access ISO 9660 discs.

With the advent of Microsoft Windows 95/98/NT, the integration of CD-ROM components within a system was made even more integral to the functioning of the system.

Resolutions of Earlier Problems

ISO 9660 solved the problems mentioned in the beginning of this section by standardizing the means by which sectors and their absolute time markers could be logically associated with the volume, directory, and file structure that applies to most operating systems. This system worked so well that it has been almost universally adopted by everyone producing CD-ROMs, and in those areas where it hasn't totally solved existing problems, committees are busy working on extensions to make it more useful. Recent developments, such as the CD-PROM, have proven that you can indeed teach an old dog new tricks. Although much of the momentum in the industry has shifted towards DVD-related issues, the huge installed base of CD-ROM drives still encourages new and innovative data storage uses that build on old formats without threatening universal compatibility. For more information on this topic, refer to *New and Evolving Standards* on page 84.

The slow CD-ROM drive performance and file-access issues were originally resolved by creating a structure where file and data links are all stored near the center of the disc in contiguous sectors. By not having to retrieve these details from all over the disc, the file-access time can be reduced considerably.

The adoption of the Universal Disc Format (UDF) for packet-written CDs and DVD discs has improved the situation further by removing the many confusing data structures and organizational issues that hindered multi-platform compatibility during the growth of the CD-ROM industry. UDF, a standard developed by the Optical Storage Technology Association, defines both the file system and the volume structure for a disc. To deal with the multitudes of existing ISO 9660 readers, a modified version of this format known as UDF Bridge offers backwards compatibility to ISO 9660 while including some of the features of UDF.

New and Evolving Standards

Lest you feel that the CD-ROM industry is standing still while the DVD-ROM industry steals most of the trade publication attention, some innovative new standards have grown new branches on the CD-ROM family tree, leading to bright new opportunities for developers and users.

CD-PROM

The CD-PROM is a recordable disc that includes a pre-recorded first session on it. Taking advantage of the flexibility of the Orange Book standard, CD-PROM provides a way to include programs along with blank media. Chapter 11 of Orange Book makes provisions for a recordable multisession disc, which they call a hybrid disc, that includes a first session that has already been mastered. The acronym stands for Compact Disc Programmable Read-Only Memory.

Taking advantage of this capability, some manufacturers, such as Mitsui, are embedding software applications in the mastered session. The Mitsui AutoProtect Demo Disc contains a flexible backup application that allows you to perform full and custom backups to blank media. Kodak was also one of the first to exploit this capability with their Picture CD, discussed in the next section.

The recordable media shares elements of CD-ROM technology (storing data on modulated grooves that resemble the pits in a conventional pressed disc) and CD-R technology (with grooves to allow laser tracking during write operations). The amount of mastered data that can be stored in the first session varies, depending on the application, according to Kodak. A stabilized cyanine dye provides the recording surface with the usual gold reflective layer that is typical of most CD-R media. The media when recorded will be readable in a standard CD-ROM drive.

Licensees of the technology, such as Rimage Corporation, are integrating CD-PROM into their publishing and duplication products, and Kodak states that the companies are working towards a system where customers will be able to create customized CD-PROMs that can perform a wide variety of applications. The goal is to enhance the system in such a way that end users will be able to write to the CD-PROM disc from the desktop.

The CD-PROM opens up a number of extremely interesting potential applications to enterprising developers. For example, specialized database applications with the database engine embedded in the first session could be useful to end users. Additional data could be appended to the recordable area of the disc over time. Multimedia tools, such as image cataloging products, could be provided in the premastered region to allow users to provide indexed access to the disc contents easily. Kodak's Picture CD also offers a fairly good proof-of-concept example.

Picture CD

Kodak's approach to storing and distributing digital images, the PhotoCD, has gained a loyal following with many professionals in the photography industry, but it never quite caught hold with the mainstream public. The newest wrinkle in image storage, the Picture CD, offers considerably more promise of widespread acceptance, including the benefits of bundled applications on the CD-PROM discs and inexpensive processing costs. The software tools included on the disc include file conversion utilities, viewing software, and image enhancement tools. Service time for processing is slated as 2 days (compared to up 2 weeks for PhotoCD turnarounds) and the image resolution (1024x1536) rivals the professional-caliber alternatives.

Kodak's approach to digital imaging employs Intel Imaging Technology, running on the Pentium-based servers that are used in Kodak's Qualex processing labs. Intel's contribution to the Picture CD also includes an Email Postcard utility and the 3D Photo Cube, two more ways that photographers enamored of digital processing techniques can distribute and display their work. Bundled software from American Greetings/The Learning Company lets computer users convert their work into greeting cards, making it that much easier to make that Christmas card of the kids and your faithful dog Sparky gathered in front of the fireplace.

From the Kodak site (www.kodak.com) you can identify a local processor who provides Picture CD service and even arrange for online delivery of images. A number of major chain retail stores will be participating in providing this service, including Walgreens, Fred Meyer, Albertson's, CVS, and target.

Unlike PhotoCDs, which store images at a variety of resolutions and can handle two or three rolls of film on a single disc, the Picture CD limits disc storage to one roll of 35mm or APS film per CD-PROM. The image resolution is also standardized at a single high-resolution maximum of 1024x1536, though this will need to be varied to accommodate the different aspect ratios available with APS images.

All in all, it looks like Kodak may have found the right mix of media and partners to bring photographs on disc to both the casual weekend photographer as well as the more serious semi-professional photographer.

Summary

Theory can be fascinating, but what you probably care most about is using the standards effectively prior to premastering a disc. Refer back to this chapter whenever you have questions about the underlying details of the different formats. The next chapter offers some insights into the standards that apply to DVD-ROMs and DVD-Videos. By the time you reach the practical applications discussed later in this book, you should be well equipped to adapt the standards to your various types of projects and choose the standard that best suits your delivery method for each project.

4

DVD Standards

The Digital Versatile Disc—DVD—fits its name very well. A veritable garden of variations have emerged from the inception of this format. By the time this book reaches bookstores, the situation may be more stable, but there is still a great deal of industry flux over the standards related to DVD. In fact, the acronym itself—DVD—is sometimes expanded as Digital Video Disc, so agreement at even the most fundamental level is still hard to find.

The good news is that there is also much stability in place, even at the moment. The discussion in this chapter will cover the variety of standards, working from the more stable to the less stable. The path roughly follows the sequence of the Books, A through E, that have been developed as the specifications for each of the formats (as discussed in the section titled *DVD-ROM* on page 88).

An Evolving Set of Standards

If the DVD standards had been developed logically and rationally by an august body of men and women, seated at a round table with all the skills of their intellect and insight coming to play, we'd probably have a more consistent framework for the growth of DVD. Instead, the standards evolution has been more like a room full of scrappy terriers trying to get the best of a fabric sock monkey and ripping it to pieces in the process. DVD has been pulled and tugged in every conceivable direction by forces representing sometimes diametrically opposed viewpoints. Some of the infighting has been over copy protection additions to discs, a vantage point made fairly ludicrous by the very quick cracking of the protection mechanism by a group of hackers interested in making DVDs operate under Linux. Some of the battles have been based on competing technologies, particularly for the rewritable forms of DVD, of which there are several variations at this moment with no sign that any one manufacturer is going to yield to the will of the others.

Nonetheless, the situation is improving. Most new DVD-ROM drives will currently read commercially pressed CD-ROMs, as well as CD-R, and some will handle CD-RW also. DVD players have reached a production volume where the prices have dropped below the $250 mark for entry level units and many of the incompatibilities exhibited by early releases of Digital Video titles have been eliminated. Consumers, initially wary of the medium from the constant flow of negative information from the press, have begun picking up players and DVD-ROM drives in record numbers and the once sluggish growth curve is now beginning to swing sharply higher. By the end of 1999, more than 5,000,000 DVD players had been sold and the unit sales continue to escalate.

DVD-ROM

The DVD-ROM is the computer data version of the digital versatile disc. As with the transition of the audio CD to CD-ROM, the DVD-ROM includes extended support for error detection and correction to allow it to be successfully applied to computer applications where a missing bit can bring down a system. Unlike the audio CD, however, DVD-ROM is considered the starting point of a succession of standards that includes:

- Book A: DVD-ROM
- Book B: DVD-Video
- Book C: DVD-Audio
- Book D: DVD-R (write-once)
- Book E: DVD-RAM (rewritable)

The DVD-Audio specification is also still in the works, with additional changes possible before it is finalized. This specification, as are all the others, will be an offshoot of DVD-ROM.

DVD-ROM

Data Storage Techniques

DVD-ROM stores data in user sectors, each consisting of 2064 bytes that are organized to support an error correction scheme. Of this total, 16 bytes are reserved for address information, error correction, and copy protection, leaving 2048 bytes for data. Data sectors are structured as 12 individual rows each consisting of 172 bytes. The beginning of each data sector contains 16 bytes of data, subdivided as follows:

- 4 bytes of identification data representing the sector ID
- 2 bytes of ID error detection data
- 6 bytes of copy protection data

The data sector is concluded with an additional code:

- 4 bytes consisting of an error detection code

The organization of the bytes within a DVD-ROM data sector is shown in Figure 4 - 1.

Figure 4 - 1   **Organization of a DVD-ROM data sector**

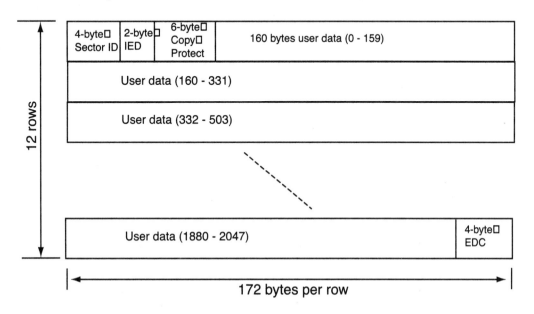

Rows of 16 data sectors are interleaved together and structured as blocks for applying error correction codes. A 16-byte Reed-Solomon code is generated for each of the 172 columns within the block. A 10-byte inner-parity Reed-Solomon code is generated for each of the 208 rows of the block. These codes are appended to the data, where they provide a flexible and robust means for detecting read errors within the data.

Through processing using 8/16 modulation, each data bytes is doubled to 16 bits, which produces a physical sector size of 4836 bytes. These bytes are generated on the disc surface, row-by-row, as channel data. As with data embedded on a CD-ROM, the Non-Return to Zero Inverted (NRZI) encoded method is used: transitions detected by the laser from a pit to a land are interpreted as binary ones; the absences of transitions are interpreted as binary zeros.

Channel data from the DVD media is transferred at the rate of 26.16M bps, which is then reduced by half by the application of the 16/8 demodulation process, resulting in a rate of 13.08M bps. After the adjusted overhead of error correction, the data transfer rate is a steady 11.08M bps. From the perspective of a DVD-ROM drive, data is transferred in logical units, each unit consisting of 2048 bytes.

UDF

One of the significant additions to the DVD standard is the widespread adoption of the Universal Data Format (UDF) as the means for dealing with files and volumes stored on disc. Although, theoretically, the data regions of a DVD-ROM can contain any type of data, most companies and organizations have followed the lead championed by OSTA, the Optical Disc Storage Association, and adhered with near-religious zeal to UDF for mapping file and volume structures.

Essentially, UDF refines a more broad framework constructed by the International Standards Organization in ISO 13346. UDF places limitations on ISO 13346, defining a structure that supports optional multivolume and multipartition divisions on a disc. This allows DVD-ROMs that include filename translations between platforms and support for extended attributes, such as the resource forks, icons, and file/creator types that are familiar to Macintosh users.

Within the UDF standard, the following platforms are supported:

- DOS
- OS/2
- MacOS
- Windows 95/98
- Windows NT
- UNIX

The ability to partition discs for different playback equipment makes it possible for manufacturers and title producers to provide content that is specific to a playback platform. In the same manner that Enhanced CDs include both audio content for a standard CD player and computer data for playback in a CD-ROM drive, a DVD disc can have player and computer partitions. The digital video content and a wide range of interactive content can be included in the partition designed for the DVD player; this can include a director's commentary on a film's production issues, alternative language editions, and so on. The DVD-ROM partition can include full interactive multimedia content, include games, screen savers, links to Web sites where additional information is available, background information on the cast and crew and design team, and similar kinds of content.

Most computer platforms also include some form of video player software for DVD, allowing films to be played on the desktop with the same crisp resolution and fluid playback that you will find on a dedicated player. The key element to making this happen is an MPEG decoder in the playback system, either implemented in hardware or as a standalone software component. Many DVD-ROM kits include MPEG decoding hardware as part of the package. However, if your DVD-ROM drive lacks this hardware, you must obtain a software decoder or you will be unable to playback DVD-Video titles. Obviously, decoders embedded in hardware relieve the playback system processor of the burden of carrying out the intensive conversion process on the fly, thereby improving playback performance, so whenever possible hardware decoding is the preferable approach. The other element to the DVD-Video playback scenario is that a hardware or software decoder must be able to handle the encryption scheme that is built into DVD-Video discs for copy protection. Xing Technologies recently had their software decoder removed from circulation because it permitted decoded video content to be captured and saved on a computer's hard disk drive. As of the end of 1999, I was unable to find any commercially available software-only decoders on the Internet, so this option may no longer be available to computer users.

DVD-Video

DVD-Video is a specialized form of DVD-ROM that is tailored to the presentation of very high quality audio and video content optimized for set-top players. This is the format that the film studios, video publishers, and consumer electronics manufacturers have been backing as the predominant delivery medium for motion pictures in the new millennium. From a hesitant beginning, the format appears to have finally caught hold solidly and, given the backing of so many of the major corporations involved in entertainment and consumer electronics, continuing success is very nearly a sure thing.

DVD-Video relies the compression capabilities of MPEG-2 to provide, minimally, 94 minutes of video playback, but up to several hours of playback using the higher capacity formats.

A key aspect of the DVD-Video format is playback on the full range of standard NTSC and PAL television displays using analog data connections, ensuring broad compatibility with the installed television sets around the world. Most current generation DVD players also include additional digital data connections, including S-Video and optical connections for those televisions that support this form of input. High-definition televisions that are beginning to appear in the market are also supported by this digital interface.

Multichannel digital audio support is also an inherent feature of this medium, allowing audio content to be played on standard stereo audio systems as well as more elaborate home theater systems. Up to eight channels of Dolby Digital audio provide the potential for excellent spatial orientation and rich, full sound to accompany videos. To achieve the best results for audio works, specialized mastering tools must be used to separate the audio tracks into individual components. If this is not done effectively, the resulting audio performance can exhibit annoying characteristics, such as drifting orientation of the dialog track or poor signal clarity.

File Formats under DVD-Video

A DVD-Video disc can contain data for both playback on a DVD player and additional data content designed for computer playback. Based on the UDF specification, a specific directory is designated to store the files, VIDEO_TS, and an informational file titled VIDEO_TS.IFO must also be present. VIDEO_TS.IFO is used to store the video manager title set, which provides the basis for the Main Menu that appears when the DVD is mounted in the player. Other title set information is contained in additional .IFO files and backup copies containing this same information are also maintained. Up to 10 video object block (.VOB) files can be created for each title that appears on the disc; these become the logical divisions by which the disc content can be navigated. Directories and files not intended for use by the DVD player must be stored after the DVD-V data; these files are typically ignored by the player.

The original UDF standard was modified with an appendix, the MicroUDF, to simplify the recommended requirements that must be met by a DVD player, in the interests of encouraging widespread manufacturing of consumer-level playback equipment. Appendix 6.9 of the UDF standard includes the following provisions:

- No multisession formats or boot descriptors are permitted on a DVD disc.

- Individual files must be contiguous and smaller than 1 Gigabyte.

- No more than one logical volume, one partition, and one file set can be included on a single-side of a disc.

- DVD players should support UDF in anticipation of the phasing out of ISO 9660.

- No more than 8 bits per character should be allocated for file and directory names.

- Aliases are not allowed for linking.

The basic file structure used on a DVD-V disc is shown in the following figure.

Chapter 4

Figure 4 - 2   **File Structure for DVD-V**

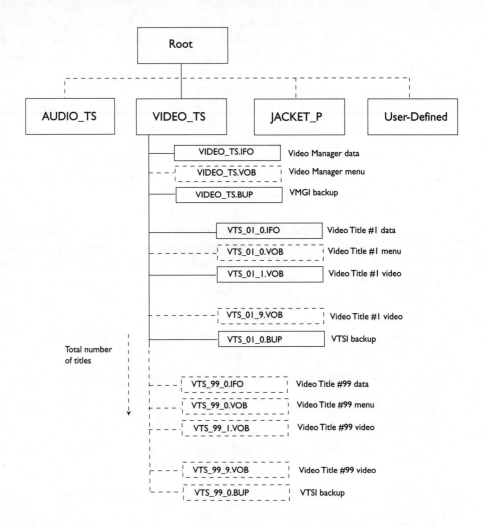

Navigating DVD-V Content

Unlike linear systems for video playback, such as film or videotape, DVD offers random access to the content on disc. Once is a disc is inserted in a player, a main menu can be accessed, which guides the user through the various types of content available on the disc. This may include a chapter view of a movie (allowing the viewer to jump to a particular point in a film), production information in video form, still images, audio content, interviews with directors and cast, and so on.

DVD-Video

This type of content is handled by a presentation engine and a navigation engine contained in the player. This provides the equivalent to a computer user interface to the viewer, allowing someone to control the selection of the different elements for playback or to activate certain features available on the DVD, such as enabling an audio commentary by the director to be played at the same time the video is being presented.

Figure 4 - 3 **Navigation and presentation engines**

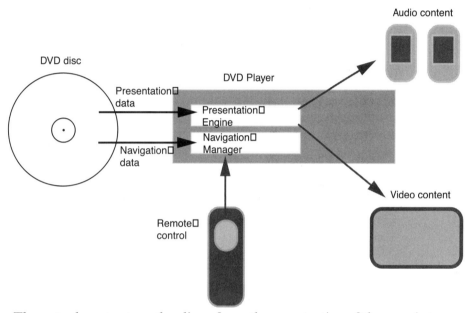

The actual content on the discs, from the perspective of the user interface, consists of:

- *Titles*, which may be films, videos, or album material
- *Parts of titles*, which may be chapters or individual songs

Each DVD can contain up to 99 discrete titles. Each title can be further subdivided into 10 chapters. For example, a disc containing the three-part television miniseries Stephen King's *Storm of the Century* might contain an individual title for each part and then 8 chapters within each title, allowing selection of any part of the series within 12 to 15 minute increments.

Optionally, a DVD can contain only a single title. For more complex projects, titles can also be nested within other titles and each title can have its own title menu. Viewers navigate between the various options using a remote control that features directional arrow keys and a select button, as well as the usual assortment of controls that one might find on a VCR remote, such as pause, scan, forward, back, and so on.

A presentation can include many different optional features that are handled by the navigation manager. For example, a project might incorporate a second camera angle that could be enabled through the menu to allow viewing of the content from a different perspective. Subtitles might be available for display when viewing foreign films. If any of these kinds of features are not present, the disc developer simply disables them while the disc is being produced.

Authoring for DVD-V

As might be expected, the kinds of authoring tools that are available for creating DVD-V content are very specialized and the professional-caliber tools on the market remain very expensive in comparison to multimedia authoring tools for producing CD-ROM content.

Encoders for producing MPEG-2 video range in price from software-only solutions suitable for home and consumer applications to elaborate standalone systems with high-speed processor tailored to providing the highest quality video at accelerated rates. Professional-caliber systems generally cost tens of thousands of dollars, although there has been a progressive reduction in cost as the industry has expanded and matured.

More information on authoring tools and techniques is provided in Chapter 17, *DVD Creation*.

DVD-Audio

DVD-Audio is being positioned as a replacement for CD-DA (Compact Disc - Digital Audio), though the launch of players and titles has been slowed by ongoing copy protection issues. A planned launch for early in the year 2000 was delayed when the DVD industry was rocked by news of a European hacker who broke the protection scheme. The spectre of widespread piracy of DVD content has been one of the ongoing stumbling blocks to the release of both DVD players and DVD-ROM drives, so it is not particularly surprising to observe the reaction of the hardware producers and industry players to this perceived threat to their copyrighted material. Anticipated release dates for new equipment have been delayed to midyear of the year 2000 by some manufacturers; others have delayed their announcements pending deeper examination of the protection issues and possible solutions.

As it stands, the DVD-Audio standard provides several significant enhancements over audio delivered via CD:

- Multichannel audio: the availability of 5.1 channels supports surround sound, encoded as Dolby Digital or MPEG-2 data
- Higher sampling rates: digital audio content can be sampled at rates up to 96K bps, producing greater frequency response
- Larger sample sizes: dedicating more bits to the sample sizes provides extended dynamic range and increased depth-of-field
- Display of album titles, lyrics, artist's names, and song titles during playback

The proposed format for DVD-Audio includes the capability of including a DVD-Video sector. This approach makes it possible to create a high-capacity DVD equivalent to an Enhanced CD, with the audio material supplemented by video material and URLs that could be linked during playback in a DVD-ROM player. Another way of looking at it is that DVD-Audio combines the features of three separate DVD formats: combining audio, video, and computer data within a single framework.

The Working Group 4 (WG-4) of the DVD Forum continues to refine the DVD-Audio standard, but the fundamental elements of this standard have at least reached the point where player manufacturers have felt confident enough to design and engineer the playback equipment, awaiting product rollout dates based on the resolution of the remaining outstanding issues (including copy protection).

Authoring of DVD-Audio content requires both a different set of tools and a different mindset for the development community. The abundant storage capabilities encourage titles that include not only audio, but video clips and computer data as well. However, as has been the case with Enhanced CDs, the extra costs associated with this type of development may limit the number of titles that include bonus content and extras. Different hardware and software tools are required to be able to record and mix the audio for 5.1 channel playback, and some competing standards, such as DTS, will vie for attention and market share with Dolby Digital.

The potential, however, for this DVD-Audio is enormous. Audiophiles and music lovers looking for a way to experience more rich and vibrant sound, with the added benefit of audio and video content should progressively be won over by the advantages as playback equipment and discs reach the marketplace.

DVD-R

If you want to burn a DVD disc designed for the widest possible playback, including current DVD players and DVD-ROM drives, DVD-R is the way to go. DVD-R is the write-once member of the writable family tree. The other rewritable formats currently have compatibility issues with much of the present generation playback equipment, limiting their usefulness to general purpose storage and archiving where there is control over the choice of readers.

The DVD Forum has defined two individual categories in the DVD-R Book version 1.9 to accommodate different uses of DVD-R:

- DVD-R for Authoring: utilizes a laser wavelength for writing of 635 nanometers. This variation is designed for authoring only.
- DVD-R for General: utilizes a laser wavelength for writing of 650 nanometers, suitable for write operations intended for general uses.

Depending on the selection of recordable media, a DVD-R can handle up to 3.95GB or 4.7GB of data per side. Double-sided media boost the ultimate recording capacity to 9.4GB per disc.

Write speeds for recordable DVD are significantly higher than most CD-R units. A 1x DVD-R unit records data at 11.08 Megabits per second compared to a 1x CD-R drive recording at 1.23 Megabits per second. Playback of recorded DVD-R discs can take place up to the maximum speed available on the DVD-ROM drive being used for playback.

The extra capacity of a DVD-R disc derives from the use of a red laser with a 635 or 650nm wavelength and an objective lens with a numerical aperture of 0.6. The smallest recorded mark possible on a CD-R disc is 0.834 micrometers, while DVD-R pits can be as small as 0.4 micrometers. This results in an increase of almost seven times the data density of the DVD-R disc over what can be stored on CD-R media.

The track pitch—the distance between two adjacent tracks on the spiral of data—is also much tighter for DVD-R than CD-R: 0.74 microns for the 4.7GB media and 0.8 microns for the 3.95GB media. All of these factors—the laser wavelength, pit size, track pitch—make it possible to achieve the high storage capacities available on DVD-R.

Playback Compatibility

DVD-R drives offer the advantage of broad playback compatibility, which includes most of the playback equipment available for DVD format discs. Discs can be recorded in either DVD-V or DVD-ROM format.

A properly recorded DVD-Video disc produced on a DVD-R machine can be played on:

- Most standalone DVD-Video players
- DVD-ROM drives installed in a host computer as long as they are equipped with either a hardware-based MPEG decoder or a software decoder that performs the same function

DVD-ROM discs that are recorded on a DVD-R drive can generally be played back on:

- A DVD-ROM drive installed in a host computer, without built-in MPEG decoding
- A DVD-ROM drive installed in a host computer with either hardware or software MPEG decoding (if access to DVD Video material is required)

Both MPEG-1 and MPEG-2 decoders are available in both hardware and software implementations. The earlier MPEG-1 encoding method was first introduced for storing video content on CD-ROM. Commercial Video CDs, including movie titles, were introduced in the market shortly after multimedia CD-ROMs became popular. The Video CD format fared more successfully in Europe and Japan than in the U.S. Many DVD players can also play back Video CDs, if you happen to have any of the released titles in your collection. Most MPEG-2 decoders can handle both the MPEG-2 and MPEG-1 data formats.

File System for DVD-ROM

As described earlier in the chapter, the file system for DVD discs is much different than the fragmented system that was designed for CD-ROMs as they evolved and grew to encompass many different types of data formats.

Instead, the more unified UDF structure was applied to all types of DVD discs and this file system is suitable for all forms of content and any type of file format designed for storage on optical disc. It is also designed to be adaptable to all the major computer operating systems.

To maintain backwards compatibility with earlier computers and operating systems that are not designed to read UDF, the UDF Bridge file system was designed. UDF Bridge is a hybrid system that includes support for discs recorded using the original ISO-9660 file system that originated with CD-ROMs. UDF Bridge also provides full support for discs containing files structured under UDF, as well.

UDF Bridge maintains an important optical disc convention that has been followed ever since the Yellow Book CD-ROM format was introduced as a means of extending the compact disc to include computer data, as well as audio data. Computer-based playback devices and decoders, as much as possible, have been designed to read all earlier formats. CD-ROM drives were designed to read audio CDs. DVD-ROM drives, from the beginning, were equipped to be able to manage CD-ROM playback. The UDF Bridge format makes it possible to produce DVDs for older PCs running operating systems that do not yet have UDF support.

In comparison, playback devices targeted for the entertainment industry and consumer use, such as audio CD players and DVD playback equipment, usually only accommodate a single format. The Enhanced CD format was designed as a hybrid format—containing both audio and computer data. This allows CD players to play the audio present on the compact disc, while the computer data files could only be read if the disc were inserted in a PC CD-ROM drive. Similarly, DVD-Video discs often contain content designed for the PC, which is essentially ignored by the DVD playback equipment. The additional data only becomes available if you insert the disc in a DVD-ROM drive.

Writing to DVD-R Media

Anyone familiar with CD recording applications and equipment will recognize the very close parallels to DVD-R write operations. The recording process is handled by an application running on the host computer connected to the DVD-R unit. For example, GEAR Pro DVD or Sonic Solutions DVD-It are two such applications. They allow the person doing the recording to select the files and organize them for the write operation. The application then manages the actual recording process, controlling the DVD recorder until the write operation is complete.

DVD-R

Like CD-R discs, DVD-R discs can be fully written in one complete operation—known as Disc-at-Once—or written incrementally over several individual sessions. Unlike CD-R discs, however, data written to DVD-R discs occurs in a slightly different sequence. The recording application first produces a lead-in area, followed by user data area, followed by a lead-out area. The lead-in and lead-out areas contain information that allows the DVD player or DVD-ROM drive to properly access the full range of data. These same two areas also appear on commercially pressed DVD discs; they are essential to playback. The user data, sandwiched in between these two regions, can vary from a 32Kb block—the minimum amount of data that can be recorded—up to the maximum capacity of the recordable media: 4.7GB.

In comparison, CD-R discs are written in a different sequence. The user data regions are recorded first. Next comes the lead-in area and the table of contents. The write operation is concluded typically by writing the lead-out area.

During write operations, the host computer must be able to deliver the data at 11.08 Megabits per second to prevent any interruptions in the data pattern being recorded to disc. Buffering is used in the DVD-R drive to compensate for any interruptions in the flow of data.

When incremental write operations are being performed, the complete file system data is not available until all of the individual writes have been finished, so the disc must be finalized before it can be read in any device other than another DVD-R drive. Finalization calculates and records the data contained in the lead-in and lead-out areas and then records this information to disc. Once a disc has been finalized, it can be read by other DVD playback devices. No further data can be recorded after a disc has been finalized.

At the recording speeds of the current generation of equipment, a 3.95GB DVD-R disc can be fully recorded in about 50 minutes. The additional recording capacity available on a 4.7GB disc requires an additional 10 minutes or so to completely finish a write operation.

Uses for DVD-R

The high cost of DVD-R recorders may dissuade many corporate and individual users who are simply looking for archival storage or backup applications. Those categories of users may find a better solution in the rewritable formats discussed in the next section.

Chapter 4

Instead, the broad compatibility of DVD-R makes it a medium better suited for high-volume data distribution, particularly in situations where there little control over the playback devices that will be used to read the discs. DVD-Video discs made using DVD-R media will be readable in the vast majority of DVD players, as well as those DVD-ROM drives equipped with the required decoding hardware. DVD-ROM discs produced using DVD-R equipment should be readable in all DVD-ROM drives.

DVD-R is also the medium of choice for testing and development applications, where it is essential to emulate the target playback equipment in preparation for releasing a title on DVD. To avoid costly errors when a DVD title is being authored, DVD-R lets developers and testing firms produce discs that can than be run in standard playback equipment—either a DVD player or DVD-ROM drive. Any inaccuracies or problems with the playback can be detected and corrected before a disc goes out for mass replication.

Producing DVD discs for presentations, particularly presentations destined for portable DVD player playback, represents another ideal use for this medium. The interactive capabilities of the DVD-Video format make it possible to author presentations that are similar to full interactive multimedia applications.

Certain archival applications, such as storing image data, audio material, motion pictures, and so on, actually favors the write-once characteristics of DVD-R. Archivists who want to record data and then ensure that it is not altered can rely on the properties of DVD-R media to protect their stored file contents. The estimated 100-year plus lifespan of the DVD-R discs also provides assurance that long-term archiving can be accomplished safely using this form of optical recording. For archival operations where data must be available for near-line access, DVD-ROM jukeboxes provide a means of storing extremely large quantities of data for convenient access. For example, a 100-disc DVD-ROM jukebox can handle close to a half-terabyte (470GB) of information.

For shorter term archiving and storage, the rewritable storage options provided by DVD-RAM, discussed in the next section, may provide a better alternative.

DVD-RAM

The DVD-RAM format was the first of the rewritable DVD formats to reach the market. DVD-RAM drives are much less expensive than DVD-R equipment, primarily because they use phase-change technology for storing data rather than the method used by DVD-R—using an organic dye to record laser impressions on disc.

The phase-change technology—which is an amalgam of the technologies used in magneto-optical cartridges, CD-RW, and Pensioned's PD devices—is also the reason that discs created on a DVD-RAM drive cannot be read in the typical DVD-ROM drive or DVD player. This significant drawback limits the utility of this particular storage method, but advances in player technology may overcome the limitation, much in the same way that early CD-RW discs could only be read in CD-RW players. Now, Multi-Read-compatible players can handle CD-RW discs and CD-R discs with equal aplomb. Similar advances may make DVD-RAM discs more widely compatible.

Data patterns written to a DVD-RAM disc are recorded to a thin film that is sensitive to laser light. The DVD-RAM write laser strikes the film surface and converts the material from a crystalline state to an amorphous state. The reflectivity of these two different states is different enough that it serves to identify bit patterns. The disc surface can be "erased" by applying a different intensity laser burst. The energy from this burst converts the film from the amorphous state back to the original crystalline state.

DVD-RAM discs employ the same kind of modulation and error-correction codes that are used for DVD-Video and DVD-ROM. This characteristic should help ensure broadened future compatibility for this media type. Both single-sided and dual-sided media are available. Each side of the DVD-RAM disc has a capacity of 2.6GB. In most regards, the physical discs have the same dimensions as DVD-R discs, but both single-sided and double-sided discs are housed in cartridges for use.

Clock data is embedded in a wobble pattern integrated into the tracks, also offering a means by which address signals can be identified by the drive.

Chapter 4

Uses for DVD-RAM

The primary appeal of DVD-RAM is inexpensive, flexible, abundant storage—this benefit comes at the loss of compatibility with the majority of players and drives in the market.

DVD-RAM is well suited to these kinds of uses:

- Short or long-term storage of press-quality images, digital audio files, digital video files, or other similar kinds of content requiring large amounts of storage space
- Local periodic backup and temporary archiving of network files, personal hard disks, organizational files, and so on
- Exchange of large-volume files between parties with similar equipment (compatible DVD-RAM drives)
- Nearline storage of mission-critical data that is important to an organization's operation, but cannot be concisely stored on the network.
- Network-resident storage for periodic system backup or personal workstation archiving, through thin server technologies

As the DVD-RAM technologies matures and the adoption of the Super MultiRead standard makes it possible to exchange disc cartridges freely, this format will become increasingly important for both large-scale storage and data exchange.

Summary

In many ways, the evolution of the DVD standards mirrors the flux and chaos that marked the course of recordable CD development, but there are also some clear advantages to the way data storage has been handled on DVD. The format is far more flexible when it comes to embedded different data types onto disc, without the necessity for creating individual formats for each of the individual data types (as can be seen on CD-ROM with Video CDs, Photo CDs, CD-PROM, CD-ROM XA, and so on). The UDF standard also intelligently handles most of the key cross-platform issues, making DVDs less prone to the kind of file translation issues and file system concerns that have complicated CD-ROM delivery. The backwards compatibility with CD-ROMs and audio CDs, which is a requisite feature of most drives and players, also makes DVD the logical successor to the CDs data storage throne.

5

Optical Recording Equipment

Selecting the right piece of optical recording gear depends very strongly on the intended use. Products in this category range from simple CD recorders that can be installed internally in the most modest PC to expensive DVD-R units designed for testing and producing discs for replication. While cost is certainly a factor when considering the best type of recorder for your applications, you should also take into account playback compatibility, media costs, the available computer hardware interface, recording performance, platform support, bundled software, and similar factors. This chapter surveys the different types of optical recording equipment and provides examples of some of the available gear.

Most recordable CD and DVD drives require a host computer. While there are a number of stand-alone recordable CD units available, primarily for use in compiling custom audio discs and producing disc masters for musicians, these units tend to be more limited in their uses than recorders that connect to a host computer. While specialized standalone recorders can be convenient and easy to operate, there are a number of advantages to being able to process and prepare files in a staging area on your host computer prior to recording. For example, if you have an interest in converting your collection of vinyl record albums to compact disc format, there are a number of excellent computer-based applications for handling noise reduction, click removal, and overall vinyl restoration. The options when capturing sound from vinyl records and rerecording to CD on a standalone recorder are much more limited; while you may be offered some rudimentary noise reduction tools, the range and flexibility of PC-based sound editing software is far superior.

Chapter 5

Beyond standalone equipment used for audio applications, a host computer is essential for most data applications. The one exception is simple stand-alone disc copiers, which are discussed in Chapter 8, Disc Copiers on page 222. These are designed to simply produce a replica of a pre-existing compact disc, so they don't require any premastering or data selection before recording.

Whether you own a Macintosh, a Wintel machine, or a Linux computer, you have many different choices available when selecting a CD recorder. The primary decisions that you need to make include:

- Choosing an internal or external unit
- Selecting an appropriate computer interface
- Deciding on the required maximum recording speed
- Deciding if you need rewritable capabilities (CD-RW)
- Determining if the CD-R capacity is sufficient (650MB) or if you need to consider DVD-RAM or DVD-R
- Deciding if you need high-speed disc duplication capabilities

The options aren't quite as numerous when choosing one of the forms of recordable DVD. The different forms of recordable DVD have only been on the market a short time in comparison to CD-R and CD-RW. Prices are higher and options are fewer. You also need to make some serious decisions as to which form of recordable DVD makes the most sense for you. Serious professionals producing discs for replication require the more expensive DVD-R devices; those with a primary interest in storing or exchanging data can choose either DVD-RAM and or one of the other emerging formats (DVD+RW is being re-engineered for a possible release in 2001).

If your needs are more specialized and they favor uses such as short-term or long-term archiving, you may want to consider those forms of optical recording such as magneto-optical drives or PD units that appeal to niche markets. A few examples of these kinds of devices are included, but this chapter focuses more heavily on CD and DVD equipment.

Selecting a Computer Interface

Like other kinds of removable storage devices, CD and DVD recorders can be installed internally within the host computer or connected externally to one of several different I/O interfaces. First-generation CD recorder products were almost invariably SCSI-based, but enterprising manufacturers have adapted the technology to almost all of the popular computer interfaces, including EIDE, parallel port, and Universal Serial Bus (USB). Some companies that once declared they would never release anything other than SCSI versions of their products have followed the trend towards inexpensive ATAPI IDE bus interfaces. This has opened up the market for wider acceptance of disc recorders, but it has also made it mandatory to carefully evaluate system performance issues and limit other operations when recording.

Before choosing an interface, you might be wondering how much of a computer is needed to effectively handle the recording process. The following section deals with the performance issues related to CD and DVD recording.

Performance Issues

Performance issues related to CD recorder operation can be deceptive. In the overall scheme of computer data transfer rates, the rates at which data gets transferred to a recordable compact disc are relatively slow. Double-speed recorders clock data at the rate of 300KB per second. More capable quad-speed recorders can double this transfer rate. As of press time, we're seeing a fair number of 8x recorders, which—if you see the pattern emerging—reach 1200KB per second. If you were to base your decisions on these numbers, you wouldn't think that you would need the latest, super-tuned Pentium-III-based machine to serve as the host computer for CD recording—if a SCSI bus can transfer data at 10MB per second, it would seem as though this level of data transfer would be adequate to keep everything flowing at the necessary rate, even if the host computer is an ancient 25MHz 486SX machine.

While a number of CD recorder applications list the minimum hardware requirement as being a Pentium processor machine or better, you need to pay careful attention to your overall system configuration to ensure that you'll be able to successfully write to disc. Even if you have more modern equipment—for example, a 500MHz Pentium-III system—you still need to be aware of a number of performance issues to maximize the success of your recording efforts. In most cases that means you will need to limit the number of processes taking place at the same time. If you try to record a disc while you're running network software, a virus protection program, and then the 3D screen saver kicks in, there is a pretty good chance your disc recording will fail. The message here is that even with the highest performance systems available, you need to carefully monitor and limit the operations that are being performed simultaneously while disc recording is taking place. In some cases, this means shutting down all other unnecessary processes.

Macintosh users have another concern. CD recording on the Macintosh used to be a nearly trouble-free enterprise, since every Macintosh on the market had a SCSI port and almost all the CD recorders were set up for SCSI. In this era of inexpensive iMac computers, you can't depend on having an available SCSI interface, so you're faced with the option of either adding SCSI capabilities (through a PC card in a laptop or a PCI circuit board in a desktop machine) or investigating one of the other CD recorder interfaces, such as USB.

The first order of business for Macintosh users is examining their system to see what interface options are available and then deciding what kind of CD recorder interface to choose. While the FireWire (IEEE-1394) standard is gaining in popularity for high-speed serial data exchange, its use so far has been more common for hard disk drives, digital video cameras, and similar equipment. Mac users should appreciate that fact that FireWire-compatible CD-R or CD-RW recorders are beginning to reach the market and these kinds of devices should become increasingly common because of the speed and reliability of this data transfer method.

As you have probably heard many times, successful CD and DVD recording depends on an uninterrupted flow of data from the host computer to the recorder. The greater the excess processing power and internal data transfer rates, the less the likelihood you will ever reach those marginal areas where your computer can't keep the data moving from disk to disc at sufficient speeds. Whenever you have a situation where a number of devices are simultaneously vying for computer resources and seeking processor attention, you have the potential for data slowdowns. There are a number of factors that can subvert the continuous flow of data, and any of these factors under certain circumstances can cause problems during disc recording. These factors are discussed in later subsections.

Interface Options

The primary interface options available for CD and DVD recorders include:

- **SCSI**: the original, predominant choice for connecting a disc recorder to a computer, SCSI is still a very popular option, particularly for professional applications. Several variations of SCSI have evolved in recent years, offering higher performance and improved behavior when servicing multiple devices. Adaptec and other companies continue to extend SCSI capabilities and introduce new host adapters into the market.

- **Universal Serial Bus** (USB): designed as a low-cost, trouble-free means of interconnecting devices—such as digital cameras and storage devices—to computers, many new computers include one or two USB ports as standard equipment. USB uses serial data transfer techniques, but achieves transfer rates close to 100 times faster than a standard computer serial port. USB-compatible devices can be *hot-swapped*, which means they can be attached or removed without shutting down the computer.

- **Parallel Port**: the chief advantage of the parallel port interface is that almost every computer has one (with the exception of Macintosh computers). Performance for the modern bi-directional parallel port (compliant with the IE-1284 standard) is adequate for operations such as CD recording, as long as other system operations are kept in balance. Enhancements to the parallel port standard, such as EPP or ECP, raise performance capabilities and may be required for some recorder connections. While not the most robust transfer method available, the parallel port is still convenient, particularly when connecting recorders to laptop computers that may not have other available interfaces.

- **ATAPI IDE**: for internal disc recorder installations, ATAPI IDE requires very little circuitry to implement and, because of this, it represents a very inexpensive interface method. Most modern computer systems can support flawless communication with recorders using this interface and setup is generally very easy. This interface option, however, is not as robust as SCSI for demanding applications in the industrial or professional sector. You are also limited to internal installations when you rely on this interface; there are no external ATAPI IDE units on the market because of the physical requirements of this I/O interface.

- **IEE-1394/FireWire**: A newcomer to the interface options, the FireWire serial interface originated on Apple computers and originally was targeted at quickly handling large volumes of data, such as encountered with hard disc arrays and digital video cameras. Like USB, FireWire is a hot swappable interface, allowing devices to be added or removed without shutting equipment down. While there are just a few disc recorders available at the moment that are designed for this interface, the benefits of FireWire may encourage more manufacturers to produce units compatible with this standard.

- **PC Card (PCMCIA)**: A few portable CD-R units, notably Hewlett-Packard's CD-Writer Plus M820e, utilize the PC Card interface (formerly called PCMCIA), which is commonly available on laptop computers and as an add-on peripheral for desktop equipment. This interface provides reasonably fast, reliable data transfer and is a good option when interfacing with portable equipment.

The available interface options are discussed in more detail in the following sections.

SCSI Considerations

The Small Computer System Interface, SCSI (pronounced "scuzzy"), is a bus standard that allows a number of different devices to communicate data over a parallel bus. Depending on the SCSI variation, the parallel bus can be 8 bits, 16 bits, or 32 bits wide. SCSI first became widely established in the Apple and UNIX marketplace, where an assortment of devices were developed to take advantage of SCSI's ability to *daisy-chain* multiple devices under a single interconnect scheme, controlled by a single host adapter. The kinds of devices that can be interconnected include:

- Hard disk drives and disk arrays
- Tape backup devices
- Scanners
- Printers
- External storage devices, such magneto-optical cartridge drives and removable cartridge drives

SCSI also supports the use of additional host adapters installed in a single system to provide multiple daisy chains of devices. This capability can prove useful in a production environment or in situations where many different types of peripherals are required. Graphics arts studios, with their needs for scanning equipment, near-line storage, and high-resolution printers, are one possible

While SCSI has made inroads in the PC marketplace, its use is more prevalent in higher end machines, including business and professional computers. The higher cost of SCSI components has led many manufacturers to use less expensive interface methods, such as USB and ATAPI IDE, to reach larger markets. Even stalwart SCSI advocates such as Plextor have expanded their drive options to include IDE equipment. The less robust device handling facilities of IDE have been to some degree offset by recent advances in microprocessor speeds. The extra processing power helps to make up for some of the bus-handling issues, minimizing conflicts when servicing multiple IDE devices.

Sometimes you can find a PC-based system that includes built-in SCSI support, but generally you have to add a circuit board to serve as the SCSI host adapter. Once you have a SCSI-capable system, however, SCSI makes it easy to add equipment—especially external equipment. Owing to the prevalence of SCSI on each of the major computer platforms, its multitasking capabilities, and its ability to maintain high data-transfer rates, many of the current CD recorders on the market use SCSI interfaces.

SCSI Connectors SCSI connectors, unfortunately, come in many different sizes and shapes. External connectors of SCSI-1 vintage usually have 50 pins and resemble Centronics printer connectors of years past. Macintosh SCSI-1 connectors, however, have 25 pins and use a D-shell type connector, similar to the connector type used for RS-232 serial connections. SCSI-2 connectors, the kind found most frequently on CD recorders, often use a 50-pin, half-pitch connector, much smaller than most connectors, with a unique clamp-in fastener to lock it to the drive or adapter. There are also a variety of proprietary connectors that inevitably require special cabling to attach to anything. Avoid these, if possible, to simplify your interconnection tasks.

The original SCSI-1 standard can support data transfers up to 5MB per second. Since recording to a quad-speed CD-R drive only requires data transfer rates of 600KB per second, SCSI-1 is sufficient for most CD-R applications. However, if a system configuration includes many different types of devices on the SCSI bus, bus traffic may create slowdowns and prevent the data flow to the CD-R from being maintained at the appropriate rate.

Most current CD recorders use the more modern variation of the SCSI standard: SCSI-2. Advances in the SCSI standard have led to the introduction of a whole series of successors, including: FastSCSI, Ultra SCSI, UltraWide SCSI, and the current speed champion, Ultra2 SCSI.

A faster version of SCSI can help reduce bus traffic concerns. The Fast SCSI version of SCSI-2 supports transfer rates up to 10MB per second over an 8-bit or 16-bit bus. Most mainstream SCSI host adapters now use 16-bit or 32-bit buses, and 64-bit buses have become available with the introduction of the PCI-X bus. Ultra SCSI and UltraWide SCSI accelerate transfers to 20MB and 40MB per second, respectively. These standards use 16- or 32-bit buses for transfers, which are becoming more common in CD-recordable equipment designs. For the ultimate in SCSI performance, the Ultra2 SCSI implementation offers data transfer rates of 80MB per second. Increasing the clock rate on the bus to 40MHz made it possible to achieve this level of sustained data transfer. This newest release is designed to be backwards compatible with all earlier SCSI devices.

Another advantage of Ultra2 SCSI is the adoption of a Low Voltage Differential (LVD) I/O interface that supports longer cable lengths by improving signal transmissions. With a simple point-to-point connection between one SCSI host adapter and a CD-R drive (or other device) the maximum cable length can be extended to 82 feet. On a fully loaded cable with multiple SCSI devices, the maximum length is 39 feet. Up to 15 devices can be interconnected on an Ultra2 SCSI daisy chain.

Adaptec is working on new versions of the SCSI standard, such Ultra160/m SCSI with a sustained transfer rate of 160MB per second. Looking ahead into the forseeable future, SCSI appears to have a ongoing role in handling high-performance I/O applications.

Rules of Thumb for SCSI Daisy Chains

A daisy chain of SCSI devices can contain up to seven units, one of which must be the host adapter. Some configuration specialists recommend connecting your CD recorder to its own independent host adapter with nothing else attached to it. In other words, keep all your other hard disk drives, printers, scanners, and backup devices on a second, separate host adapter. Other specialists and engineers suggest that there is no particu-

SCSI Considerations

lar advantage to using two adapters, and in many cases it just adds configurations problems to your system.

However you choose to implement your devices, you should ensure that you don't string two many devices off a single adapter—two, possibly three, is the limit to the number of hard disk drives that should share the daisy chain with your CD recorder. Furthermore, keep noisy devices, such as scanners, off the chain when doing recording; these devices can generate spurious signals even when they are not actively being accessed. Keep your cables lengths short, and use double-shielded cables in environments where stray signals may be a problem.

Termination

SCSI is a fairly simple standard for equipment use, with a few simple implementation rules that nonetheless seem to cause recurring problems for people interconnecting equipment. The first rule involves *termination*: The first and last devices in a SCSI chain must be terminated. Terminators, or termination resistors, are designed to damp signals traveling through the cables so that the signal does not reflect or "ring." Ringing results when the signal reaches one end of the chain and a portion of the signal is reflected back in such a way that the original signal is distorted. Lack of termination can interfere with the communication of devices on the bus and make the entire SCSI chain unusable.

Termination resistors come in a number of different forms. Your CD recorder may have a switch on the back panel that enables or disables termination, or you may have to add a terminator plug (which looks like a connector without a cable extending from it) to the unused connector on the back panel. Host adapters need to be terminated if the system has only an internal driver, or if the system has one or more external drives and there is no internal drive. In both these cases, the host adapter represents the end point in the chain. The termination is usually in the form of a jumper somewhere on the host adapter, or a series of resistor packs that plug into sockets on the adapter circuit board. Check your host adapter documentation for positive identification of the termination technique in use.

If the host adapter is in the middle of the chain—in situations where you have both internal and external devices in the system—you should *remove* the termination (by changing the appropriate jumper or removing the resistor packs from the circuit board).

Sometimes the lack of termination resistors on a SCSI chain won't stop the devices on the chain from working completely, but will cause sporadic data errors and mysterious system failures. If you experience these kinds of problems on a system with SCSI devices, always check to see first that the terminators are installed properly. Newer SCSI devices are equipped with active termination, consisting of electronics that automatically detect when a device is at the end of the chain and, if so, enable the necessary resistors. With active termination of this type, you can avoid the perils of SCSI bus disorders; this feature is a very desirable addition to a SCSI configuration.

SCSI IDs The other common SCSI pitfall involves IDs. Each of the devices on a SCSI daisy chain must maintain its own unique identification number, from 0 through 7. If two or more devices share the same number, they will attempt to respond to SCSI commands issued to them. The result will be complete failure of communication on the SCSI bus. ID numbers are assigned to CD-ROM drives and recorders through jumpers, rotating switches, or other mechanical switching techniques. As you add any new SCSI device to a chain, check to see what numbers are assigned to the other existing devices, and then assign a unique number to the device you are adding. The SCSI host adapter itself has its own ID, which may vary from manufacturer to manufacturer, so check your computer specifications or SCSI host adapter data to determine what ID has already been assigned to the host adapter. Avoid using this ID number for any connected devices.

Short Cables One final bit of advice for avoiding SCSI hassles: Keep the lengths of SCSI cables short. The total length of the SCSI bus, including all of the internal and external cabling in the chain, should not exceed 10 feet. If it does, you begin to see occasional failures, more data errors, and perhaps a collection of ambiguous and confusing system problems. Use the shortest cables you can find that still allow you to set up equipment around your computer conveniently, and—if you do run into problems—recheck the cable lengths to make sure that you haven't exceeded the 10-foot limit.

You can mix SCSI and non-SCSI devices in your recording configurations. For example, you can use an IDE hard disk drive to communicate data to your SCSI recorder, and in most configurations this won't result in any problems. In the interests of configuration simplicity, however, some engineers suggest that using SCSI versions of hard disks helps to ensure the consistent system interactions that are needed to keep a balanced flow of data from drive to recorder.

Portability of SCSI Drives

Because many CD recorders rely on a SCSI interface, and SCSI is well-supported on DOS, Macintosh, and UNIX platforms, the recorder itself—if an external unit—can usually be freely transported from computer to computer. In situations where you would like to get maximum utility from a disc recorder, this easy portability can be a major factor. You do need the necessary driver software and a premastering application resident on the host computer to record, but many manufacturers routinely include both Mac and Windows applications with their SCSI disc recorders.

SCSI Connectors

If you are considering sharing a disc recorder between different computer platforms, pay careful attention to the type of SCSI connector included on the recorder. The four most common types include:

- HD50 (50-pin SCSI-2 or MicroSCSI)
- 68-pin Fast/Wide SCSI
- Centronics 50-pin
- DB-25 (25-pin D-shell)
- 30-pin PowerBook connectors

There are many different types of cables available to bridge the variations between the connector available at the SCSI host adapter and the connector type installed in the disc recorder. High-quality SCSI cables tend to be expensive, ranging from $20 to $75, so consider this factor in the cost as you are investigating equipment for purchase.

Platform-Specific Issues

There is no particular advantage to choosing one platform over another unless you are a developer planning to create applications for a single-platform audience (such as owners of Sun SPARCstations). In such a case, you may want to maintain the file system in the native structure and use the same file system on your mastered CD-ROM. To take an example, for a SPARCstation-only audience you could use Young Minds' *CD Studio* (described later in this chapter) to create a disc using the UNIX file system (UFS); the resulting disc could not be played back on most Macintosh or DOS computers unless the UNIX operating system was installed on each machine.

Similarly, if your audience is limited to the spectrum of worldwide Macintosh users, you may want to master discs using the native Macintosh file system, in which case you would want a Mac as a host. However, some of the mastering applications let you produce CD-ROMs with Macintosh file systems (or with hybrid systems that are half-Mac, half-ISO 9660) from a DOS or UNIX platform. The tendency for the newer applications is to provide maximum flexibility in terms of your output format. In actual practice, however, in most cases if you are mastering file systems for a particular platform, it will be easier to do it from a host computer that runs the native environment that you are targeting. You will also want to test your completed CD-ROM under the native environment, which suggests using the same computer your audience will be using.

One of the chief advantages of CD-ROMs, however, is their ability to break down barriers between different computer platforms. If you maintain ISO 9660 file conventions, your resulting work will be available to anyone with access to a CD-ROM drive. You lose some flexibility in file naming and file organization, but you gain a much wider audience. You can create ISO 9660-compatible discs from DOS, UNIX, or Macintosh platforms. You may want to choose a more powerful computer if your mastering plans involve use of a 8x or 12x recorder, but for 4x applications you can often rely on a typical workhorse business machine.

ATAPI IDE Considerations

Expense has always been a consideration when designing I/O interfaces for personal computers and the IDE bus was devised with ease of use and simplicity at the forefront. The first variation of this standard, Extended IDE (EIDE) added features that improved the behavior of the bus when multiple devices were attached. The ATAPI standard (AT Attachment Packet Interface) extends the capabilities of IDE to encompass certain kinds of peripheral devices, including CD-R and CD–RW drives. Under the right conditions, ATAPI drives perform quite successfully, even on older systems. However, this interface method is not nearly as robust as SCSI for more demanding applications. Bus performance can be seriously degraded if two active devices are sharing the data lines for operations.

On a positive note, ATAPI IDE drives have brought CD recording within reach of almost anyone with a PC. CD-RW drives now sell for approximately what a state-of-the-art CD-ROM drive sold for two years ago. Many new systems now come with built-in CD-RW drives, offering convenient storage and data distribution for both general purpose and specialized uses.

The successor to EIDE, Ultra DMA (UDMA), is sometimes referred to as generic disk I/O. It is a new extension of EIDE that includes a maximum throughput of 66MB per second, although this throughput can't be achieved if two devices are using the bus at the same time. Bus support under UDMA is designed to be backward compatible with earlier EIDE devices.

As with standard EIDE, UDMA is designed for internal devices only. Unlike SCSI, UDMA does not allow overlapping operations to be performed by two devices sharing the same channel. A maximum of two devices per channel are allowed. Typically, a PC will have two UDMA channels internally. The maximum cable length is 11.8 inches.

For simple applications, the ATAPI IDE interface offers an inexpensive means of connecting internal CD-R and CD-RW drives. For more professional applications and high-volume production use, you would be better served by equipment designed for SCSI or FireWire interconnection.

Optical Disc Recording Issues

If there is a single rule for success when you are recording compact discs or DVDs on a desktop system, it is "Keep the data streaming." Once a CD-R or DVD-R recorder starts firing laser pulses, creating impressions on the recordable medium, it can't go back and erase the data. The term *write-once media* tells the story. Data flows from the host computer through the SCSI bus or EIDE bus or FireWire interface to an internal buffer inside the CD recorder. If the data flow is interrupted, the recorder starts using the data stored in the buffer. If the buffer runs out of data and the laser beam stops recording transitions in the middle of a sector, in most cases you cannot go back and correct the mistake. In the colorful parlance of the industry, the disc becomes a coaster—useless for reading or writing data but very handy if you don't want your cup of hot cider to make a ring on the end table.

The specifications of Orange Book bring us back to a few simple inescapable facts when working with recordable compact discs, as discussed in the following subsections. As you work on CD-R or DVD-R projects, these fundamental points will affect many of your decisions and strategies when you are determining the best ways to transfer data to disc.

Chapter 5

Non-Erasable Media

The medium originally known as *Compact Disc - Write Once* (*CD-WO*), is not erasable; you can write once and only once. If you interrupt the recording flow in the middle of a sector, the current state of the technology does not let you go back and start recording transitions from the point where you stopped. Some of the newer recordable CD archiving applications create the illusion of being able to erase files, but what you are actually doing is changing the contents of the disc's directory structure. The file is still on the disc—you just eliminate the pointer to it. Similarly, DVD-R is designed for recording the data only once.

The non-erasable nature of recordable compact discs makes it important that the information written to the disc be exactly right. Since a typical recording session can last anywhere from 8 minutes to half an hour or more and involve hundreds of Megabytes of data, all the system components must work together flawlessly. The computer system, hard disk drive, SCSI host adapter or other I/O interface, software driver, and disc recorder must all perform their role in the data transfer process without error for the duration of the operation.

This characteristic is both a strength and a weakness. If your goal is to maintain an unchangeable copy of a set of data files or a graphics library or a technical document, the optical disc will store your files for up to 100 years without the risk of overwriting the data or accidentally erasing the files. If, however, you are looking for a temporary place to keep large quantities of data, the rewritable optical disc formats are more appropriate, as described in the following section.

Rewritable CDs and DVDs

Recording to rewritable discs does not require the same attention to uninterrupted data flow as does writing to write-once media. Data written to CD-RW and DVD-RAM media can be corrected by rewriting. Performance issues become less important since mistakes can be remedied. The benefits of packet writing also increase the utility of rewritable media, since the overhead associated with storing data that has been written incrementally is large eliminated.

The widespread acceptance of *CD-RW* (*Compact Disc - Read/Write*) provides an alternative to the write-once approach of conventional CD-R systems. CD-RW media uses a six-layer system that includes two dielectric layers and a recording layer that captures phase changes between a crystalline state and an amorphous state (of reduced reflectivity). The laser used in this technology pulses between two different power levels—one to record data and the other to return the media to an unrecorded state. MultiRead-capable CD-ROM players can read the CD-RW discs, making the technology useful for both in-house backup and archiving of files, as well as distribution of data. Similarly, the Super MultiRead specification is under design for DVD media, which would allow rewritable DVD+RW discs to be read in specially equipped mainstream players and DVD-ROM drives. As of press time, this specification is still being finalized.

The rewritable formats open up optical recording to more conventional storage uses, making discs a practical alternative to storing large amounts of data on hard disk drives or magnetic tape. The primary limitation to this form of storage is that playback is restricted to a smaller range of devices than the full universe of CD-ROM drives or DVD player/DVD-ROM drives. You also cannot use rewritable media as a method of producing one-off discs for replication purposes. Replication facilities receiving data on CD-RW or DVD-RAM have to pull individual files from the disc and premaster to tape, CD-R, or DVD-R before a glass master for replication can be produced. The extra step can add a great deal of unnecessary expense to a replication project and adds the potential for possible file errors during data transfers and restructuring of disc contents.

Disc Formatting Considerations

Formatting of a compact disc is irregular in comparison with a formatted hard disk drive. The physical contents of a hard disk drive are laid out in concentric tracks, each utilizing an identical sector format. The hard disk's logical file system contents are periodically updated and maintained in a file allocation table of some sort. Bad blocks on the hard disk surface—due to surface irregularities—can be marked as non-writable areas and excluded from the data map.

In comparison, a single-session compact disc has one table of contents, written once. Also, unlike the regular, preformatted sectors on a hard disk, each compact disc sector can contain a different number of data bytes depending on its contents. Error-correction techniques and embedded codes vary by type of sector. Mixed-mode discs can contain purely audio information following a track of computer data. Multi-session recording builds an ongoing structure that references the contents of earlier recorded files. As recording is taking place on a compact disc, the system is building a series of roadmaps and pointers to encompass the data. If a roadmap points in the wrong direction when a disc is being read, the data that it was supposed to identify becomes inaccessible.

Multi-Session Recording and Packet Writing

Multi-session reading and writing is covered in Orange Book, but the means of execution is not specified in detail. Because of this, the original implementations varied depending on each manufacturer's interpretation. Incremental write operations are subject to control by the CD recorder's feature set, by driver software, and by the premastering application. Some applications are more lenient than others in regard to interruptions of a write operation. Most current implementations have turned to *packet writing* as a technique for recording smaller quantities of data at a time and for being able to bridge interruptions in the data flow without ruining the media. There are still variations from manufacturer to manufacturer as to how this process is carried out, and most companies include a software driver that is used as a bridge between the computer operating system and the packet-written data on disc. For example, Adaptec's DirectCD program is an example of a tool that supports seamless reading and writing of data on rewritable and recordable discs.

For packet writing to be supported, the CD recorder must be equipped with hardware that is compatible with this process. A firmware upgrade cannot be accomplished for a CD recorder that lacks the appropriate hardware components to make it capable of packet writing. When using packet writing, data is written to disc in small increments—predictably called "packets"—that can be as small as a single file. Unless using conventional multisession write operations that require as much as 23MB of overhead for creating a single session, packet writing makes more efficient use of the disc storage space, adding minimal information to the disc to support file storage and retrieval.

Packet writing opens up recordable-CD applications to many mainstream uses, such as simple file storage and distribution within organizations, or network backups that are performed by administrators. Depending on the package and the implementation, packet writing can be applied to both CD-R and CD-RW applications. Those applications that are designed for CD-RW typically use the UDF standard for the file system. CD-R implementations often employ ISO 9660, for more universal playback.

All of the packet writing solutions require that a disc be formatted before use—whether you are writing to CD-R or CD-RW media. The formatting prepares the disc for identification and recording. Most applications of this sort automatically bring up the formatting utility when they detect that blank media has been inserted in the drive.

PacketCD

PacketCD is a method for incremental disc writing designed by CeQuadrat. PacketCD installs under Windows 95/98/NT and equips the system to provide drive-letter access to the disc recorder. Drag and drop file operations from the desktop are supported, or the standard "Save" commands from within an application can be used to write files to the recordable media.

PacketCD utilizes the Universal Disc Format (UDF) for its active file system, which requires that a UDF reader be available to the operating system for accessing the files. Both CD-R and CD-RW media can be used for packet writing. When write-once CD-R media is used, deleted files are not actually removed from the disc—instead, they are logically erased. With CD-RW media, files can be physically erased and the space reclaimed and reused.

PacketCD includes a defect management utility that can track and report the condition of CD-RW media. Since CD-RW discs can only support a finite number of erasures, defect management provides assurance that deterioration of media surfaces will be detected before problems develop that could result in data loss.

A number of drive manufacturers bundle PacketCD with their CD-RW units, including Verbatim, COMPRO, and Ricoh. CeQuadrat also offers software solutions for disc recording.

DirectCD

Adaptec's DirectCD software supports incremental write operations to CD-R or CD-RW media. When performing write operations to a disc, DirectCD normally stores the files using the UDF format. You have the option of closing the disc to ISO 9660 format so that it can be read back under the majority of CD-ROM drives (except for those running under DOS or Windows 3.1).

As with other software in the packet-writing realm, DirectCD lets you treat your CD-R or CD-RW drive as if it was a hard disk drive, accessible by a drive letter, on your system. You can drag and drop files from Windows Explorer, save files from within an application, or delete files from the desktop. DirectCD performs the necessary operations transparently, whether you're working with write-once media (CD-R) or rewritable (CD-RW).

DirectCD has versions for both Windows and Macintosh, and discs created on one platform can be readily exchanged with the other, even if the disc has not been closed. The formatting of a disc to support packet writing requires some overhead. For example, a 74-minute CD-R disc formatted under DirectCD, offers 621MB of data storage. A CD-RW disc formatted under DirectCD provides a total storage capacity of 531MB.

Two forms of packet writing exist: *fixed* and *variable*. Under fixed packet writing, the packet size written to disc is always the same. Using the Macintosh version of DirectCD, only fixed packet writing is available. Variable packet writing allows the packet size to be adjusted as required to suit the nature of the data being written. The Windows version of DirectCD handles both fixed and variable packet writing. Since this approach provides more efficient use of the disc storage space, it can be a useful means of archiving or backing files in small batches, unlike standard CD-R multi-session write operations, which are not practical for small file backups.

DirectCD is bundled with many different CD-R/CD-RW recorder packages, including products from Pinnacle Micro, Ricoh, and Hewlett-Packard Company. Direct purchases and upgrades are available from the Adaptec Web site (*www.adaptec.com*).

CD-R FS Packet Writing

Sony introduced their variation of CD-R packet-writing software as a package that runs as part of the operating system on Windows 95/98/NT, Macintosh, and UNIX systems. The software extends the operating system capabilities by allowing the Save As command, provided by most applications for saving files, to direct the file output to a specially formatted recordable disc. Recorder support includes Sony's Spressa drives, as well as drives from other manufacturers.

As interim recordings are made, the only machine that can read the files on the compact disc is the CD recorder itself, until the point at which the *final* disc contents are recorded, after which the data can be read by any CD-ROM drive. This approach is typical for multi-session CD-R writing: the disc must be finalized with an overall table of contents to provide universal playback.

Standard multi-session recording uses a *link block* to unite the individual sessions. This link block is created each time the recorder is stopped and then restarted. In Sony's packet-writing implementation, the overhead requirement is greatly reduced (compared to the approaches taken by other methods), which relieves the need for reserving large areas of the disc to coordinate the packet-writing operations.

By integrating CD-R write operations into individual operating systems, recording to disc becomes a seamless, transparent operation that can be performed from applications ranging from graphics programs to database applications. To the user, the operating system behaves as if it is performing a simple file-write operation to a hard disk drive (although the operation may take a bit longer to complete).

Selecting a Host Computer

Unless you are a professional musician or a studio engineer, you will probably want to configure your recordable CD environment in the conventional manner: with the CD recorder connected to a host computer through a SCSI interface.

For musical applications there are some interesting stand-alone CD recording devices. The Marantz CD-R610mkII lets you record 74 minutes of Red Book audio from a variety of digital sources and does not require a host—it is a fully self-contained CD recording system. However, such a system limits you to recording only Red Book audio and is proportionally expensive in relation to computer-based configurations. Even for musicians and musical applications, connecting a CD recorder to a suitable host computer provides the most flexibility and the most value for your money.

Minimum System Requirements for CD Recording

When CD recorders were first introduced, a high-performance system was essential to successful recording. A 486 machine on the PC side might do the job if configured carefully, but a Pentium-class machine was preferred. A Macintosh 68040 could generally be coaxed to record reliably under the right circumstances, and the introduction of the PowerPC guaranteed seamless recording operations.

Modern personal computers with processing power rivaling early generation supercomputers can typically handle disc recording tasks with grace and ease. If you have a relatively modern computer purchased since

1998, odds are you will not encounter any performance difficulties when recording as long as you don't attempt to operate multiple devices or run multiple processes during a long recording operation. Anyone in a production environment running CD duplicator equipment should probably invest in a dedicated system to handle write operations. Anyone else should do fine for occasional recording as long as they follow the guidelines presented in this chapter.

As a yardstick for comparison, many manufacturers list hardware requirements in terms of the operating system running on the computer (perhaps assuming that if the computer can run the designated operating system, it is fast enough to handle CD recording). For example, Yamaha lists the requirements for their CRW6416sxz drive (6x CD-R write speed, 4x CD-RW write speed, 16x read speed) as follows:

PC Computers Require   IBM-PC compatible computer running Windows 95/98 or Windows NT 4.x to run supplied software; SCSI adapter card required. Computer with fast hard disk: <19-millisecond access time with a 700KB per second or better transfer rate.

Macintosh Computers Require   Macintosh running System 7 or System 8; available SCSI-2 controller port; SCSI cable. Computer with fast hard disk: <19-milliseconds access time with a 700KB per second or better transfer rate.

Other manufacturers frequently use similar requirements, since the computer processing power is less commonly a critical factor in ensuring the appropriate performance requirements for recording.

Distributing Files to Replicators

If you plan to distribute files to a CD or DVD replicator on media other than a one-off CD-ROM or DVD-ROM, you will need some form of high-capacity data storage device connected to your system with removable media. Tape is popular for this purpose. Digital Linear Tape (DLT) is still the submission medium favored by replicators for DVD data. DLT cartridges have a 40GB native capacity and can handle data transfer rates of 360MB per minute. This is enough to get you through any conceivable DVD replication project.

Some CD/DVD recorder applications, such as GEAR Pro DVD, directly support data output to a number of different tape drives. Earlier nine-track tapes can be used at some replicators, though these are gradually being phased out throughout the industry. More recent 8mm EXAbyte or 4mm Digital Audio Tapes are supported at many locations.

Earlier forms of storage can sometimes be used by consolidating the data on several cartridges. For example, magneto-optical cartridges or SyQuest cartridges in various sizes can sometimes be used for distributing data to a replicator. Since these media usually cannot store the complete contents of a CD-ROM and certainly not a DVD-ROM, you need to carefully document the file contents for the replication service. Prepare a comprehensive list of source files to help the replicator assemble a CD-ROM image, DVD-ROM image, or DVD-Video image during premastering.

Other Hardware Considerations

In setting up a workstation to perform CD-R or DVD-R recordings, you may want to consider the hardware-related information in the following subsections.

Uninterruptible Power Supply

To ensure constant operation for maintaining data flow during disc recording, there is one more item you should add to your developer's workstation. An Uninterruptible Power Supply (UPS) large enough to handle both your host computer and the disc recorder can provide the extra insurance that short power interruptions or glitches in the line voltage will not disturb the integrity of the recording. Power interruptions that might otherwise go unnoticed, because they might not be long enough to cause your system to shut down, or to cause unrecoverable data loss to occur in normal transfer operations, could conceivably ruin a CD or DVD recording.

A UPS ensures a steady flow of power to your computer and recorder for the 15 minutes to an hour that it takes to complete a recording. If you notice during the course of a day how many times the lights flicker off and on or dim, you are witnessing the potential for ruined media at any time recording is taking place. A capable UPS large enough for a computer and recorder may cost from $250 to $600. For any kind of professional application or in a production environment where lost time is critical to your business, we recommend that you buy one and use it.

Chapter 5

Hard Disk Considerations

The combination of increased processor performance and advanced hard disk drive designs has reduced the concern that used to exist over data transfers from hard disk to recorder. Drives that are capable of average seek times from 15 milliseconds down can typically retrieve data fast enough to keep most disc recorders operating smoothly.

If you are using an older drive for disc recording applications, seek out an A/V model. A/V (audio/video) drives became a popular add-on peripheral for use with CD recorders in the late 1990's. This drive type originated to solve a common problem with audio and video capture, such as when performing digital recording to disk or when using a video capture board. As with CD recording, the string of data being sent to the hard disk needs to be continuous. Any interruptions will result in an audio track that misses a beat or a video sequence that drops several frames.

A/V drives typically solve this problem in two ways. One involves thermal calibration. The other involves the manner in which data caching is handled. Although these two features were designed in response to the needs of audio and video capture, they suit the requirements of CD recording perfectly.

Thermal Calibration Thermal calibration is a dynamic, periodic fine-tuning operation that high-capacity non-A/V drives perform in response to changing internal drive conditions. The gradual rise in temperature inside the drive requires adjustments to compensate for the relative locations of the drive read/write heads and the platters. This adjustment occurs during a thermal calibration cycle, sometimes called a T-cal, which occurs on a periodic basis without regard for the current computer or disk operation being performed. A/V drives circumvent this problem by rescheduling the thermal calibration for less critical times, periods when no significant data transfers are taking place. While a T-cal cycle on a non-A/V might interrupt data flow just long enough to disrupt a write operation to a compact disc (ruining the disc), an A/V drive does not enter a T-cal cycle during a write operation, and thus ensures disc data integrity.

Data Caching Caching of data is another area where A/V drives differ. The most common types of computer hard disk activities involve frequent small transfers of data, and the disk cache is tailored to these kinds of operations. In comparison, disk caches on A/V drives are optimized for large, continuous transfers of data, such as take place when a video stream is being digitized or 600 Megabytes of data is being output to a CD recorder. With an A/V drive, the transfer of data occurs at a predictable, steady level, and data transfer rates do not suffer under different access conditions.

Virtual versus Physical Images

Earlier recording technology often required that the files to be recorded to disc first be copied to an ISO 9660 image on a dedicated hard disk. Without this image, adequate performance could not be maintained throughout the recording process.

More-modern hard disks with faster performance and improved premastering software have generally eliminated the requirement for creating an ISO image before recording. Instead, a virtual image can be constructed during premastering, an image composed of pointers to the files required for mastering and a sequence to follow for transfer. The success of this approach depends on files being accessed quickly enough to keep the cache buffer loaded with data as the recording is taking place. Even for recording at 8x speeds or above, however, most systems can support use of a virtual image. The system configuration must be able to access files quickly, on the fly, from different parts of the disk and transfer them to the structure required for the CD recording. This feature reduces your system configuration needs; in most cases, you don't need an additional, dedicated hard disk drive to contain the large image file, since files can be accessed dynamically from your primary drive. Defragmenting the hard disk drive before recording is a common technique to make sure that performance is not hindered by trying to re-assemble files that have been stored in several different non-contiguous locations on disk.

In some situations where you are preparing a set of files for transfer to a replication facility, and your transfer medium will be tape or data cartridge, you may want to create an ISO 9660 or UDF image to simplify the interaction with the replicators. This requirement will vary from replication facility to replication facility, so check before submitting material to see if there is an advantage to creating an ISO image of your files.

Selecting a CD Recorder

Recordable CD technology is still specialized enough that only a handful of hardware manufacturers compete for market share. A number of other companies are repackaging drives from these primary vendors of CD-R equipment and bundling additional software—such as MPEG compression tools or software for cataloging media assets—and sometimes additional hardware to offer all-in-one premastering workstations. This

Chapter 5

section discusses the features available on the current crop of CD recorders and suggests some of the considerations that should guide your purchase decision.

Pricing of CD Recorders

The entry level for obtaining new recordable CD equipment is—at press time—around $149. Factory refurbished CD-R and CD-RW drives can sometimes be obtained for under $100. The least expensive drives generally use the ATAPI IDE interface and are limited to internal installations. External drives primarily employ SCSI interfaces—the addition of a cabinet and SCSI interface hardware can add an additional $50 to $150 to the price of a drive. You can also obtain external drives using the USB port as an interface; these are sometimes slightly less than comparable SCSI-based external drives. Special-purpose CD-R and CD-RW units exist that can switch among several I/O interfaces, such as SCSI, USB, and parallel port. These tend to be more expensive than single-interface drives.

On the high end, if you want a recorder that can produce several discs at once, perhaps over a network connection, the price of entry quickly rises to $5000 and above. Drives in this category, commonly called CD Duplicators, are discussed in Chapter 8.

Because this market is still growing and competitive pressures continue, prices may continue to decline for the near future. Larger manufacturers may attempt to boost their market share by aggressively pricing recorders at the lowest possible costs.

Recording Speed

Single-speed recorders have not been available for several years. You may still see older 2x recorders at discount outlets and liquidation stores. The older CD-R gear typically weighs more, takes up more desk space, provides less universal support for the various CD-ROM standards, and generally suffers from a small buffer capacity. Older units often have problematic operation quirks that will plague you during recording. Unless you find a genuine bargain on one of the more reliable units with some assurance of future support, we suggest you steer away from the earlier recorders and look instead to more recent equipment.

Most baseline units on the market today are 4x recorders, capable of mastering a 650MB disc in about 16 minutes. The data transfer rate for these units during recording equates with typical 4x players, 600KB per second.

Selecting a CD Recorder

You'll pay a bit of a premium for the faster recorders—those capable of 6x and 8x recording speeds. Because of improvements in manufacturing design and hardware circuitry, the disparity between the base level 4x recorders and faster 6x and 8x recorders is much less than in the past. For an additional $75 to $175 over the cost of a 4x CD-R unit, you can generally gain the extra speed offered by 6x and 8x rates.

The fastest CD recorders currently available can write at 12x rates (1800KB per second); these units require careful system tuning to ensure that data can be transferred fast enough during the recording process.

Onboard Buffers

Buffer sizes for CD recorders, like minimal memory configurations for PCs, have risen steadily over the last few years. All CD recorders have some amount of built-in onboard buffer space, generally located on the circuit board that houses the drive electronics. The buffer serves as a storage repository for data being transferred to the laser write head for recording. The buffer compensates for small interruptions in the data flow from computer to CD recorder. As long as data in the buffer does not get drained during the recording process, the recorder can burn a continuous stream of pits into the disc and maintain the absolute integrity of the recording. Clearly, all other factors being equal, a larger buffer provides better insurance against the data being totally emptied from this temporary storage area and the recording being interrupted.

Buffer sizes on modern CD-R and CD-RW equipment typically range from 1MB to 4MB. Buffer upgrades can increase the size to as much as 32MB. Manufacturers, attentive to the feedback from those early adopters of recordable CD technology, have moved towards 2MB as the prevailing standard for most equipment. Other factors can influence the success of the recording process and buffer size alone does not guarantee that data flow can be successfully maintained.

For most purposes, look at the buffer size as just one important factor in the suitability of a recorder. Larger is better, but a large buffer alone should not be your sole consideration.

Software Support

With only a few hardware manufacturers, producers of mastering applications can adequately ensure support for most of the CD recorders present on the market. Indeed, this is clearly the case for many of the current applications, such as Adaptec's Easy CD Creator Deluxe or GEAR's

Pro DVD. As new recorders are introduced by additional manufacturers, however, the waters may become muddied and the one-to-one correspondence between mastering application and drive may become less assured. If you have a favorite mastering application, do make it a point to check the compatibility with that application in the CD recorder's specifications.

Read Speed

A disc recorder can easily serve double duty, reading CD-ROMs as well as writing CD-R blanks. If you intend to frequently use a disc recorder to read commercial discs (as well as recorded discs), the maximum read speed may be a factor in your purchase decision.

Many disc recorders offer 20x read speeds, which is comparable to performance levels of low-end CD-ROM drives at this time. A few units achieve read speeds as high as 32x. A few only manage to reach 16x speeds. If this is a critical factor in your purchase decision, check the specifications before buying to determine the maximum read speed of a disc recorder.

Easily Upgradable Firmware

Computers and many peripheral devices conventionally contain some form of Read-Only Memory (ROM), which generally contains low-level, machine-specific instructions that allow the computer or device to interact with an operating system, host adapter, or bus controller. ROM chips containing this type of firmware on a computer motherboard or on a circuit board internal to a peripheral can be updated only by physically swapping the outdated ROM with a newer version containing more recent firmware. The recordable CD industry in particular had numerous problems with the firmware included in early units, many of which contained ROMs with built-in bugs and defects.

A recent trend in current-generation recorders is the incorporation of Flash ROMs in place of ROMs for firmware storage. This features provides chips that can be rewritten by software, making the upgrading of firmware a far simpler task. With an upgrade diskette from the manufacturer or a downloaded upgrade file, you can reprogram the device from your computer without ever removing the CD recorder cover. Several manufacturers have signed on to this method of upgradability, including Plasmon, Pinnacle, and JVC. Check the specifications for any recorder that you are interested in to see if it supports firmware upgrades through software. This capability is a definite plus. With the changing nature of

Selecting a CD Recorder

optical recording standards, upgrade capabilities also provide an avenue for you to add new features to your disc recorder, as well, as the industry adopts new standards and protocols.

Direct Overwrite Feature

In the case of CD-RW, the write operations are also affected by another factor. The capability of performing a direct overwrite allows a CD-RW drive to erase previous data and record new data in a single pass. If a drive cannot perform direct overwrite operations, a separate pass must be performed to erase the earlier data on the disc before the new data can be written. This, of course, becomes a performance issue, increasing the amount of time that it takes to perform write operations to the media. Many specifications for CD-RW equipment detail whether direct overwrite is supported. Look for this feature if you want to gain the benefit of the performance improvements that it offers.

SCSI Version Supported

Many mainstream CD recorders are SCSI devices, but these are still divided evenly between SCSI-1 and SCSI-2 implementations. Besides the faster data rate offered by SCSI-2 (which becomes more important if your CD recorder is sharing a daisy chain with other SCSI devices), one of the main differences in these two forms of SCSI is the connector type. SCSI-2 connectors utilize a unique half-pitch 50-pin connector with a thumb-operated snap-on latch. You may have to scramble to find SCSI cables for chaining the newer SCSI-2 connectors with earlier SCSI-1 that may be present on your old SyQuest drive or tape backup device, but fortunately SCSI-2 connectors that adapt to varying configurations are becoming more common.

Mixing and matching certain kinds of SCSI devices in this manner may present problems in some configurations. Scanners on the same daisy chain as CD recorders often cause problems. Be equally cautious of any SCSI devices that are likely to generate sporadic noise on the SCSI bus; physically disconnect or power down any unnecessary equipment from the chain before beginning a recording.

Laser Power Calibration

To generate pits on recordable media in a precise enough manner that the recorded disc can be read back easily by conventional CD-ROM drives requires accurate calibration of the laser power used by the recorder. Orange Book specifies a region on a blank disc that must be used for laser power calibration. Unfortunately, as critical as this feature is, not all recorders support power calibration, particularly earlier units. An excess of power during recording, or a lack of sufficient power to the laser, results in pits that are either too long or too short. Discs produced by non-calibrated recorders can present problems when being read; if the disc is to serve as the master for large-scale replication, this problem can be perpetuated through thousands of manufactured discs. Performing a thorough disc analysis before committing to replication can avoid this potential problem.

To ensure the highest accuracy of disc recording, make sure the recorder that you purchase is designed to perform a power calibration cycle as part of its normal operation. There are many other features that contribute to overall data integrity and broad media compatibility, but you should actively seek power calibration in any drive that you purchase.

Running OPC

Another power calibration feature is called Running Optical Power Calibration (OPC). Recorders equipped with Running OPC continuously adjust the power of the laser beam during writing to adjust for variations in the surface of the disc (including fingerprints or other irregularities). The laser power is increased slightly to compensate for anything that obstructs the write operation. Reflected light is used as the basis for this surface check. Drastic changes in the level of the reflected light can also trigger a read verification cycle to ensure that the data has been written correctly to disc.

Running OPC is included as a feature on a number of CD recorders from Philips, Sony, Hewlett-Packard, Ricoh, and others. Having this feature available adds an extra level of data integrity to your recording efforts, which should definitely be considered a bonus.

Track-at-Once or Disc-at-Once

The majority of CD recorders and supporting software applications use a technique known as Track-at-Once when writing to disc. During Track-at-Once write operations, each time a track completes, the laser is paused and two run-out blocks are added to the disc. When recording restarts, an additional link block and another four run-in blocks are recorded to provide continuity between the tracks.

The problem with these added blocks between tracks is that they cause clicks and pops when encountered by a typical CD players, so if you are recording audio for Red Book playback, you create a very unprofessional sounding glitch between the tracks of your CD. (These additional blocks don't cause any problems for conventional CD-ROM uses involving data; they are only a noise problem when recording audio CDs.) Most replication facilities will not accept as masters discs that have been recorded using the Track-at-Once technique.

Disc-at-Once recording provides start-to-finish recording in one pass without halting the laser. No additional run-in or run-out blocks are required. The disc is closed (fixated) at the end of the recording.

If you plan to use your CD recorder to produce master audio discs for replication, make sure that the CD recorder and associated software support Disc-at-Once recording operations. Most current generation recorders support both Disc-at-Once and Track-at-Once, but it doesn't hurt to check the specifications if you think you might be creating one-offs for replication at some point.

Additional High-End Features

Once you get away from the economy recorders, there are a number of features that can add some very interesting capabilities to your compact disc recording ambitions. Drive units designed to be installed on the network can bring the benefits of disc recording to everyone in a workgroup. These types of drives often require specialized software for operation and are often sold in a hardware/software bundle. CD Duplicators, discussed in Chapter 8, support the creation of multiple discs. Combined with robotic arms and automated printers, these devices become compact CD production facilities, capable of churning out hundreds of discs at a time and running unattended for hours. If your requirements include occasional needs for more than a dozen or so discs, you might find the features offered by the CD duplicators as extremely appealing.

At the high end of the disc recorder family are the DVD-R units. Engineered to produce discs that are readable in DVD-ROM drives, DVD-Video players, as well as CD-ROM drives, these devices, available only from a few manufacturers, have undergone significant price reductions over the last 18 months. One of the primary problems at the moment for these units is the short lifespan of the recording heads, which require replacement after burning only a few hundred discs. After purchasing a $5000 DVD-R device, you probably don't want to be changing the recording heads after few months, so this area of the design still requires some improvement before the equipment takes hold for mainstream business and corporate uses.

Current Examples of CD Recorders

The following subsections provide some examples of equipment that is available in the disc recorder market. This is by no means an attempt to provide a comprehensive list, but it should offer you good representative examples of the kinds of products that are currently available. Most of the units are manufactured in Europe or Japan and, as such, are subject to changes in international currency rates. You can expect that the list and street prices described will have changed by the time you read this. We hope, however, that providing typical prices will give you a relative basis for comparing the various units.

Sony Spressa Professional (SCSI) CRX140S/C

As one of the originators of the compact disc format, Sony has also been an active player in the market and their recording gear is consistently well conceived and executed. The CRZ140S/C is an internal CD-RW drive upgrade kit that uses a SCSI interface and offers 8x recording speeds for CD-R media. As a reader, this unit can handle 32x rates and it can also record to CD-RW at 4x speeds. It requires an open 5.25-inch drive bay, a SCSI host adapter, and, at a minimum, a Pentium 233MHz PC with 32MB RAM.

Quick Specs

Table 2: Sony Spressa Professional specifications

Parameter	Value
Interface	SCSI-2 (single-ended)
Buffer size	2 Megabytes
Applicable formats	CD-ROM, CD-R, CD-RW, CD Video, CD-DA, VideoCD, CD-ROM XA (Mode 2, Form 1 & 2), PhotoCD, CD-Bridge
Recording formats	CD-R, CD-RW

Current Examples of CD Recorders

Table 2: Sony Spressa Professional specifications

Parameter	Value
Recording method	Disc-at-Once, Track-at-Once, Multisession, Fixed and variable packet writing
Average seek time	150 milliseconds
Burst transfer rate	Synchronous: 10MB per second; Asynchronous: 5MB per second
Sustained data transfer rate	4,800KB per second 32X reading CAV; 1,200 KB per second, 8x CD-R write/read; 600KB per second, 4x CD-RW write/read.
Capacity	650 Megabytes
Loading mechanism	Motorized tray

HP SureStore CD-Writer Plus M820e

Hewlett Packard's entry into the recordable CD marketplace was a cause for celebration by many of those involved in this industry. HP's reputation for quality, reliability, and design excellence has helped gain broad acceptance for CD-R technology in the marketplace. HP has also entered into partnering agreements with a number of other companies to provide unique combinations of matched hardware and software for specialized uses.

The HP CD-Writer Plus M820e is a very small, highly portable CD-RW drive designed to connect to the PC Card slot of a laptop computer. Weighing less than a pound, the M820e is actually less than an inch thick and could make an ideal companion to your traveling PC for creating presentations on the road or generating custom CD one-offs containing marketing literature, databases, large PowerPoint files, or similar kinds of material suited for the capacities of CD-ROM.

The M820e diminutive dimensions don't restrict its performance. It can perform write operations at 4x speed and read discs at 20x, extremely respectable performance for such a small package. As a backup device for your laptop or a nifty data distribution tool, the M820e combines the best features of CD-RW in a package that tucks neatly into a briefcase.

Chapter 5

With the solid support of HP and a respectable feature set, the SureStore CD-Writer M820e should satisfy the requirements of CD-ROM developers, audio recordists, and corporate users.

Quick Specs

Table 3: HP CD-Writer M820e specifications

Parameter	Value
Interface	PC Card Type-II (PCMCIA) or, optionally, SCSI-2
Buffer size	2 Megabytes
Applicable formats	CD-ROM, CD-R, CD-RW, CD Video, CD-DA, VideoCD, CD-ROM XA (Mode 2, Form 1 & 2), PhotoCD, CD-Bridge
Recording media	CD-R, CD-RW
Recording method	Disc-at-Once, Track-at-Once, Multisession, Incremental packet writing
Logical recording format	UDF and ISO-9660
Encoding method	EFM
Average seek time	150 milliseconds (one-third stroke)
Burst transfer rate (min.)	Synchronous: 10MB per second; Asynchronous: 5MB per second
Sustained data transfer rate	3,000 - 1,500KB per second, 20x CD-ROM read; 600 KB per second, 4x CD-R write/read; 300KB per second, 2x CD-RW write/read.
Capacity	650 Megabytes
Loading mechanism	Motorized tray

Current Examples of CD Recorders

APS DVD-RAM External SCSI Drive

Specializing in storage equipment of all types, APS Tech offers an extremely wide selection of CD recorders, tape drive units, high-capacity hard disk drives, and similar kinds of equipment. Quick to introduce equipment whenever new storage devices begin to appear in the market, their DVD-RAM drive offers good value in an emerging market niche.

As with other drives of this type, the DVD-RAM can store up to 5.2GB on double-sided media, using cartridge-based discs. When reading CD-ROMs, the drive is equivalent to a 20x CD-ROM drive. When writing recordable media, the recording speed is 2x.

Quick Specs

Table 4: APS DVD-RAM specifications

Parameter	Value
Interface	SCSI-2, Fast SCSI
Buffer size	1 Megabyte
Applicable formats	CD-ROM, CD-R, CD-RW, CD Video, CD-DA, VideoCD, DVD ROM, DVD-Video, PD Cartridge, DVD-RAM
Recording formats	DVD-RAM, PD cartridge
Average seek time	95 milliseconds
Burst transfer rate	Synchronous: 10MB per second; Asynchronous: 5MB per second
Sustained data transfer rate	DVD-RAM: 1.385MBps; DVD-ROM: 2.770MBps; CD-ROM: up to 3.0MBps; PD cartridge: 1.141MBps
Capacity	5.2 Gigabytes
Speeds	2x20

APS CD-RW 8x4x32 FireWire

This APS unit is one of the first to adopt the IEEE-1394 FireWire interface in a disc recorder. FireWire supports speedy data transfers as well as providing hot swapping of units. FireWire interfaces and I/O boards are now appearing on PC computers, as well as Macintosh equipment, so users on both platforms can take advantage of this flexible interface option.

Quick Specs

Table 5: APS CD-RW 8x4x32 FireWire specifications

Parameter	Value
Interface	IEEE-1394 FireWire
Buffer size	1 Megabyte
Applicable formats	CD-ROM, CD-R, CD-RW, CD Video, CD-DA, VideoCD,
Recording formats	CD-R, CD-RW
Recording modes	Disc-at-Once, Session-at-Once, Track-at-Once, Fixed and variable packet writing, Multisession
Average seek time	95 milliseconds
Capacity	680 Megabytes
Speeds	8x CD-R; 4x CD-RW; 32x read speed

Yamaha CRW6416sxz External CD-RW Drive

First on the block many years ago with a quad-speed recorder (with the venerable Yamaha CDE100 II), Yamaha consistently rates near the top of the pack with reliable, high-performance CD-R and CD-RW recording equipment. The CRW6416sxz uses the SCSI-2 interface and includes software for both Windows and Macintosh applications. With the 6x recording speed for CD-Rs, this tray-loading Yamaha drive can burn a CD in approximately 12 minutes. CD-RW write operations can be carried out at a healthy 4x speed. The 2MB onboard data buffer ensures uninterrupted recording under a wide variety of conditions.

The bundled recording software includes Easy CD Creator for Windows users and Adaptec Toast for Macintosh folks. A version of Adaptec

Current Examples of CD Recorders

DirectCD for packet writing operations is also included for both Windows and Mac users. Additional software includes Adobe PhotoDeluxe, Adobe PageMill, a clip art library, Graphic View 32, a poster and sign making program, and software for generating jewel case inserts.

Quick Specs

Table 6: Yamaha CRW6416sxz CD-RW Drive specifications

Parameter	Value
Interface	SCSI-2
Buffer size	2 Megabyte
Applicable formats	CD-DA, CD-ROM, CD-R, CD-RW, CD Video, CD-ROM-XA, VideoCD, PhotoCD, CD-I, CD-Extra
Recording speeds	6x, 4x, 1x
Access speed	160 milliseconds
Capacity	680 Megabytes
Read speeds	16x, 10x, 6x, 4x, 2x, 1x
ReWrite speeds	4x (requires 4x capable media), 2x
Disc loading	Tray
Media	CD-R Orange Book Part II; CD-RW: Orange Book Part III
Max. data transfer rate	2400KB per second
Recording modes	Disc-at-Once, Session-at-Once, Track-at-Once, Fixed and variable packet writing, Multisession

Chapter 5

Young Minds, Inc. CD Studio and DVD Studio

Young Minds, Inc. provides turnkey solutions to CD-R and DVD-R recording, specializing in UNIX applications and network capable recording systems. Their founder, Andrew Young, earned his stripes as the author of the Rock Ridge Interchange protocol, which extended the capabilities of ISO 9660 to include discs designed to run under UNIX.

CD Studio

Young Minds uses existing hardware as a springboard to systems development, adding additional value in the form of software innovations and integration of components into smoothly operating systems. Their CD Studio is a full desktop CD-ROM production system that runs under a variety of operating systems, including Windows NT and UNIX. You can select a recorder that fits your budget and your requirements.

Young Minds' approach to CD recording also includes network capabilities. Their premastering software, MakeDisc, operates as a background process, letting you launch a recording from anywhere on the network without consideration for where the actual files reside. The application locates and assembles the files and then performs the recording of the compact disc without any need for devoting a dedicated workstation to the task.

Since they started in 1989, Young Minds has focused on UNIX as their operating system of choice, and, in some specialized environments, their solutions are the most flexible and practical for the UNIX platform.

Young Minds also offers a unique approach to the data security issue by incorporating an NSA-approved SCSI media encryptor into their enhanced product, Secure CD Studio. You have the option of creating fully encrypted, partially encrypted, or unencrypted discs. Classified material encrypted on a CD-ROM is considered unclassified for shipment and distribution, allowing top secret material to be distributed with the convenience and advantages of the CD-ROM media.

The UNIX development expertise of Young Minds proved useful in their recording support for UNIX/POSIX file systems. ISO 9660 places serious restrictions on file naming and directory organization, restrictions that are particular impediments to UNIX systems which traditionally present robust and complex organizational frameworks. The Rock Ridge extensions encompass ISO 9660 standards while still providing access to UNIX/POSIX information, but Young Minds implementation offers several different choices to UNIX developers. CD Studio produces discs using the native UNIX file system (UFS), ISO 9660 for full cross-platform compatibility, ISO 9660 with translation tables (giving UNIX functionality without requiring Rock Ridge drivers), and ISO 9660/Rock Ridge (which does require UNIX drivers). With this range of choices, the UNIX CD-ROM developer can choose between maximum cross-platform compatibility or native UNIX support (or several positions in between). If your CD-ROM development plans are strongly positioned towards UNIX delivery, CD Studio may be your mastering package of choice.

DVD Studio DVD Studio extends Young Minds' turnkey approach to systems design into the realm of DVD-R. DVD Studio provides the hardware and software for creating DVD-ROMs, including a DVD-R drive that can be installed as a network accessible device.

The MakeDisc for DVD premastering software relies on a Java interface to set up the disc file structure in preparation for recording, or you can use a command-line interface to direct operations in batch mode. DVD images are tested and verified by an intelligent controller without impeding the very high throughput rate realized by this system. The system can be extended to include a CD-R recorder to complement the DVD-R recorder, providing the capability to take advantage of CD-R's low media costs whenever appropriate.

DVD Studio can create fully cross-platform compatible DVD discs from your choice of platforms: UNIX as implemented by Sun, Linux, IBM, HP, and Digital, or Windows NT 4.x. The system is designed to be completely scalable and easily upgradable as production needs or network requirements increase.

Guidelines for Hardware Installation

While installation procedures will differ to some degree for each of the disc recorders, the following general guidelines may help you avoid problems during the installation of your disc recording equipment.

- Many disc recorders include an external chassis grounding screw. You should run a wire from this screw to the chassis of your computer (you can often connect it to one of the screws that secures the computer case to the frame). Secure grounding can prevent a number of signal level problems that may affect your recording.

- If using a SCSI recorder, keep the total SCSI bus length less than 10 feet, if the chain contains all external drives (including both hard disk drives and CD recorders). Some SCSI experts recommend that you don't mix internal and external drives on the same host adapter. Others recommend that if you do mix external and internal drives that you keep the total SCSI bus length less than 7 feet. This is more conservative than the actual specification, but it may reduce the chances of running into perplexing SCSI problems.

- Use active termination on both ends of your SCSI chain. Active termination ensures that the signals along the SCSI bus will be balanced and stay within acceptable amplitude levels (ensuring data integrity throughout the bus). Active terminators are built into many newer drives or can be added as a terminator plug (available from companies such as APS Technologies at www.apstech.com).

- Avoid having extraneous devices on the SCSI bus, such as scanners or printers. These devices can often introduce signals onto the bus any time they are powered on even if inactive. Any erroneous data that appears on the SCSI bus can potentially disrupt the recording process, which, of course, can ruin the media being recorded.

- Disconnect network interface cards in the CD recorder host computer prior to recording and disable their network drivers. Network activity in the midst of a recording can cause problems (unless you're running applications that are designed specifically to perform recording operations in the network environment.

- Disable or disconnect fax/modem boards installed in your computer that may be set up to respond to incoming faxes or communications from other computers. If the host computer tries to respond to modem communications while recording a CD, it will probably ruin the recording.

- Disable any TSR applications running on your host computer before starting to record a CD. Screen savers, timers, alarms from personal manager programs, and similar memory-resident applications can disrupt a recording operation if they are active.

- Once your hardware is installed, you can generally test its proper operation by inserting blank media and performing a "test" write operation (which moves the laser record head and simulates the data transfer to see if the equipment and host computer can successfully write to disc). If the test operation doesn't work at all or the CD recorder application cannot identify the attached recorder, you need to recheck each aspect of the hardware installation (power, cabling, driver installation). If the test write operation can be performed at 2x speed but not at 4x speed, you may need to consider performance improvements to your host computer (simplifying the system configuration or adding a faster hard disk drive).

- Keep in mind that if you plan to perform audio recording using the Red Book standard, data transfer rates must be slightly higher than when performing data recording. This is because Red Book audio adds fewer correction codes to the data stream (something normally done by the recorder firmware just prior to writing) and therefore must keep more data moving along through the interface. For example, at 1x recording rates, the required data transfer rate for computer data is 150KB per second; audio data transfer rates must be 172KB per second. At 2x recording rates, the audio data transfer rates must be 344KB per second, as compared to 300KB per second for computer data. If you're doing audio recording, your system performance must be more finely tuned.

From Hardware to Software

Once your hardware is successfully installed, you can turn your attention to selecting a premastering application to use with your disc recorder. You may have already received some type of application with the recorder; the question is: does it have the features necessary to accomplish the kinds of disc recording in which you are interested? Low-end products can serve for file archiving and distribution, but usually deliberately hide disc formatting and file system controls from the user. Higher end applications require a greater degree of knowledge about the placement of files, the implementation of each of the different CD-ROM formats, and other standards considerations. The next chapter discusses the major software applications available for recording CDs and the benefits and tradeoffs associated with different approaches.

6

Disc Recording Software

Premastering is the act of selecting and structuring files in preparation for creating an optical disc. A premastering software application supports this creation process and typically produces a physical image or virtual image of the file structure and associated formatting, which can be saved to hard disk, tape, or a recordable optical disc. The term *premastering*, however, suggests that a disc or other media is being prepared for mass replication (as a preliminary step to creating a glass master). In reality, many discs are produced in small quantities on CD-R or DVD-R equipment for independent distribution, without any consideration of volume replication. This chapter uses the term *disc recording software*, referring broadly to those software applications that control current generation CD and DVD recorders. You will sometimes see these programs referred to as premastering applications in other references. The terms are used interchangeably in this chapter.

Unless you own one of the standalone disc recorders, with its simple controls and small LCD interface, the disc recorder application will typically be running on a host computer to which the disc recorder is connected. Some network applications also allow individual workstations to select files and initiate disc recording operations. This chapter examines the different kinds of software applications that let you create optical discs, provides some examples of these programs, and offers some guidelines for evaluating and selecting an appropriate application for your uses.

Chapter 6

Evolution of Recorder Software

Recordable-CD software applications had a rocky start—plagued by unreliability, confusing interfaces, limited hardware support, and other problems—but most of the software producers have learned from the experience. Recorder software shipped with current generation CD-R or recordable DVD equipment (or purchased separately) shows significantly improved design. The more technical aspects of the disc recording process are concealed beneath a drag-and-drop veneer. Often, even professional-caliber applications submerge the more involved control settings and format options at deeper levels of the application, where they are accessible if needed, but out of the way if they are not.

These days, premastering applications rarely cause major system conflicts and crashes. CDs or DVDs can usually be burned successfully on the first attempt. Most programs also include wizards to guide you through the file selection and formatting process. As a part of this process, the wizard will often run a complete performance test on both your system and your disc recorder to ensure that uninterrupted recording can take place without data throughput problems. If problems are encountered during the test recording, the wizard often prompts you with a recommendation to resolve the difficulty. For example, if the wizard detects that a screen saver is active on the host computer, it may prompt you to disable the screen saver before proceeding with the operation. This kind of practical feedback eliminates many of the difficulties often associated with disc recording, particularly with novices who are burning their first discs.

Early Travails

Today's situation is very different from the trials faced by early adopters of CD recording technology. Typically, after you added a CD recorder to your system and installed the recorder software, you could expect to spend days reconfiguring the system to optimize performance, troubleshooting to eliminate recording glitches, or debugging endless driver problems. Macintosh users generally fared better in terms of consistency and reliability when recording CDs than did Windows 3.X users. Fortunately, Windows 95 helped resolve many problems, primarily by providing more stable memory management and better integration and handling of system components. The operating system architecture established under Windows 95 was better suited to maintain the data transfer rates necessary to keep a CD recorder buffer from emptying. One feature in particular, *preemptive multitasking*, ensures that unless you're concurrently running two or three demanding applications, there is little likelihood of normal system operations stealing vital processor cycles and interrupting the data stream en route to the CD recorder.

With support for multithreaded I/O, data transfers flow uninterrupted while CD recording is taking place. Windows 98 refined these techniques further, and Windows 2000 promises to offer the most stable CD recording environment yet. The larger percentage of the software for CD premastering continues to be produced for the Windows platform, but Macintosh, UNIX, and Linux users also have a reasonable number of applications from which to select.

Development Platforms

Also noteworthy is the fact that the Macintosh continues to be well represented in the DVD marketplace with many tools for both producing DVD-Video and DVD-ROM content. The stability and high performance of the Mac architecture based on the G4 microprocessor, the first microprocessor with a performance level worthy of the label *supercomputer*, has furthered the Macintosh presence in the important developer marketplace. While the creative community still bears a strong allegiance to the Macintosh, a number of high-end DVD authoring packages have been optimized for Intel architecture and operation under Windows NT.

The clearcut selection of one of these platforms over the other for content creation and premastering is not so easy. Game development, broadcast quality 3D animation, digital video production, and similar tasks have fostered the design of many fine production tools, some of which are more clearly associated with the Mac, others which have been specifically designed for the Windows platform. In the end, most developers will choose and work with the tools that run on the platform with which they are most comfortable (or that offers them a profit advantage). With the speed at which advances and changes take place in the computer industry, it generally doesn't pay to predict what tools and technologies will prevail in the marketplace. By the time this book reaches bookstores, BeOS may be the dominant platform.

For many different types of specialized applications, recordable CD modules are being embedded into other types of applications. Particularly in vertical applications such as medical imaging and database storage, the perplexities of CD formats are made transparent to the end user. As the last step in using the application, you can often burn a set of files to disc. Typically, the software producers who adopt this approach license the

Chapter 6

CD-recordable module from one of the major CD software developers, so these add-on modules are usually very reliable and bug-free.

DVD premastering applications have built on the successes of CD recording development, but the range of programs and the complexity of the recording process have made this technology less accessible than CD recording. Steep entry prices for professional-caliber applications have also created a barrier for those interested in becoming involved in title development. Some of the commonly used programs are discussed at the end of this chapter.

All in all, this chapter surveys the different types of recorder applications and itemizes the important features to look for when purchasing this type of software. I've also tried to provide an arm-chair view of common activities that are encountered when you are running these programs. This will, hopefully, provide a sense of how the typical interface is designed and the types of decisions that you need to make when preparing to create a disc.

While the ever-changing nature of this marketplace makes it impossible to provide exhaustive coverage of the software packages available, you should get some valuable insights as to the types of programs that are available and what you need to know to use these programs. The CD-ROM included with the book features trial versions of a number of disc recorder applications; you can explore the interfaces of these programs and try burning some discs from your own computer. If you don't yet have a CD or DVD recorder, many of these applications let you create an image file on disk for later transfer to a replication facility. You can also simply save a file to later burn a disc when you do have the necessary equipment.

Terminology of Recording Software

Some of the terms associated with CD and DVD recording tend to create doubt in the end user's mind about what process is actually taking place. What is the difference between premastering and authoring? Is burning a disc different than pressing a disc?

Essentially, a premastering application is a piece of software designed to produce a set of files formatted to create a CD-ROM or DVD-ROM. The files themselves may be outputted to DLT tape, a CD one-off, or a DVD-ROM, but the key characteristic is that these files are structured in a format appropriate for producing an industrial master in preparation for replication. Despite the fact that many CD-ROMs and DVD-ROMs will be produced for distribution with no intention of replicating hundreds or thousands of discs, the nature of the premastering or disc recorder software ensures that the formatted disc meets the requirements to be played back on whatever platforms have been selected.

Authoring is the act of preparing files and designing content in one of many digital formats for playback on one or more computer platforms. For example, you might use Macromedia Director to author QuickTime content for playback under the MacOS or Windows 2000. The sum total of the files containing this authored content can then be transferred to optical disc using a premastering or disc recorder application.

The term *burning* a disc is generally applied to the physical process of writing to a recordable disc using a disc recorder. The laser essentially burns a pattern into the dye layer of the recordable media. The discs used are recordable media, such as CD-R or DVD-R discs.

Pressing a disc, on the other hand, typically refers the replication process where manufacturers uses stampers to impress a data image into molten plastic. Although a laser is used to create a glass master from which the stampers are produced, the actual manufacturing process does not employ lasers.

Types of Disc Recorder Applications

The currently available disc recorder software products fall into four basic categories, generally sorted as to their intended use. As you might expect, some of the categories overlap in different products.

- **Backup and archiving utilities**: provide a means to store large volumes of data for short-term or long-term archiving where random access is an advantage. This includes many typical business archiving operations, as well as individual backup requirements.

- **Basic disc recording tools**: support the creation of basic optical disc formats, such as audio CDs, Yellow Book CD-ROMs, or DVD-Video discs. Generally designed for less experienced users, these tools often simplify the file selection and formatting process and include wizards to guide novices through fundamental tasks.

- **Professional premastering applications**: support the full range of disc formats for all applications, including specialized formats for limited or vertical markets. Professional-caliber applications pro-

Chapter 6

vide access to the lower level details of each format, allowing developers to control and define each individual element, if necessary.

- **Disc recording components**: provide a disc-recording add-on to a specific kind of application, such as an MP3 music player that supports the creation of recordable discs if a CD-R unit is available. Specialized components, such as medical imaging software with disc recording capabilities, are also included in this category.

Backup and Archiving

Disc recorder applications designed for ease of use generally hide underlying formatting details from the user. The goal is typically to make the disc recorder just another device identified by drive letter or name to which files can be transparently copied. Network variations on this theme include products such as SmartCD from Smart Storage, Inc. that extend the file drag-and-drop approach to CD-R jukeboxes installed on the network. Some products, such as Adaptec's DirectCD, offer CD-R access from any application, so that the user can write files to disc through the Save command in the application being run. Often, both CD-R and CD-RW media are supported. Other applications in this category are simple backup tools that copy files to a recordable disc and maintain a history of file operations.

Using non-erasable CD-R discs for backup and archiving provides a guaranteed file history or audit trail, an especially important characteristic for many types of organizations that could be put at risk losing data from erasable media. Although the applications often provide the illusion that a file is being deleted (removing it from view in a displayed directory), the earlier version of the file will still be on the CD, but its pointer will be removed.

Examples of programs in this category include Backup NOW! by NewTech Infosystems, Inc. and Adaptec DirectCD.

Basic Disc Recording Tools

Applications in this category provide simple premastering capabilities, including the necessary tools to create CD or DVD one-offs, which are single discs that can serve to create masters for replication. For example, a typical program such as Adaptec's Easy CD Creator 4 supports a range of the CD-ROM formats, including creation of Audio CDs, Mixed Mode discs, CD-ROM XA, and hybrid discs for individual platforms.

Some control of disc geography (physical placement of files on the CD) is sometimes included as a part of this type of product, although this feature is becoming less common— higher performance playback on 8x to 12x CD-ROM drives makes precise file placement less of a concern. Examples of products in this category are: HotBurn by Asimware Innovations, Inc., DVDit! by Sonic Solutions, Toast DVD by Adaptec, and WinOnCD by Adaptec, Inc.

These mid-range applications are suitable for inhouse preparation of multimedia training materials, all forms of corporate electronic publishing (portable document format files, infobases, technical support libraries, and so on), as well as personal applications (creating PhotoCD family histories, recording original music for demo or promotional purposes, archiving research material from the Internet). Many of them do have the necessary features to produce master CDs or DVDs to submit for large-scale replication, although if this is an important consideration for you, read the fine print in the features list carefully, particularly the disc formats supported.

Professional-Caliber Applications

Applications designed for professional use provide the most control over disc formatting, with full control of disc geography and support for all of the standard and many of the less common CD-ROM or DVD formats. The most versatile products in this category can produce disc images for any type of disc, regardless of platform, including some of the lesser-used game formats, Enhanced CDs, and other less commonly used formats. Professional-caliber applications should be able to handle any format or specialized use that you require. Two examples of programs in this category are: GEAR PRO DVD from Command.com Software and Buzzsaw 4.0 by ISOMEDIA, Inc.

Obviously, the primary uses in this category will be producing master discs for replication or high-volume distribution. Some applications are not sufficiently precise when handling the low-level details of CD-ROM or DVD formats. Non-standard disc one-offs, not suitable for replication masters, can be produced by low-level deviations in link block handling, certain obscure multisession constructions, and the inability to manage Disc-at-Once recording to those disc recorders that support it. If an application cannot produce discs to be used at a replicator to cut a glass master, it does not fit into the category of professional applications.

At a minimum, a professional-caliber application should be able to produce a disc usable as a master, but, ideally, it should be able to handle all of the industry recognized CD-ROM or DVD formats.

Disc Recording Components

This is a fairly recent category borne out of a wave of applications designed to produce electronic publications, infobases, or data collections accessible through search engines. Often the final step in the publication process is to output the data files to CD-R, so these applications include the required software module to seamlessly handle the process. The electronic publications also may be designed for World Wide Web or intranet distribution; some of these applications have facilities to simplify the creation and maintenance of links with Web sites directly from the CD-ROM.

The primary application in this instance is generally the creation of corporation information databases, infobases, or portable document format libraries. An example of a product that uses this approach is Sonic Foundry's Siren, which allows CDs to be burned from downloaded audio content.

Bundled CD Recorder Applications

Disc recorder applications are commonly bundled with CD and DVD recorders, particularly by value-added resellers who are integrating components from other manufacturers. Smart and Friendly, Inc. is one company that integrates high quality hardware and software—such as recorders from Sony and software from Adaptec and Macromedia—and offers it to consumers at value prices.

This approach is almost universal throughout the industry. It is unusual to find disc recorders that are sold without any kind of supporting software application. Another benefit of having the software applications included as part of a product bundle is that you have some assurance that the hardware and software will work well together. This is not always the case if you purchase the hardware and software separately, although software compatibility with a wide range of recorders is more common than in the past. As a general rule, the larger software producers, such as Adaptec, will have more complete support for the full range of recorders simply before they have more resources for development and testing. If purchasing a disc recording application from one of the smaller software producers, check first to be sure that it supports your disc recording equipment.

If your intended applications are not too rigorous, bundled applications may comfortably satisfy your requirements, but rarely are the high-end professional-caliber applications included in bundles. If you have a clear sense of your objectives in recording and know which of the CD-ROM or DVD formats you want to be able to create, these factors will help guide your selection of an appropriate disc recording application.

Disc Recording Software Features

How much control do you need over the creation of disc formats? If you know what you're doing, you can create discs that full advantage of the most advantageous characteristics of the media. For example, you can create hybrid discs that allow Macintosh or UNIX users to use the full capabilities of their native file systems, without the confining restrictions of ISO 9660, but also allow full file access to Windows 2000 users, as well. You can fine tune the performance of a multimedia CD-ROM title by tweaking the positions of the files on disc in relation to their importance in running the application, placing the critical files near the center of the disc. You can create specialized discs in some of the lesser used formats, such as Video CD, Enhanced CD, or CD-I. Newer packages, such as Command.com's developer-oriented Gear DVD Pro, also include support for emerging formats, such as DVD.

To get the widest possible range of support for CD and DVD formats, you generally need to invest in one of the high-end, professional-caliber applications. These packages usually range in price from around $250 to upwards of $1000. You can often achieve similar results while sacrificing some features and flexibility with a mid-range applications, which start at about $50 and range up to $250.

On the other hand, if optimizing playback performance of the CD-ROMs you burn is not important, and the only file system you care about is ISO 9660 (the universal system that offers platform independence), many inexpensive applications will let you freely copy files to recordable CD or DVD media for archiving, storage, or distribution. Some of these applications have overcome the costly overhead usually associated with multisession recording (which can be as much as 9MB of overhead per recorded session). Such approaches, including the several different techniques clustered under the title "packet writing," permit files to be copied to CD in smaller batches, either fixed or variable sizes.

For business uses where the emphasis is on simplified, high-volume data distribution, applications that focus on backup and storage are often ideal for workgroup use. Possible applications include the archival storage of high-resolution graphics files or digital audio files, periodic backup of workgroup member data on networks, and near-line storage of research material collected from the Web or other digital sources. Since CD-ROMs and DVD-ROMs offer random access to stored data, unlike linear tape backup systems, this method of storage and archiving can be much more convenient than any tape-based system. Optical discs also have about 20 times the lifespan of the typical magnetic tape cartridge, so if long-term storage of data is required, CD-ROMs and DVD-ROMs win hands down over any form of magnetic media.

If your requirements involve support for disc-recording towers or autoloaders, the supporting applications tend to be more expensive and more complex. At the highest level of complexity are applications that both support network access to recording equipment and offer compatibility with Towers and Autoloaders. Packages such as ISOMEDIA's Buzzsaw and some of the software packages from SmartStorage fit into this category.

Getting the Most Benefit from Recorder Software

Rarely are recorder applications married to a particular piece of hardware. More commonly, each application is capable of supporting a range of equipment, sometimes extending beyond simple recorder support to include DLT drives, DAT, and other forms of media. For applications where you are creating masters for replication, access to some of these popular tape formats can be useful in certain situations. Because new recorders are frequently being introduced, most software manufacturers provide a means of updating the drivers that operate with their product to allow customers to gain the benefits of the newest equipment on the market, without having to release a new version of the software.

Hardware capabilities also vary and it makes sense to ensure that the software you purchase supports most of the features offered by your hardware. Professional-caliber software generally supports the widest range of hardware features, allowing developers to control every aspect of the formatting and exercise every disc option available.

For example, you may have a software application that can perform packet writing, but unless the connected CD recorder includes support for incremental packet writing, the feature is useless. Similarly, to consistently produce audio CDs that are probably formatted for submission to a replication service, both the CD recorder and the CD-R software must be

Disc Recording Software Features

capable of performing Disc-at-Once recording. While almost all CD-R applications support Disc-at-Once recording, the feature cannot be used unless the hardware also can handle this method of recording. You can sometimes circumvent the limitations of earlier hardware by performing a firmware upgrade of the equipment (either by inserting an upgraded ROM chip into the recorder or transferring the firmware upgrade directly into recorder's Flash ROM—which is supported on some recent recorders). Many of the CD recorder manufacturers include extensive information about available upgrade paths on their Web sites.

When evaluating the purchase of a disc recording application, keep this factor in mind: Any high-end features that are beyond the capabilities of your disc recording equipment won't do you any good, and you'll most likely pay more for the kinds of applications that offer these extra features.

This section discusses and describes the features common to different disc recording applications.

Interface Considerations

A well-designed interface can contribute greatly to productivity when working with a disc recording application. Learning the program initially is much easier if the interface makes sense and the program makes the most commonly needed features easily accessible. Less-needed features should be placed where they can be quickly found, but they intrude on the interface.

Since a large part of the disc-recording process consists of selecting and organizing files, many applications rely on interfaces modeled after the Windows 98 Explorer. File selections are typically made from a collapsible directory tree. This approach usually uses a pair of windows—one representing the disc image and the other showing the system resources through a directory tree view, often actually using File Manager or Explorer as the vehicle for locating and selecting files and directories. Using drag-and-drop techniques, files are selected and copied to the CD image window. This technique best fits a model where the primary goal is data distribution and placement of files geographically on the disc is less critical.

Other disc recording programs support file selection through a browse-and-click techniques, where you navigate through your system resources and select items to add to a list composing the CD image. Some software builds the disc image geography based on the order that you select individual files and directories (usually placing the files selected first closest to the center of the disc where they can be most rapidly accessed during playback). In some cases, you can drop down to the sector level and specify precisely the physical arrangement of the files on the recordable disc. Other interface approaches exist as well. The simplest is based on setting up the disc recorder as a drive letter (or by name in Macintosh environments). You transfer files to disc by copying them as you would to a hard disk drive. The best interfaces combine simple access to the disc recorder on one level, but also offer the capability of dropping to a lower level and performing placement of data very precisely.

Progress gauges and visual displays of disc organization are included in many applications. For example, Adaptec's Easy-CD Creator includes a capacity bar graph at the bottom of the display that shows the percentage of CD-ROM space consumed by files you have selected during premastering. When the bar graph reaches the top, you've reached the capacity of the CD-ROM (so it's time to stop adding files to the image).

Demos of CD recorder applications can help you decide if the look-and-feel of a program suits your tastes and requirements. Since everyone's tastes are different, an interface that pleases one user might be totally inappropriate for another user. Professionals may be annoyed by novice-level features that can't be hidden or wizards that can't be easily turned off.

To help you evaluate the differences in look-and-feel between applications, we have included trial versions of several different disc recorder applications with this book. If you haven't yet purchased a disc recording program, or you're looking for a replacement for your bundled disc recording software, we strongly encourage you to sample these applications.

File Format Support

CD-ROMs have their own file system, ISO 9660, designed to facilitate exchange of data in a platform-independent way. You would have to do some heavy duty searching to find a CD recorder application lacking support for ISO 9660. Only applications created prior to the adoption of ISO 9660 in 1987 are likely to be deficient in this area. The predecessor to ISO 9660, the High Sierra file structure, is often included as an option in CD recorders applications, even though the use of this file system is essentially obsolete. We're not exactly sure why manufacturer's think it is important to give you the option of recording in an obsolete format, but almost all of them do.

Beyond ISO 9660, many CD recorder applications include support for native file systems, particularly those available in the Macintosh and UNIX environments. Macintosh and UNIX users lose much of the flexibility and power of their native file system when restricted to ISO 9660 conventions. For this reason, many of the CD recorder applications support different forms of hybrid discs that combine characteristics of two different file systems while still permitting ISO 9660 data interchange.

In comparison, the various forms of DVD all use the UDF file system. This simplifies the creation of DVD images, since the developer can consistently rely on one file system to bridge all platforms and formats. The UDF file system is also used to support the incremental forms of recording that are part of CD-RW. This provides a consistent and reliable means of distributing files among all readers that support the MultiRead format; CD-RW discs, however, cannot be used as masters when submitting files for replication.

ISO 9660 Issues

ISO 9660 includes three interchange levels. Level 1 is commonly used for CD-ROMs that are targeted towards playback on all platforms because it supports the lowest common denominator: playback on MS-DOS computers. Software packages have various ways of ensuring that the files that you move to an ISO 9660 fit into the strict naming conventions, limiting the characters to A through Z (all uppercase) and the numerals 0 through 9. For variety, you can use the underscore character (_), but any other characters (such as # or $ or &) are forbidden under ISO 9660.

The most common approach to this in software applications is to let you choose to enforce or over-ride the ISO 9660 restrictions. If you enforce them, the program generally will display any files you have selected that violate the convention and allow you to rename the file before you create the disc image. Of course, some programs or routines on the CD-ROM may be looking for a particular file and changing a single character may cause that program to malfunction. Use care whenever renaming files to meet ISO 9660 requirements. It's generally better to modify your source files (whether a database, multimedia application, HTML document, or portable document format file) to comply with ISO 9660 than to try to change names on the fly when preparing files for CD mastering. Over-riding the file naming conventions is also an option if your intended audience does not include DOS computers (for example, Window95/98/NT, UNIX, and Macintosh).

Ideally, a good software application should have an intelligent method for screening filenames during mastering, offering you a reasonable solution to convert filenames, if necessary, and provide support for ISO 9660 Level 2 and Level 3, as well.

There are tradeoffs to creating hybrid discs and this approach is usually taken only when the benefits for users to work with files in an accustomed manner are compelling. For example, some approaches to creating hybrid discs that include the Macintosh HFS file system divide the data regions into two separate areas on a disc, one for the Macintosh and the other for all other platforms. This approach effectively cuts the data storage capacity of the disc in half, since many of the data files are duplicated. However, certain software applications let you use longer Macintosh filenames and create icons for files and still share the single ISO 9660 data storage region. If maintaining a comfortable working environment for Macintosh users is among your CD-ROM title development goals, make sure you select a disc recording application that supports a workable solution to multiple file systems.

Flexibility for implementing UNIX file structures in native and hybrid forms is less widespread, although many CD recorder applications can handle the Rock Ridge extensions, a set of rules governing creation of ISO 9660 discs with more UNIX capabilities. Only applications designed for use in the native UNIX environment, such as Young Minds CD Studio, seem to provide complete flexibility in this area. CD Studio supports the creation of a variety of different hybrid discs with varying approaches to the UNIX file structure, including full native support.

Because the CD-ROM serves as a great leveling medium and it effectively breaks down barriers between different platforms so effectively, most CD-ROMs are created in ISO 9660 format, allowing the widest possible distribution. Many of the applications that also support DVD-ROM recording let you create CD-ROMs on a CD-R drive so you can reach the widest possible audience.

Recorder and Tape Support

The greatest disc recording application in the world is useless to you if it doesn't support the disc recorder that you plan to use. While there are less than twenty disc recorder manufacturers competing in the market,

Disc Recording Software Features

some disc recording applications do not support the full range of recorders. As a general rule, those companies that have been actively developing software during the full period of recordable CD evolution have the broadest support for hardware. Command.com with their GEAR software includes support for almost every available disc recorder. Adaptec's Easy-CD Creator also includes a long list of supported hardware devices. More recent software products may only offer support for the most common recorders or most recently released recorders. Some specialized applications may be designed to only work with a few recorders.

Because there may be situations where you want to copy a premastered disc image to tape, direct support for tape drives from within the disc recording application can be very useful. Most replication facilities accept incoming data on 8mm Exabyte tape or 4mm Digital Audio Tape. Those replication facilities designed for DVD replication generally accept Digital Linear Tape (DLT) for submissions.

Support for some of the earlier tape formats, such as 9-track and Quarter-Inch Cartridge (QIC), also exists at some replication houses, but these tape formats are rapidly being superseded by more modern Exabyte, DLT, and DAT formats.

Support for interconnected tape drives varies from program to program. If this feature is important to you, check carefully to ensure that your first choice for a disc recording application offers this support.

Simulation

Simulation features come in a couple of different varieties:

- Simulation of the disc recording process prior to actually burning a disc
- Simulation of the performance of a CD-ROM or DVD-ROM title from a stored physical image

Each of these processes is important in its own way, as explained in the following two sections.

Recording Simulations

Almost all of the disc recording applications—with the exception of the simplest backup programs—let you "test" the recording process prior to burning a disc. The essential feature was born out of the ease with which early CD-R systems destroyed media during the recording process. When the media costs were in the $12 to $20 range, you could easily burn up a $100 in discs during a morning of failed recording. Simulation before recording is also important in DVD-R recording, where media costs are much higher than for CD-R.

To be effective, the simulation needs to actually access data from the source storage media and transfer it across the SCSI bus (or other I/O bus) to the disc recorder. The testing should also exercise the laser write assembly as it would if recording was taking place. The realism of the test stops at actually energizing the laser, so the process remains a simulation—although you may see the red recording light illuminate on the front of your disc recorder (if it has one), no bits are burned and no media is altered.

The better disc recording applications can detect possible performance problems that may interrupt your data recording and ruin the media. These kinds of warnings can alert you to the fact you may need to further optimize your system by setting up additional caching, connecting a faster hard disk drive, or installing a dedicated SCSI host adapter for the recorder.

Sometimes you can sidestep performance difficulties that are identified during a simulation by recording from a physical CD image rather than performing on-the-fly recording from a virtual image. In fact, if a recording simulation fails on the first attempt, you should reduce the recording speed from 8x to 4x or from 4x to 2x. If that attempt fails, try recording from a disc image (a file recorded to your hard disk drive that contains the exact contents of the disc ready for transfer). If neither of these techniques improves matters, you probably have some fundamental problems with your system setup that need to be corrected before you will be able to record. Simulation is an excellent way for detecting these kinds of problems without burning up a lot of recordable discs.

Performance Simulations

To help you predict how your completed disc presentations and programs will perform when released, some disc recording applications also let you run simulations from a hard disk drive directly from the physical image in which the files are stored.

For this approach, the application generally lets you set up a virtual disc driver that introduces the equivalent access delays that you would encounter if you were transferring data from a pressed disc. Performance problems can be identified and perhaps corrected by repositioning files in the disc recording application so they will appear closer to or further from the center of the disc.

While simulation of this sort can only approximate real-world performance, it can be a useful aid in refining the performance of a CD-ROM or DVD-ROM title. Modern disc drives have reduced the need to go to extreme measures to achieve satisfactory playback, so this type of simulation becomes important primarily in cases where you're trying to reach the highest level of performance for a game or for database access or for training simulations, including videos or similar materials. For routine CD-R or DVD-R applications, the performance simulation is probably overkill.

Simulation can also be an important tool for detecting logical errors that exist within a CD-ROM image. One recurrent problem is caused by the renaming of files during premastering to fit ISO 9660 conventions. If an application or search engine used in a CD-ROM title requires a file of a certain name for operation and that file has been renamed during conversion to ISO 9660, the application may not run. It's best to identify and correct these kinds of errors before burning a disc. To avoid this kind of error, when you're creating an application, use ISO 9660 names during development so that no conversions are necessary.

The task of running through the full contents of a CD-ROM image and testing all the various permutations of hyper-link paths or file accesses can be extremely time consuming and it can require a very methodical approach to avoid missing potentially serious, disruptive errors. This feature can be a valuable testing and debugging aid for any CD-R or DVD-R user; those software packages that support performance simulation will reward you with smoother running, trouble-free discs if you take advantage of this option.

Multisession Support

One of the handicaps of early CD recorders and applications was the inability to record in any mode other than Disc-at-Once. A single-session recording, even if it only transferred files that occupied a small percentage of the total storage capacity of the disc, closed the disc to any further write operations. Users either had to save up files for storage and archiving until they had enough to justify copying them to CD, or they had to be content with storing lesser quantities of data.

Ironically, as many manufacturers adopted support for multisession recording, a contradictory problem arose. Hardware and software that didn't support Disc-at-Once recording could not produce CD-ROM masters that could be taken to a replication service and reliably deliver properly structured discs. Track-at-Once recording caused low-level link information to be embedded in the data, ruining the one-off disc for replication purposes.

The trend now is for software and hardware packages to provide full multisession writing capabilities, including incremental write sessions using packet writing, but also to be able to switch to Disc-at-Once recording mode in order to support production of audio CD masters or other CD-ROM formats intended for mass replication.

Multisession writing provides the ability to create a number of individual sessions, recording each at separate periods of time, with the option of fixating the disc when full. Fixating completes the recording process by creating a single table of contents to access the files in individual sessions. Fixating also makes the disc accessible to a standard CD-ROM drive, allowing anyone to read it. Prior to fixating, only the CD recorder or CD-ROM drives operating under specialized drivers can read the file data.

Multisession support carries an overhead consisting of up to 20MB of data to support the initial session and 9MB for each individual session. Manufacturers have taken different approaches to handling session creation and linking, so the exact overhead varies from one implementation to another.

Among the different approaches to multisession recording, you will find each of the following.

Kodak Multisession

Kodak's approach to multisession recording, particularly as implemented for the PhotoCD standard, has been adopted by other manufacturers as well. One of the ideas behind this approach was to allow consumers to bring in their photographic film for processing and to have the images recorded to CD-R. While getting film processed onto CD-ROM has not caught on quite as vigorously as Kodak had probably hoped, the technique is extremely valuable to graphic arts professionals and organizations that deal with massive numbers of images (such as NASA's libraries of photos from space explorations).

Sometimes termed "last-session-first" recording, the PhotoCD recording technique allowed Kodak to add images to a recordable CD in several different sessions. This allows the full capacity of the CD to be used instead of closing it out after the first writing. As each new session is recorded, the directory information from previous sessions is merged with the new directory, making the CD appear as a single, cohesive unit. This approach easily adapts to data as well as images. Kodak Multisession refers to directories constructed using ISO 9660 conventions and also implies CD-ROM XA disc formatting. Hardware and software designed only to meet Kodak's definition of multisession often can not read CD-ROM Mode 1 multisession discs, as defined by Orange Book.

Hardware Multisession

From the perspective of many hardware manufacturers, recording a session consists of writing one track or a group of tracks to disc at a time, regardless of whether the data is presented in the ISO 9660 standard. Each session consists of a Lead-In and Lead-Out area with the Open New Program (ONP) bit set to indicate fixation. This definition is much broader than the Kodak Multisession approach, but it still assumes that new sessions have an awareness of earlier sessions.

ECMA 168

Members of the European Computer Manufacturers Association have agreed on a set of guidelines defining a session as an integral file system incorporating ISO 9660 and Orange Book conventions. Within these guidelines, new sessions can be appended to a recordable CD in a standardized manner. This approach is fundamentally the same as the conventions expressed by the Frankfurt Group. Both CD-ROM XA Mode 2 and CD-ROM Mode 1 disc formats are included.

Autonomous Multisession

Based on Orange Book standards, autonomous multisession recording assumes that each session on a disc is isolated and independent from every other session. There is no attempt to extend the directories in more recent sessions to include the information in earlier sessions. This approach has particular value in storage or archiving situations, where the software accessing the CD-ROM recording over several sessions has the necessary driver to jump from session to session to retrieve the data. In most instances, discs recorded using this approach can't be played back in a standard CD-ROM drive without a separate specialized driver. Typically, the first recorded session will be available, but none of the others will be visible.

Chapter 6

Online Assistance

To lessen the learning curve for disc recording applications, many of the software companies have incorporated an abundance of online information in the form of wizards, guides, and extensive online help. The better implementations of this approach can have you recording discs within a few minutes of firing up the application program, even if you don't have any previous experience with this type of media.

For example, Adaptec's Easy CD Creator employs a Disc Wizard that is automatically activated when the program starts. You progress through the process by filling in the edit boxes in a succession of dialog boxes and then selecting audio or data files to add to the CD. By the time you've advanced through the steps presented by the Wizard, you're one button press away from burning your first compact disc. Experienced users can bypass the wizard and make their selections through the options in the main program window.

CeQuadrat's WinOnCD ToGo! (now owned and distributed through Adaptec) follows a similar strategy in the form of a Guide. Selecting the Guide from the program's main application window takes you through all of the necessary dialog boxes, from hardware setup to file selection, for initiating the recording process. The guide can be enabled or disabled as needed at any time you are running the program.

A program's ease of use can be greatly enhanced by online procedures and guided sequences. If you are new to disc recording, you might want to select a recording application that provides a generous amount of online assistance.

Using Wizards

The use of wizards has lowered the barrier to mastering disc recording. Wizards are similar to online tutorials, guiding new users through the process of setting up files, selecting formats, and recording discs. Unlike most tutorials, however, wizards accomplish useful work and if you follow the process through to the end, the result is a completed disc recording.

Hewlett-Packard CD-Writer Plus Wizard

Hewlett-Packard did much to legitimize CD recording when they entered the market with their SureStore recorders several years ago. Besides significantly lowering the entry price for disc recording, with the first significant unit that sold for under $1000, the abundance of support from HP and the features designed for ease-of-use helped CD recording achieve broader acceptance in the mainstream computer market. Typical computer users—not just professionals, developers, and musicians, began to adopt the technology—and CD recording became much more commonplace.

Hewlett-Packard continues this tradition with their well-designed CD-Writer Plus series, which includes numerous features to guide the setup and recording process. The following material shows how the wizard handles the task.

The wizard is activated when you select *Create a CD* from the HP CD-Writer Plus Start Menu option. As you can see from Figure 6 - 1, the first question, What do you want to do?, prompts the user through the full range of possibilities for disc recording. Each prompt provides details as to the nature of the operation. In this case, you can choose to:

- Make a music CD (Red Book audio)
- Copy an existing CD
- Make a data CD through drag-and-drop selection
- Make a data CD for wider distribution
- Create CDs and a diskette for disaster recovery
- Define a backup schedule using HP Simple Trax

Chapter 6

Figure 6 - 1   **Wizard lets you select the task to perform**

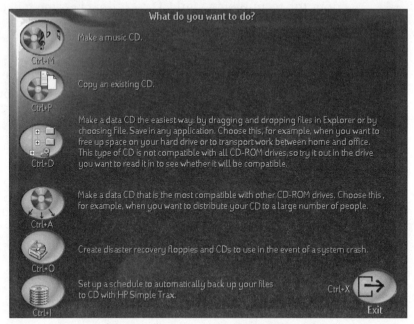

Your selection at this point launches the appropriate application or a brief recommendation message. For example, if you choose the option for creating a data CD by drag-and-drop, the wizard displays the message shown in Figure 6 - 2.

Figure 6 - 2   **Recommendation for disc type**

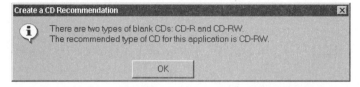

As you can see, the wizard recommends the use of CD-RW as the media for the selected operation. If you click **OK** to proceed, the wizard launches the appropriate application, in this case, the Adaptec DirectCD Wizard, to continue the process.

Figure 6 - 3   Welcome to the Adaptec DirectCD Wizard

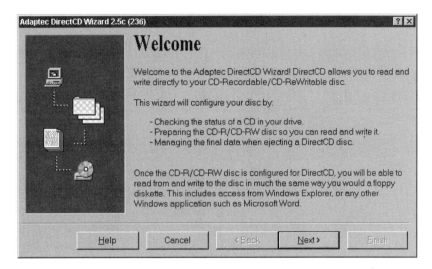

The DirectCD Wizard explains how it will configure the disc recording by:

- Checking the status of the blank disc media
- Formatting the media for the write operation
- Structuring the data to fit the selected format

Clicking **Next** continues the operation. The wizard identifies the connected CD-RW drive, the HP CD-Writer+ 8200. The type of media installed in this case is a blank CD-R disc.

Chapter 6

Figure 6 - 4   Identifying the Drive Information

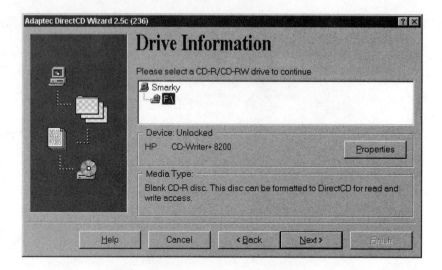

DirectCD can format both CD-R and CD-RW discs for simple drive-letter access. Of course, only the CD-RW discs can be erased and overwritten. Formatting of CD-R discs takes only a few seconds of preparation. Formatting of CD-RW discs, since they require more elaborate addressing information, can take a few minutes.

Once a disc has been prepared for use through DirectCD, it is in an interim state where it can only be read through the DirectCD driver. If you attempt to eject it while it is in this state, it displays a screen prompting you as to whether you want to make the disc accessible for use in other CD-ROM drives or keep it in its current state for future use through DirectCD. This is illustrated in Figure 6 - 7.

When the initial formatting operation is complete and the CD is ready for drag-and-drop file operations, the wizard displays the completion screen, as shown in Figure 6 - 5.

Using Wizards

Figure 6 - 5   **CD ready for read and write operations**

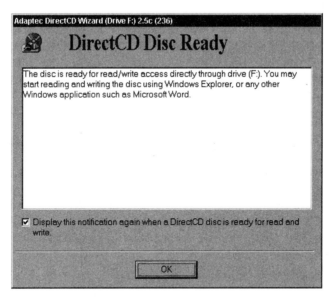

When the disc is prepared in this manner, you can open up the window for the disc recorder and simply drag files and folders to it. Windows will treat the operation as it would any file copy operation to a hard disk drive, Zip cartridge, or diskette. For example, Figure 6 - 6 shows the appearance of the disc contents window at the completion of the write operation.

Figure 6 - 6   **Files and folders copied through DirectCD**

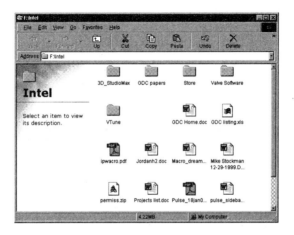

Chapter 6

If you attempt to eject the disc in its current state, the DirectCD wizard prompts to see if you want to close it, so that others CD-ROM drives will be able to access it, or to leave it so that it can be reused through drive-letter access for future DirectCD operations. If you don't close the disc, it can only be read in CD-R/CD-RW drives that are operating under DirectCD.

Figure 6 - 7   **Ejecting a disc through DirectCD**

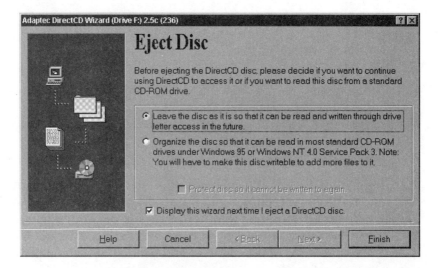

If you write to a CD-R disc in this manner and you continue adding files, you will eventually run out of space, since CD-R discs can only be written once. Although DirectCD lets you delete files (by removing the pointers to them), it cannot reclaim the space that has been used for storage.

In comparison, CD-RW media can be rewritten many times. For this reason, CD-RW media makes more sense for temporary storage and short-term archiving.

As you can see from this short example, this wizard made the process of recording to a CD very simple and not much different than formatting and writing to a diskette. The wizards for other operations, such as creating audio CDs and performing periodic system backups, are not much more difficult. Wizards exist for nearly every type of disc recording operation and they can definitely shorten the learning curve for this technology and help novices achieve successful results.

Examples of Disc Recording Applications

The following subsections highlight some of the more capable disc recording applications from those currently available in the market.

HP Disaster Recovery Wizard

CD-R and DVD-R applications designed primarily for backup and archival data storage generally lag behind their tape-based counterparts. The clear advantages of recordable discs as a backup medium include longevity, stability, reliability, accessibility, and platform independence. Anyone who is ever sorted through the file directory catalogs for three months worth of backup tape cartridges knows how much easier it is to slip a recordable disc into a drive and locate any required files from a directory tree.

This situation is changing, however, and many software producers are designing specialized utilities to handle archiving and backup operations in a simple and straightforward manner. These applications are designed to take full advantage of the capabilities of disc recorders and media, supporting such operations as the chaining of multiple sequential discs to enable several Gigabytes of data to be backed up as part of a recovery set.

The worst case scenario that a backup operation prepares you for is a catastrophic hard disk failure. The HP Disaster Recovery application, bundled with many of the recorders in the CD-Writer Plus series, creates a disaster recovery set consisting of diskettes to reboot your computer and CD-RW discs to restore your data and applications. By running the recovery command after a hard disk failure, you can restore your system to its prior state.

A wizard guides the setup process. When you select CD-Writer Plus from the Start Menu, the initial wizard screen appears, as shown in Figure 6 - 8.

Chapter 6

Figure 6 - 8   **Disaster Recovery wizard**

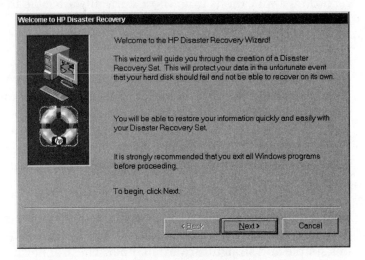

The application calculates the quantity of files that need to be backed up in preparation for disaster recovery and indicates the system status. In this example, two diskettes are needed to create a Windows 98 startup disk used by the recovery application followed by five CD-RW discs for the actual data backup.

Figure 6 - 9   **Storage requirements for backup**

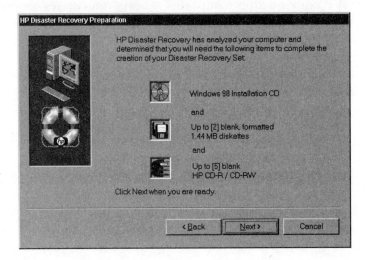

Examples of Disc Recording Applications

When you click **Next**, the program prompts you to insert a diskette for the startup disk creation, as well as the Windows 98 CD-ROM. Typically, one or two diskettes are created, depending on the current configuration of your system. The disaster recovery utility then customizes this startup disk to work in combination with the CD-RW discs that will contain the backup data.

Figure 6 - 10   **Copying data to recordable discs**

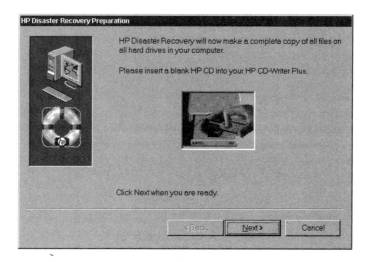

If you haven't formatted the CD-RW discs for disaster recovery, the program formats them at this time. This can be a fairly tedious operation requiring up to 40 minutes to complete. The disaster recovery program proceeds to format each disc, as required, and then copy all of the data from your hard disk volumes to CD in a sequential set.

The program prompts you to insert new blank media as it fills up each disc in the set. An overall progress bar indicates the progression of the backup operation. You can stop the process if necessary by clicking Cancel; otherwise, the program continues until it has stored each of the files on your system onto CD-RW discs.

Chapter 6

Figure 6 - 11   **Formatting a disc for recovery**

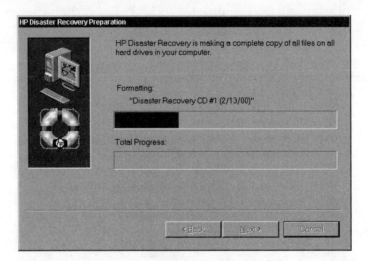

Once done storing your files on disc, the program gives you the opportunity to test the recovery process. You can proceed as if you're performing a complete recovery, but stop short of actually overwriting any files.

This type of backup offers an ideal use for CD-RW. You're safely protected from any crashes that damage your hard drive and you can reuse the media as necessary to create new recovery discs whenever your system changes significantly. This type of backup is comparable to the Complete option that is offered by many tape backup systems; it can be supplemented with an incremental backup of files that change periodically. The HP Simple Trax program, described in the next section, offers an automated way to maintain historical copies of files, a feature that can be extremely useful if you are working on a project involving many different files and you need to archive and track the progression of the project changes.

Automated Backup with HP Simple Trax

The HP Simple Trax program provides a means of tracking the contents of CDs, even if they are not inserted in your system. It also lets you copy specified files to recordable media on a regular schedule. In other words, the program provides both indexing and automated copy operations.

Examples of Disc Recording Applications

Simple Trax can be configured either through a wizard or by means of a control panel that lets you specify schedules and select files for automated copying to disc. Since we've already provided a couple of examples of wizards in action, let's look at the control panel features. As shown in , the control panel includes tabs for **Selection**, **Schedule**, and **Event Log**.

Figure 6 - 12   **Simple Trax control panel**

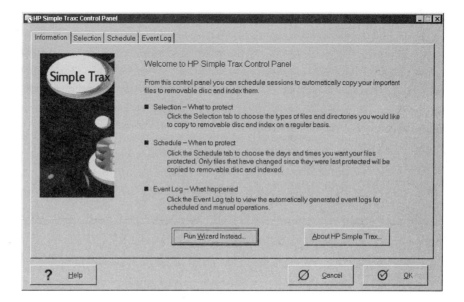

The **Selection** tab lets you check the types of files that you want to copy automatically and also check the directories that you want to include in the operation. You can limit the operation to simply data files, such as Microsoft Word or Adobe Illustrator files, or you can select any conceivable file on your system. Once you have determined the range of files to include, you can proceed to the next tab and schedule the copy operations.

Chapter 6

Figure 6 - 13   **Simple Trax Selection tab**

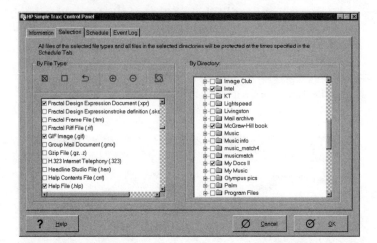

The **Schedule** tab offers considerable flexibility for automating copy operations. You can set up multiple days and multiple times for copy operations by checking the appropriate boxes. Simple Trax will then carry out the automated tasks according to the schedule indicated.

Figure 6 - 14   **Simple Trax Schedule tab**

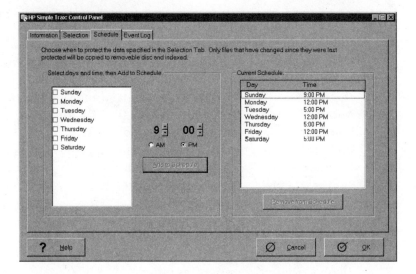

Examples of Disc Recording Applications

The operations that have been performed are visible as events that appear on the **Event Log** tab. Once files have been copied, you can view them through Windows Explorer. If the disc containing a file is not in the drive when you select it, Simple Trax prompts you to insert the CD containing the file. All the files that have been copied appear in the index and Simple Trax records and maintains a list of the file changes over time.

Figure 6 - 15   **Simple Trax Event Log tab**

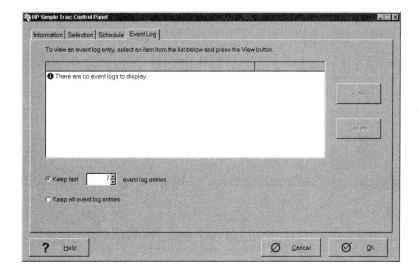

Simple Trax is not sophisticated enough to serve as a network backup utility or serious archiving tool, but it is capable and flexible enough to handle the requirements of individuals who want a failsafe way to maintain backup copies of important files necessary for their work.

Adaptec Easy CD Creator

Easily the most ubiquitous disc recording program in the industry, Adaptec Easy CD Creator has evolved through four versions over the last decade. With each iteration, it has been trimmed and streamlined for easier operation and wider acceptance in the mainstream market. Adaptec has forged bundling arrangements with many of the disc recorder manufacturers; Easy CD Creator is included with recorders from LaCie Ltd., Pinnacle Micro, Hewlett-Packard, Smart and Friendly, Ricoh Corporation, as well as others. At one point in the history of CD recording, Corel was winning accolades for the ease-of-use designed into

Chapter 6

their CD Creator product. With its test routines, guided configuration process, wizards, and well-designed interface, CD Creator took much of the difficulty out of disc recording. Adaptec purchased the product from Corel and integrated the best features into Easy CD software (which they had acquired from Incat Systems). The merged result, Easy CD Creator, combines a fair amount of control over the low-level aspects of recording with one of the more refined user interfaces in the CD recording world. The program handles audio discs and most formats of data discs with drag-and-drop simplicity. It also lets you layout jewel case inserts and design the actual disc layout to be printed via an inkjet printer or any other printer capable of printing to recordable media.

For the uninitiated, the program opens the Easy CD Creator Wizard at startup (shown in Figure 6 - 16) by default. This well-organized sequence leads novices through the process of recording their first disc and removes much of the confusion from the decisions that must be made along the way. Once you're more comfortable with disc recording, you can deselect an option in the program and open immediately to the main application window of Easy CD Creator.

Figure 6 - 16   **Easy CD Creator Wizard**

From the main application window, shown in Figure 6 - 17, Easy CD Creator present four panes—two that let you navigate through your hard disk contents and two that show the current structure of the disc content you are organizing.

Figure 6 - 17   **Main application window with its four panes**

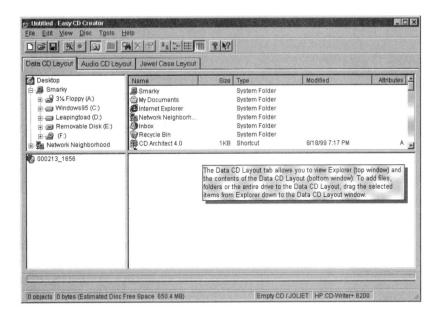

As you navigate through the disk volumes that appear in the upper left pane, you make selections by dragging files and folders to the lower right pane. As you do, the program fills out the hierarchical structure and displays it in the lower left pane. If you happen to select one or more files that does not meet the criteria for the selected disc format, the program displays a warning message and lets you rename the specified files. In the example shown in Figure 6 - 18, the filename exceeded the 64-character limit of Joliet.

Chapter 6

Figure 6 - 18 **Filename changing on the fly**

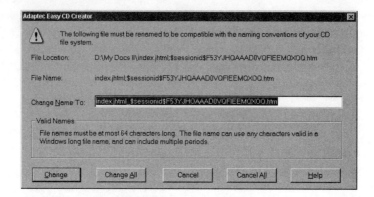

As the available space is filled on the blank media, a progress bar, shown along the lower edge of the window, indicates how much recording space remains, as shown in Figure 6 - 19. You can rearrange the content until you have it to your liking and then simply click the **Record** button to initiate the disc write operation.

Figure 6 - 19 **Capacity bar indicates how much potential space remains**

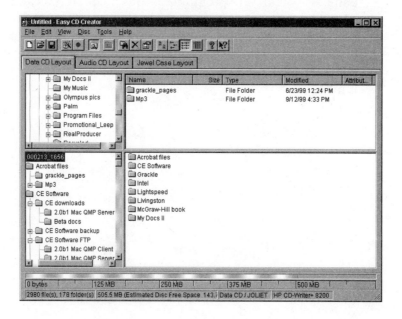

180

Examples of Disc Recording Applications

Easy CD Creator employs a very simple approach to handling the Disc Layout and various format options. You can make some simple selections that control items such as the file format (switching between Joliet and ISO 9660). You can maintain a modest degree of control over the file ordering and positioning using the options for reordering files for disc speed or for disc space, or you can preserve the ordering originally specified. These options are illustrated in Figure 6 - 20. If you select the ISO 9660 file system, you can choose to make the disc bootable or create a CD Extra disc.

Other settings on the neighboring tabs control the inclusion or exclusion of particular file types, assign your own volume label, and identify an audio disc title and artist for players that can read this information. All in all, the low-level controls are not as extensive as other disc recording packages, but, of course, this typically works in the program's favor for less experienced users—there are less ways to make a mistake when preparing files for recording.

Figure 6 - 20   **CD Layout Properties**

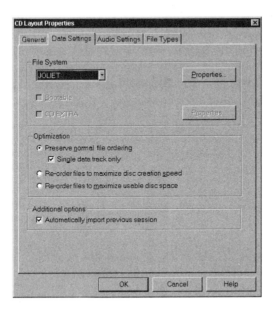

Another nice feature of this program, a carry-over from CD Creator days is a design tool for laying out the jewel case insert and the CD design. Text and graphics can be added to each of these templates and then printed out to an appropriate inkjet printer or other media printer.

While the artistic options are somewhat limited, this feature is easy to use and a good means of adding some polish and professionalism to a completed disc being distributed to a customer or colleague.

Figure 6 - 21 **Printing to discs**

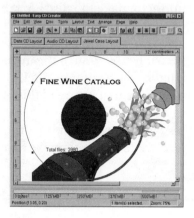

Test Utilities A suite of system test utilities is available from the **Tools** menu. These tests let you evaluate the performance and operation of your system components to ensure that successful recording can be accomplished. The overall transfer rate of the system can be tested and you can also test the connected CD recorder, through simulated write operations using a blank disc, to ensure that performance is adequate at the highest available speed of the recorder.

Testing provides reasonable assurance that your system can handle the demands of the recording process before you launch a project and start burning discs. Any problems encountered during testing can be corrected prior to attempting to write data to disc. Keep in mind that not all applications work with all hardware. Test failures reported by Easy CD Creator may not be recreated under different applications.

When you have creating a CD layout in the program, and you are preparing to burn a disc, the program also offers you the option to simulate the recording process first. You can access this function by checking the **Test only** option. You can also choose **Test and create CD**, which proceeds with the recording if the test is successful. If you know your system and you're confident enough to just start recording, you can choose the **Create CD** option.

Easy CD Creator offers a reasonably complete collection of CD recording tools, smoothly handling the creation of data discs (including mixed mode), audio discs, and some of the special-format discs, such as CD Extra. The program handles both CD-R and CD-RW media without difficulty.

Utilities such as the jewel case layout editor and a simple printing utility for disc designs complement the overall program functions. Easy CD Creator has just enough complexity that business people, students, and other who have just begun using disc recording should find enough depth to meet their needs for some time. Features such as the Easy CD Creator Wizard make the program highly accessible for novices and business users, and accelerate the learning curve for more experienced users.

Adaptec, Inc.

801 South Milpitas Blvd.
Milpitas, CA 95035
Tel: (408) 957-2044
Fax: (408) 957-6666
URL: *www.adaptec.com*

HyCD Publisher

ISO 9660 succeeds very nicely in bridging various platforms and creating media that can be distributed without concern for the playback equipment. The limitations, of course, are that the file naming restrictions and other conventions are quite restrictive. Macintosh users don't like losing their icon-based work environment. UNIX users resent losing long filenames, deep directory depths, and permissions.

HyCD, Inc. (formerly Creative Digital Research) addressed this problem with their product HyCD Publisher, a CD-R premastering application that produces multi-platform hybrid discs. Discs burned in the hybrid format supported by this program can be read by machines under Windows 95/98/NT, DOS, the latest versions of UNIX (including SGI IRIX 6.5 and Sun Solaris 2.6), and the MacOS, while still retaining the familiar operating environment of each platform. In other words, the disc format allows the simultaneous use of Joliet for Windows 95, plain vanilla ISO 9660 for Windows 3.11 and DOS users, ISO 9660 with the RockRidge extensions for UNIX, and HFS Macintosh complete with desktop icons.

HyCD, Inc. has focused their attention on developing a product that spans platforms, so it's a natural conclusion that the premastering application, HyCD, would have versions that support major platforms (including Silicon Graphics workstations, Sun Microsystems machines, Windows 95/98/NT. Macintosh support is suspiciously absent, but this omission is probably forgivable since the Mac native format is supported in the hybrid approach. HyCD, Inc. has designed their software to include most of the multimedia formats, including CD Plus (Enhanced CD), Mixed Mode, CD-ROM XA, Video CD, and Red Book audio.

The program also includes a number of features that are sometimes only seen in high-end professional caliber applications. The ability to control individual file placement on disc is a welcome feature for developers who are constantly tweaking disc geography to gain playback advantages. Support for autoloaders and robotics lets the HyCD Publisher be integrated into production environments. The support for CD-RW drives extends the usefulness of the product to backup and archiving, as well as premastering applications.

HyCD, Inc. also claims full support for the most recent disc recorders on the market. HyCD publisher is also bundled with disc recorders from some manufacturers, including JVC and Ricoh. Software upgrades are available through their Web site (*www.HyCD.com*), as well as their new products, such as HyCD Play&Record v3.0, an audio tool that can burn MP3 files directly to disc (among many other audio features). Integration of the Fraunhofer MP3 audio compression software into the upgraded version of this product provides powerful encoding capabilities beyond the more pedestrian MP3 encoders.

Different manufacturers have different solutions to hybrid disc creation. If this aspect of CD recording is important to your end application, you will probably want to thoroughly investigate the approach taken by each software producer. HyCD provides a practical solution to developers who want to release titles that encompass the full range of platforms. Among a number of competitors, HyCD excels at hybrid creation.

HyCD, Inc.

5300 Stevens Creek Blvd., 4th Floor
San Jose, CA 95129
Tel: (408) 244-7007
Fax: (408) 244-7077
URL: *http://www.HyCD.com*

GEAR PRO DVD

Ideally, professional-caliber applications should let you do anything that you could conceivably need to do when it comes to disc recording. As might be expected, these are the most expensive applications, but many of the ease-of-use features common to more basic applications also appear in these high-end counterparts. No longer is it necessary to struggle with command-line syntax or difficult format issues to premaster a CD-ROM or DVD-Video disc. GEAR PRO DVD evolved from the GEAR product line introduced by Elektroson, a Netherlands-based CD-R pioneer.

Elektroson originated many of the processes that have become common in the recordable CD field, such as mastering from virtual images and simulation of CD performance. They also produced some of the earliest versions of premastering tools, and the product line of succession led directly to GEAR. Early versions of GEAR could handle premastering on a wide variety of computer platforms, including Windows, Macintosh, and UNIX. GEAR is noteworthy for the level of precision and control it permits over the recording process, with functions that manage interleaving of different data types, sector-level layout, individual sizing of sectors, and other low-level controls.

Now distributed by Command.com, a software firm based in Jupiter, Florida, GEAR PRO DVD includes the capability of premastering to CD-R, CD-RW, or DVD-R media. File system support runs the gamut from ISO 9660 and Joliet to UNIX with the Rock Ridge extensions and version 2.0 of UDF.

Upon launching the program, GEAR opens to a conventional application window, shown in Figure 6 - 22, that displays the drive volumes in two upper panes and the selections for disc recording on two bottom panes. Files and folders are dragged from directory tree above to the lower right pane. The resulting file structure is displayed in the two lower panes. File information, including the starting sector number where the file will be written to disc, appears to the right of each filename.

Figure 6 - 22   **GEAR PRO DVD main application window**

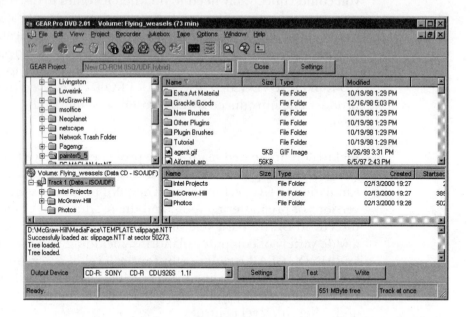

You start by defining a GEAR Project, which can consist of any of the following:

- DVD-ROM (ISO/UDF)
- DVD-ROM (ISO)
- DVD-Video (ISO/UDF)
- CD-ROM (ISO/UDF hybrid)
- CD-ROM (ISO)
- CD-ROM XA
- CD-Audio
- CD-ROM mixed mode (Mode 1 or Mode 2)

For CD-ROM projects, the volume can be based on any of the existing blank media sizes: 18 minutes (158MB), 63 minutes (553MB), 74 minutes (650MB), or 80 minutes (703MB). For DVD-ROM and DVD-Video projects, the images sizes supported for one-off creation range from 428 minutes (3.95GB) to 509 minutes (4.7GB). DVD projects aimed at utiliz-

Examples of Disc Recording Applications

ing the larger capacities of double-layered media can be outputted to DVD premaster tapes, allowing you to create 8.5GB up to 17GB projects.

Support for tape output in GEAR is quite flexible. For CD mastering, the developer has control of many aspects of the tape file strategy, as shown in Figure 6 - 23. Tapes can be created that place tracks in separate files, all of the same type tracks in a single file, or that consist of one contiguous image file. Tape blocking factors can be adjusted and options for specifying the sector size and pregap index are included in the settings options.

For DVD mastering, the DDP format can be used when outputting to tape or an individual tape file strategy can be selected. For DVD-ROM Mode 1 applications, the sector size and tape blocking factor can be specified.

Figure 6 - 23   **Tape settings for CD mastering**

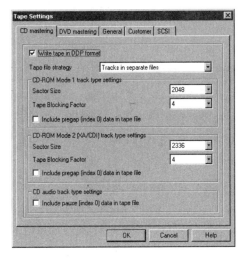

ISO Filenames   GEAR offers a variety of options when it come to translating filenames to meet ISO 9660 requirements. You can choose your level of translation and the program can either prompt you before changes or automatically make any necessary character substitutions. The following options are available:

- Do not translate names, length up to 64
- Translate to uppercase, length up to 64

Chapter 6

- Translate to ISO names, length up to 64
- Translate to ISO names, truncate to 8+3

Selection of the appropriate UDF version to apply is also a useful addition to the Preferences options. Individual settings are available for the UDF version for CD-ROM and DVD-ROM, as well as the UDF version for DVD-Video. Support for ISO/UDF hybrid discs also makes it easier to handle file system issues as UDF support continues to become more prevalent.

Test Runs

Once you've set up the ISO/UDF track contents to your liking, the Test Run icon lets you perform a simulated recording operation, as shown in Figure 6 - 24. You can choose whether to continue with an actual recording or stop once the test run has completed. Once you're assured of your systems ability to output successfully to disc, you can initiate a recording to the selected output device. Data verification can take place after the recording has completed.

Figure 6 - 24 **Performing a test run**

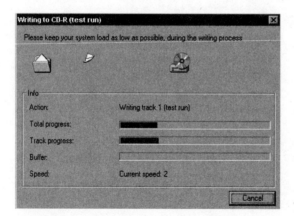

One particularly nice feature of GEAR is a log window that is displayed on the lower portion of the main application window. The log window provides the current status during any operations. It gives a running commentary on each action performed and these step-by-step details can be extremely helpful if you need to troubleshoot a malfunctioning recording system or if you just want to monitor the flow of events while recording or testing is taking place.

If you've selected the XA format for a track, GEAR provides two options for controlling the interleaving of data, including the capability of specifying the interleaving through manual means through the COPMMF command. This isn't a task that will normally be performed by anyone other than a developer trying to optimize the performance of a multimedia CD-ROM XA title, but it's nice to know the feature is there if you need it. GEAR consistently has all of the features and options you need tucked away in different parts of the program, like tools on a Swiss army knife, ready to click into place and perform some useful task.

Premastering to Tape GEAR's built-in support for tape systems is more extensive than most other products in this class. Some replication services still prefer tape to other media for creating glass masters for large-volume replication (although most services now accept CD-R and DVD-R one-offs). The program supports many common SCSI-based tape systems, including 8mm Exabyte, 4mm DAT, and Digital Linear Tape (DLT). A previous staple in the industry, M4 9-track, is rarely encountered these days. Two common output devices supported by the software that are generally accepted by replication facilities are the Exabyte tape subsystem and the Hewlett Packard HP35470A DAT drive.

As with premastering to CD-R, premastering to tape can proceed from a virtual image of the files. GEAR checks the details in the volume during downloading to tape to ensure that the actual file sizes and time stamps match the values recorded in the volume data. To ensure data integrity, GEAR can also check the validity of the files recorded to tape against the original volume data.

API Toolkits GEAR Software is firmly committed to extending CD-R and DVD-R technology to third-party application developers. Increased focus on optical disc recording applications in digital imaging, COLD applications, multimedia tools, MP3 audio utilities, and high-capacity storage devices has led GEAR Software to make their APIs for disc recording available for developers. The API tools make it possible to integrate the core technologies provided by the GEAR software into a variety of applications. Already, a number of products are appearing on the market that incorporate burning a recordable disc as the last stage in data collection and organization. These types of products will continue to proliferate as disc recording devices fall in price and the technology becomes more widespread in businesses and organizations

Chapter 6

GEAR Summary Over the years, GEAR has consistently rated very high in third-party reviews and evaluations. The range of UNIX platform support is exceptional, providing extensive coverage of various UNIX machines. GEAR provides genuine professional-level control over the disc recording process. Much flexibility is also available for file naming, including ISO Level 3 support (allowing you to use full Windows 95/98/NT directory filenames, up to 37 characters).

Handy features such as the real-time log window give you a handy visual display of events during recording and simulations and also offer valuable insights if the need to troubleshoot a recording system should arise.

All in all, GEAR PRO DVD offers a nearly perfect balance between low-level control over disc recording processes and the ease of use that comes from a carefully designed user interface.

GEAR Software USA

1061 East Indiantown Road, Suite 500
Jupiter, FL 33477
Tel: (561) 575-3200
Fax: (561) 575-3026
URL: *http://www.gearcdr.com*
Email: sales.us@gearcdr.com

ISOMEDIA Buzzsaw CD-R Recording Software

If your disc recording requirements include multiple recorders embedded in Tower units and shared through a network, Buzzsaw v4.0 provides a number of professional-caliber features that can simplify your workload. Available in both a Standard and a Pro version, Buzzsaw works under Window 95/98/NT. A flexible upgrade system allows you to start with the most basic version and then perform upgrades to gain access to more recorders to add the professional features.

The professional version of this software lets you record multiple unique images to as many recorders as you have connected to your system (up to a maximum of 32). Some applications designed for tower systems only support sending a single image to all of the recorders, much as a copy machine operates. This is the approach used by the Standard version of Buzzsaw.

Because of the flexibility of its Professional version, Buzzsaw easily supports interdepartmental disc recording requirements, where individual departments may be collaborating on the components for a project and need frequent access to a number of different recorders as workgroup members produce and modify content.

The network capabilities of Buzzsaw make it especially appealing for workgroup applications. Disc recording can be initiated from images stored on network directories or from pre-existing discs. An elaborate queuing system handles up to 500 jobs at a time forwarded from individual workstations on the network. The administrator of the queue can adjust priorities, set up batch operations, and manipulate duplication tasks in a number of ways. Images premastered by other applications, such as Adaptec Toast or Easy CD Creator Pro, can be recorded to disc through Buzzsaw, either as single images directed to multiple recorders or as multiple images directed to multiple recorders. This multitasking feature allows up to 8 unique 650MB images to be recorded simultaneously under Windows NT when retrieved from a single SCSI hard disk drive (rated for A/V operations; see Hard Disk Considerations on page 126 for more details). RAID drive systems can make even higher levels of multitasking possible.

Hardware support under Buzzsaw includes a number of 4x, 6x, and 8x CD-R recorders. Among the drives supported are:

- Yamaha 400, 4260, 4416
- Panasonic 7502-B
- Plextor PXR 820
- Teac 56S

A Bit-for-Bit Verification feature ensures the data integrity of each disc recorded by Buzzsaw. A flexible and configurable caching system allows the administrator to set the cache to an appropriate depth to prevent interruptions to the recording. Premastering to industry conventions includes Disc-at-Once closed sessions with the following formats supported:

- ISO 9660 Level 1
- ISO 9660 DOS-compliant Level 2
- ISO 9660 with full ASCII CD-ROM discs and UDF (including Rock Ridge encoding)

Audio CDs can be produced from existing .WAV files and also by extracting the necessary files from an existing audio CD.

Buzzsaw also has support for a number of autoloaders, making it possible to set up network-resident disc production systems. Current autoloader support includes:

- MediaTechnics
- CD-Robotics
- TS Solutions
- Cedar
- CopyPro
- Champion
- Rimage

Refer to Chapter 8, *CD Duplicators*, for more information about the operation of autoloaders.

ISOMEDIA, Inc.

2457 152nd Avenue NE,
Redmond, WA 98052
Phone: 800 468 3939
Fax: 425 869 9437
URL: *www.isomedia.com/ps/cdr-buzzsaw.htm*
Email: forder@isomedia.com

Deciding on an Application

Choosing the best disc recording application for your purposes from a group of descriptions is not easy, especially when the competitive features and pricing of these applications is so closely aligned. We hope that this chapter has at least given you a solid basis for comparing feature sets and learning a bit more about what is involved in the disc recording process.

We've bundled demonstration versions of several CD recorder applications on the CD-ROM included with this book. We invite you to sample these programs in the hopes you'll gain additional insights about the look and feel of programs in this category. Most of the demonstration programs allow you to create CD images—sometimes putting a limit on the size of the image—and perform actual recordings if you have a CD recorder available.

The best CD-recorder applications tend to be expensive, so it is generally wise to spend some time investigating the various features, the hardware support provided, the methods for creating images, the overall interface design, and other related features. You may find that the best solution is to select one package tailored to backup and archiving to CD-R for everyday file storage and a more professional application for your multimedia development or electronic publishing requirements. As next-generation CD software based on full drive-letter accessibility becomes widely available, programs that permit rapid file transfers and storage will also become useful additions to the toolkits of CD-ROM developers.

7

Recordable Media

While recordable discs bear a superficial resemblance to manufactured discs, under the surface the differences are considerable. Most forms of recordable disc rely on the light sensitivity of a dye layer to capture the patterns made by a tightly focused laser beam. Many different dye formulations have been developed to respond the varying conditions encountered during disc recording, such as the varying disc rotation speeds, the variety of disc recorders—each with its own laser write assembly and power calibration method, the temperature and humidity at the recording site, and many other factors.

Early problems with inconsistent dye formulations, media that would work on one brand of recorder and not another, and similar problems have been largely eliminated. While media quality and integrity are not an inconsequential consideration, the selection and use of recordable media is not as critical a concern as it was in the early days of disc recording. There are still frequent debates over the projected lifespans of recorded discs and indications that media test results may be available at some point, but, by and large, if you purchase a branded disc from one of the large media manufacturers that has been recommended for use at the maximum speed of your disc recorder, your chances of encountering media problems are quite minimal.

This chapter examines the different types of media available and investigates their individual properties and characteristics.

Chapter 7

CD-R Media

Recordable compact discs have gained widespread acceptance throughout the computer industry—hundreds of millions of units of blank media are sold each year. This large-scale acceptance has led to significant reductions in prices, making it possible for small production units utilizing CD duplicators to be competitive with replication services when quantities are in the hundreds. With the current CD-R disc costs at $1 to $3 per blank, this media is significantly less expensive than other form of random access data storage—with the least expensive discs, the calculated costs are less than $.0015 per MB.

Dye-Polymer Variations

The most noticeable visible difference among the various recordable-CD brands is the color of the dye that is used in the sub-layer—the material that reacts to the laser beam. Several different formulations have been developed and introduced into the market over the last few years. While their characteristics vary somewhat, each type functions reliably in the majority of recording applications. Higher speed recording, however, tends to favor the more recent dye formulations, including phthalocyanine and metal azo. Cyanine-based media, the first dye formulation that was used for CD recording, appears to have greater sensitivity to UV light exposure (suggesting shorter lifespans) and less responsiveness when used for high-speed recording. Even within the cyanine family, however, metallic materials added to the dye greatly improve the resistance to casual UV exposure, so as new formulations continue to be developed, it is difficult to make any generalizations on this subject.

Independent media tests have never been performed on a widespread basis, and testing by the individual manufacturers lacks the objectivity necessary for non-partial evaluation. Proponents and detractors of the various dye types can be readily found throughout the industry, but, unfortunately, most of their judgements are based on anecdotal evidence and individual experiences.

In the early days of recording, certain brands of recordable media consistently worked more reliably with certain recorders, due to a number of factors in recorder characteristics, laser assemblies, calibration procedures, and so on. Modern recorders and most media are more consistent and recorder features, such as Running Optical Power Calibration, overcome many of the small variations from one media type to another.

As a general rule of thumb, seek out a media brand that works reliably with your recorder and stick with it. This is not an absolute guarantee for recording perfection, since dye formulations can vary from batch to batch and sometimes branded media comes from more than one manufacturing source. However, until independent media testing is performed on a large-scale basis, it's the only recommendation that can be reasonably made.

The most common dyes types are:

- **Cyanine**: the original organic dye used for CD recording; cyanine media can be identified by its greenish hue. Among the manufacturers of this type media are Ricoh, TDK, and 3M.
- **Phthalocyanine**: identified by a yellow-gold or greenish-gold reflective surface, phthalocyanine was introduced after cyanine. Primary manufacturers of this media type are Kodak and Mitsui Toastsu.
- **Metal Azo**: with their silver reflective layer and metal azo dye, recordable discs of this type exhibit a bright blue recording surface. The primary manufacturer is Mitsubishi Chemical Company and the media is marketed through Verbatim.

Licensed Dye Formulas

Because the dye formulations used in CD media are patented, they have remained under the control of the six original Japanese companies who have been the primary media producers for over 10 years. This list includes:

- Pioneer
- TDK
- Taiyo Yuden
- Ricoh
- Verbatim/Mitsubishi
- Mitsui Toatsu

As the growth of CD recordable technology increased sharply in the late 1990's, a number of other companies tried to jump into the media production game, some of them playing fast and loose with the patent laws as they formulated dyes that replicated or closely matched the patented formulas on the market. Generally, the six original companies have not been overly anxious to license their formulas to other media producers. Because of the dynamic flux in this industry, and the fact that relabeling and branding media from the six main manufacturers occurs regularly, tracing the origins of some of the media that appears on the market can be difficult.

Recording Process

The heat energy generated by the laser beam striking the recording surface of a blank disc is the mechanism that writes the data patterns. The dye coating the sublayer of the recordable disc is a photo-absorption surface. The microscopic beam of the laser, producing 4 to 8 milliWatts of power, generates a pulse that heats the dye to about 250° Centigrade. In response to this energy pulse, the substrate layer expands into the absorption layer, the polymer combines with the decomposed dye, and an area of reduced reflectivity is formed on the surface—a pit. On any write-once media, such as CD-R, this mark is permanent—it is essentially burned into the substrate and cannot be erased. The mechanism is quite different for rewritable media, as explained in *CD-RW* on page 199.

The pit formed by the melding of polymer and dye produces substantial changes in the reflectivity of this area. For an 11T pit, this can be as much as a reduction from 75% to 25%. This represents enough of a change that the laser read assembly of a conventional CD-ROM drive or CD player can detect the difference.

Media that is designed for high-speed recording applications much maintain a tenuous balance between a dye formulation that can respond quickly to laser pulses during more rapid disc rotation speeds (up to 12x speeds) as well as create appropriate pit sizes at much slower rotations speeds when performing low-speed recording (at 1x or 2x speeds). For more details on this topic, refer to *High-Speed Recording on Low Quality Discs* on page 204.

Extended Capacity Discs

Some manufacturers, including Verbatim and Kodak, have released recordable CDs that offer additional minutes of recording space. Verbatim's Extended-Capacity CD-R80 Discs provide an extra 6 minutes (approximately 50MB) over the normal 74-minute storage capacity of a CD. This technique is achieved by adjusting the wobble frequency that is built into the pregroove running through the disc's spiral data track. The disc, while being recorded, is rotated more slowly, which allows more laser bursts to be packed into a tighter area. This increases the data capacity.

Kodak distributes a high-capacity disc called Ultima 80 Media that also provides 80 minutes of recording space (700MB). The Kodak disc uses a reflective layer composed of a combination of silver alloy and gold.

Extended storage media of this type may not work with all CD recorders. Manufacturers typically publish a compatibility list on their Web site or through another source to indicate those recorders that can access the extra recording area. For example, Kodak lists qualified recorders at: www.kodak.com/global/en/service/tib/tib4187.shtml

If your goal is to achieve maximum compatibility for a recorded disc you are distributing, you should avoid extended capacity media. They are handy, however, for gaining extra storage space if you own a compatible drive and you're not worried about distributing the discs you burn elsewhere.

CD-RW

The rewritable version of the CD-ROM format, CD-RW makes it possible to expand the format into applications such as short-term archiving and data exchange.

Orange Book Part III, authored by Sony and Philips, outlines the standard that elevates the compact disc from a write-once medium to a rewritable form of storage. CD-ReWritable (RW) has already had a big impact on the industry, despite the fact that the initial versions of this product did not enjoy universal playback on all CD-ROM system. The reflectivity of discs recorded in CD-RW drives can be up to 25 percent lower than commercial CDs or CD-R discs. This difference in reflectivity unfortunately makes the CD-RW unusable in the original 100-million CD-ROM drives that flooded the market; newer drivers, however, incorporate additional laser circuitry and electronics to bridge this difference. A CD-RW disc resembles a conventional CD-ROM in size and weight, but the layers contained beneath the disc surface are significantly different than recordable CD discs, as shown in Figure 7-1.

Chapter 7

Figure 7 - 1 **Layers in a CD-RW disc**

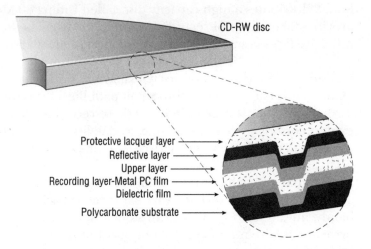

In a conventional recordable disc, the recording layer is an organic dye; once the dye is altered by a laser beam, it can't be restored to an unrecorded state. The recording layer of a CD-RW disc, however, consists of a crystalline layer composed of an alloy consisting of silver, indium, antimony, and tellurium (shown as the Metal PC film in Figure 7 - 1). In its unrecorded state, as a blank disc received from the factory, the crystalline layer presents a highly reflective surface. The laser beam, at a specific power setting, creates pulses the convert the crystalline layer to an amorphous state characterized by low reflectivity. This combination of low reflectivity and high reflectivity regions on the disc surface become the equivalent to the pits and lands on a CD-R. The laser beam effectively causes the recording surface to undergo a phase change. The recording surface can be changed back to the equivalent of an unrecorded state by firing another laser pulse at a lower power setting. When the recording layer cools, it returns to its original crystalline structure.

As the laser beam traverses the CD-RW disc, it alternately changes power settings to erase data or write new data. In other words, to write new material to the disc, the laser must first overwrite earlier data and return the recording surface to its crystalline state. Like a CD-R disc, the polycarbonate substrate of a CD-RW disc includes a pre-groove to guide the laser. Additional recording information is embedded in the disc in a specific data region, including recommended erase and write power settings, recording speed calibration, and individual formatting details.

CD-RW

To read discs created in a CD-RW drive, CD-ROM drives must contain additional circuitry allowing Automatic Gain Control functions to be used with the laser read head. While not an expensive modification to CD-ROM drive hardware, it is unlikely that earlier drives can be inexpensively retrofitted for this feature. Many of the CD-ROM drives currently on the market are equipped with MultiRead capabilities, allowing them to effectively read the contents of CD-RW discs.

CD-RW was initially more important from a data archiving and storage point of view, but the more universal compatibility of these discs makes it possible to use them for every type of application for which CD-R is used. There is, however, one noteworthy exception. CD-RW media cannot be used to produce CD one-offs that are to be the basis of preparing a master for mass replication. The method of formatting and data storage is significantly different than the method used for a CD-R disc written using Disc-at-Once recording (which is the appropriate method to use when creating discs for replication masters).

Media costs for CD-RW discs were initially about five times as expensive as standard CD-R discs, but volume manufacturing and consumer demand have successfully brought prices down. Current prices for CD-RW discs are approximately $5 each. Bargain brands purchased in quantities of 50 or more can sometimes be found for about $2.50.

Formatting CD-RW

One of the disadvantages of using CD-RW media is that the discs must be formatted for use in packet-writing applications—the incremental write operations normally used to store data in UDF on CD-RW. Depending on the speed of the host computer to which the CD-RW drive is connected, as well as the speed of the CD recorder, the formatting operation can take anywhere from ten minutes up to an hour.

The fixed-length formatting method that is used in this approach divides the storage space on the CD-RW disc into logical sectors of an identical size (unlike the variable-length packet writing used for CD-R). The CD recorder is then able to perform write operations to individual sectors and to erase data, similar to the approach that is used for a hard disk drive or a diskette. Without the capability to perform a random erase of data stored in individual sectors on the disc, the storage space will be eventually consumed. Early approaches to CD-RW let you delete a file, so that it would no longer appear in the disc directory, but the actual file remained resident on the disc. There was no way to reclaim the storage space without actually reformatting the CD-RW disc.

If you don't have the patience to wait for this lengthy formatting to take place, you have three options:

- Use an application that performs the formatting as a background operation, such as Adaptec's DirectCD. Discs can be formatted while you're working on a Word file or crunching numbers in a Quattro Pro spreadsheet.

- Purchase pre-formatted CD-RW discs, which are just becoming available. The first product release of this type was from Verbatim, who introduced pre-formatted 4x CD-RW media in November of 1999 as part of their DataLifePlus Media product line. Other manufacturers are following suit.

- Verbatim offers a Sequential CD-RW formatting option that they refer to as *quick formatting*. This method prepares the disc for random CD-RW packet writing and consumes less space for the formatting overhead (approximately 120MB less). Instead of taking 20 minutes or more, quick formatting takes 30 seconds. This approach maintains 100 percent compatibility with the UDF standard. By using this approach, however, the random erase capabilities of CD-RW are eliminated. Files are deleted by removing references to them. The actual files are only removed when the disc is fully erased.

As you can see, you have different formatting options available when using CD-RW. To achieve the random erase capability, you sacrifice storage space, allowing only 535MB of data on the disc. To gain up to 649MB of storage, you can use Verbatim's quick formatting approach, but you lose the capability of physically deleting individual files (although they can still be logically deleted). Verbatim's quick formatting is one of the features available through CeQuadrat's *just!burn* software, which also includes PacketCD (their implementation of packet writing).

Limited Rewriting

Unlike hard disk drives or other forms of magnetic rewritable media, CD-RW media can only be rewritten a certain number of times. Depending on the quality of the media being used, a typical disc can be rewritten anywhere from over a thousand times (the minimum required to meet the CD-RW standard) to over ten thousand times. For most practical applications, this should be sufficient to enable a disc to best used for several years without difficulty, or much longer if the media is only used for periodic archiving or backup.

The material used for the recording surface can only be converted from a crystalline to an amorphous state a certain number of times before it begins to lose the capacity for undergoing this transition. Data errors will begin to appear as the reflectivity changes in the material no longer achieve the levels that allow a CD-ROM laser read assembly to detect the data. If you are using CD-RW for mission-critical applications, you might want to mark the disc with the date that it was first put into service and then retire it at some predetermined point—maybe 24 or 36 months in the future. Performing a reformatting operation of the disc can also provide indications of possible media surface problems if errors are generated by the formatting application.

MultiRead Compatibility

The MultiRead standard makes it possible for CD-RW discs to be read effectively in compatible CD-ROM and DVD-ROM drives. This standard is necessary because CD-RW media reflects less light than other forms of pressed or recorded compact discs. MultiRead-compatible drives can adapt to the variations in recorded data from one type of media to another, primarily by adjusting the intensity and focus of the laser read assembly when CD-RW media is detected. The MultiRead standard also specifies the interface requirements to the host computer and ensures compatibility in this area.

An extension of this approach called the Super MultiRead standard is being developed for rewritable DVD media (including DVD-RAM and DVD+RW) to provide similar compatibility for data recorded on rewritable DVD discs and played back in Super MultiRead compatible DVD-ROM drives and DVD-Video players.

Hybrid CD-R

A new and interesting innovation fostered by Mitsui adds a mastered session to a blank CD-R, providing a means of combining content distribution with the recordable capabilities of the media. Sporting the label, Mitsui AutoProtect Demo Disc, these discs feature a backup application known as AutoProtect, which was produced by Veritas. This backup application can guide you through the backup of selective files or a full backup of your system—multiple discs can be used if you overflow the capacity of a single CD-R. File compression is used to gain maximum use of the available disc storage space.

Chapter 11 of Orange Book, which describes the recordable forms of disc, defines a hybrid disc variation that consists of a mastered first session and a second recordable session. In Mitsui's approach, the first session contains an AUTORUN.INF file that will launch the installer to put the backup program driver on your system (if the auto-insert notification feature is turned on in the Windows operating system). Once the driver is installed, inserting a disc automatically brings up the backup program, allowing you to access all of the backup and restore features of the program. The producer of the backup application, Veritas (*www.veritas.com*), also offers other backup applications, some of which are easily adaptable to optical disc recording.

Kodak has taken this same principle—the ability to generate a multi-session recordable disc—and created a product called CD-PROM. The original discs employ a recorded first session, in addition to the recordable areas of the disc, but Kodak's plans are to incorporate mastered first sessions. The first session of this type of hybrid disc could contain many different sorts of applications. Cataloging and indexing programs would be especially useful to anyone archiving different types of data to recordable disc. Encoders for MP3 music or compression utilities for video or graphics could be handy. The hybrid CD may become an important vehicle for adding extra value to recordable media—one more means for a manufacturer to separate themselves from the competition.

High-Speed Recording on Low Quality Discs

As CD recorder speeds continue to ramp up, the issue of media responsiveness to high-speed recording becomes more critical. Many of the specifications in the Orange Book were designed in the days of low-speed recording and the issues become somewhat different when 8x and 10x recorders come into play. The introduction of 12x recorders makes the issue even more important.

Media that have been tested for high-speed recording speeds are crucial to recording integrity. In particular, a dye formulation that can meet the demands of the higher recording speeds is key to the process. Thinner dye layers that have been optimized for high speeds must also be able to operate effectively when recording speeds drop down to 1x levels. Check the recommendations included in your CD-R or DVD-R operator's manual for guidelines as to what will work and what won't. Many of the recorder manufacturers will list one or two recommended media types, giving you some assurance that they have actually tested the media and found it compatible with their equipment.

Some manufacturers have deliberately engineered their media to work with the most recent and fastest disc recorders. For example, Verbatim introduced the first 12x-compatible CD-R media as part of their DataLifePlus media family. These discs use a metal azo dye formulation as the recording surface. Kodak offers a high-speed recording line called Kodak CD-R Gold Ultima Recordable. This particular media incorporates a dye that has been formulated for improved interchangeability among the full range of readers and recorders. The supported recording speed range is 1x to 8x.

If you attempt to use inexpensive discs for high-speed recording you may be disappointed in the results. Odds are that unless the media has been deemed compatible with the higher recording speeds by the manufacturer, you will generate multiple write errors when attempting to write to it using an 8x, 10x, or 12x recorder. Most media will clearly state on the package whether it is been designed to work with various speed recorders. Without this assurance, your recording efforts become a very large gamble.

DVD-R Media

Recordable DVD media uses principles similar to recordable CD media for storing data. The technique relies on a dye layer which is affected by laser light. The dye has been spin-coated onto one side of a clear polycarbonate substrate; on the other side, a thin layer of reflective metal is laid down. As with CD-R media, the spiral (which will serve as the continuous data track) consists of a groove that helps position the laser beam as write operations are taking place. The groove—which is referred to as a pregroove—contains an undulating surface molded into it that is used to generate a signal. This signal synchronizes the speed of the spindle motor in the DVD-R drive as recording occurs. Additional information is added to the area between the grooves to provide addressing data; these Land Pre-Pits (LPP) are unique to DVD-R media.

For dual-sided recordable media, this same combination of materials is used on the opposite side of the disc so that two full recording surfaces will be available for use. Single-sided media often employ the non-recordable side of the disc for a silk-screened image or printed label, but dual-sided media must be kept free of labels or markings so that when flipped in the playback device, both sides can be read by the laser beam.

Data is recorded by focusing short bursts from a red laser onto the dye layer. The write laser generally has an intensity of 8 to 10 milliwatts and the duration of the on-and-off periods, known as the write strategy, is varied according to the speed of the disc rotational speed and other factors to produce marks of an appropriate length. The intensely focused laser beam causes a permanent alteration of the dye layer, creating microscopic impressions—tiny marks that serve as the equivalent to the pits that are embedded in a commercially pressed disc.

When playback takes place, the read laser focuses on the disc surface and detects the pits, which do not reflect light back to the optical head, as opposed to the lands—the unmarked portions of the surface, which reflect most of the light directed at them. This information is then decoded and transferred to the host computer, where it becomes available as the familiar bytes, words, and double-words of computer data.

How Long Do DVD-R Discs Last?

Clearly, one of the primary prospective uses for DVD-R media is archiving and storing of large amounts of data. Many manufacturers have performed their own testing and produced estimates of the lifespans of the media by subjecting discs to a variety of conditions—including intense light, heat, humidity, and shock. As is the case with recordable CD media as well, little independent testing has been performed, so interpreting manufacturer's test results requires a slightly jaundiced approach and some healthy skepticism.

Some manufacturers of DVD-R equipment, such as Pioneer, who may brand recordable discs from other OEMs, rate media lifespans as better than 100 years. Others project lifespans as being in the range of 80 to 100 years. Either way, the media itself will probably outlive the practical lifespan of the playback equipment, which may be obsolete and impossible to find in as little as 20 years. Archiving programs that involve digital media should include a system for updating the stored contents to more recent media types on a periodic basis. Fortunately, since there is no loss of data integrity involved when transferring volumes of digital information, data can be easily re-archived on new media and re-stored an infinite number of times. As long as there applications available that can read the original file formats, and equipment to accept the media, the preserved data can be read.

DVD-RAM Media

DVD-RAM media—like magneto-optical discs—comes in a cartridge, sealed against the possible data corrupting influences of dust and fingerprints until it is inserted in the DVD-RAM drive. The principles used for storing and retrieving data from DVD-RAM are similar to those adopted for CD-RW. Like CD-RW, DVD-RAM discs must be formatted prior to use. Some formatted DVD-RAM disc options exist. Plasmon IDE has introduce pre-formatted DVD-RAM media to the market, citing a higher degree of reliability to offset the somewhat higher price. Since the pre-formatting process helps weed out media that is defective in some way or that does not meet close specifications, this claim is very likely true.

With a number of manufacturers producing DVD-RAM products for different equipment in different capacities, the media availability situation is not exactly crystal clear. Kodak has released a 5.2GB DVD-RAM disc cartridge, as has TDK and Maxell. Maxell also offers a 2.6GB DVD-RAM disc that can be purchased in individual jewel cases, in bulk on a spindle, or in a removable cartridge. As new DVD-RAM equipment is introduced, the requirement for keeping the disc within a cartridge may become less important and more media options may become available.

Magneto-Optical Media

In many ways, magneto-optical recording equipment helped establish the market for CD-R and DVD-R applications. Early 128MB and 230MB 3.5-inch magneto-optical cartridges became one of the first high-capacity storage devices that became widely accepted throughout the industry for exchanging graphics files, program source files, digital audio material, and even files for preparing a CD image for replication. This format never achieved the same level of universal data exchange as did the compact disc, particularly because of the reliance of playback devices specific to one platform or another.

While magneto-optical (MO) recording does employ laser optics in the recording process, it also relies on magnetic materials for actually storing the data. For this reason, the process is sometimes referred to as optically assisted magnetic recording.

On the media surface of a MO disc, the magnetic particles are aligned perpendicularly to the surface plane of the disc, following a pre-grooved recording pattern. Focused through an objective lens, the laser beam strikes the surface of the media and heats it to a temperature known as

Chapter 7

the Curie temperature—a temperature at which the magnetic characteristics of the medium can be readily affected. The oxide particles at that spot on the disc are altered magnetically by a recording coil. The precision of the laser beam in heating a precise point on the disc allows much more data to be compressed into the recording area than could be accomplished by magnetic means alone, such as a conventional magnetic recording head.

Data is read optically taking advantage of the Kerr effect—a phenomena that causes a small divergence of the reflection of polarized light from magnetized materials. This slight difference between the light reflected from the reverse-polarized regions compared to the non-polarized regions is read optically and converted to data. The media can be erased and rewritten simply by reapplying the laser beam and the magnetic field through the recording coil to any region on the disc.

MO drives and data cartridges come in a number of different sizes and data capacities, from higher capacity 5.25-inch discs to the 3.5-inch discs that became popular storage devices for personal computers. Capacities range from 128 Megabytes up to several Gigabytes. One recent version of the MO media, referred to as Optical Super Density (OSD) media, has a storage capacity in excess of 40GB.

Another form of the media exists for archival storage applications where it is important to avoid any tampering with the originally recorded data. Financial, government, legal, and medical applications often require that data security and integrity are better protected than in the rewritable forms of MO storage. Companies such as Verbatim offer Continuous Composite Write-Once Read-Many (CCW) media; these discs detect and record any attempts to alter data that has been written.

MO Media Lifespans

MO media, because it is based on a composite magnetic-optical recording method, lasts much longer under frequent use than rewritable media such as CD-RW. A typical MO disc can be rewritten millions of times. Estimated lifespans for the an MO cartridge are in the range of 30 to 40 years, if stored in an environmentally benign location.

Printable Media Surfaces

With the increasing popularity of disc printers, most media manufacturers have introduced blank media with coated surfaces for thermal or inkjet printing, as well as other forms of disc printing. Coated media of this sort is generally more expensive than blank discs without a prepared surface or discs with an area surfaced for labeling with a permanent marker. Don't spend the extra dollars on a printable surface unless you really need it for disc printing.

As an example of the range of surfaces available, Kodak offers these blank disc variations:

- **Inkjet Printable White**: for the purest color renditions and most saturated hues, white surfaces work best for disc printing. Kodak offers this bright-white coated media that is optimized for inkjet printing.

- **Inkjet Printable Gold**: gold surfaces will somewhat affect the colors overprinted on them, but the gold can be effectively integrated into the printed design for striking effects.

- **Thermal Printable White**: the overcoat on this disc surface is designed for CD thermal transfer printing. The bright white surface provides sharp image quality and vivid colors.

Other variations are available from other manufacturers, including printable silver. You can also purchase silk-screened media for recording, based on a custom design or commercial design. Some of these are designed so that they leave an area free for overprinting so that a title or additional information can be added after the disc has been recorded.

Poor Man's Labeling

If you're not ready to invest in a $1200 disc printer, you may want to label your recorded discs manually using a permanent marker. Avoid the temptation to use a ball-point pen—even with a light touch you run the risk of deforming the reflective surface in the substrate and making portions of the disc unreadable.

Some sources advise against using any kind of a marker containing solvents. I've found that quick-drying ultra fine tip markers work fine and I've seen no evidence that any damage to the surfaces underlying the writable disc surface occurs.

Two possible solutions for markers are:

- **Dixon-Ticonderoga Redi Sharp Plus**: offers sharp, permanent labeling with a choice of seven colors and three different tip sizes, including fine, extra fine, and ultra fine. The ink is water-based and should not affect the disc coating or leach through the lacquer.

- **Sanford Sharpie Permanent Marker**: this does contain solvents, but the ink dries very quickly and, in my experience, doesn't cause any disc damage when you limit your writing to the areas designed for labeling on the disc surface. The ultra fine point works best for titling.

8

CD Duplicators

CD-Recordable technology introduced the idea of desktop disc mastering, but it wasn't until CD duplicators were introduced that efficient production-level desktop electronic publishing on CD could be realized.

Disc copiers were the first machines to be introduced in this class, but these are only capable of low-volume production since they usually include only a few recorders, from one to three or four, and they are hand-loaded.

Towers came next, with a larger number of recorders, but they were originally handicapped in the maximum number of recorders that could be fit into a computer case, and the limitations on the number of bus slots on a motherboard. Today's tower enclosures can hold up to thirty recorders, but, because their recorders are hand-loaded, when you are recording discs with small data sets, setting up the operation could take longer than the actual recording.

The next advance came when robotics were added to disc recording on multiple recorders. With this strategy, manufacturers made it possible to finally perform true production-level disc publishing with recordable discs. There are, however, other applications of robot handling—when combined with disc recorders—besides simply churning out multiple copies of a title.

An auto-loading duplicator in an enterprise environment can provide productivity and product quality enhancements that far outweigh the initial investment cost. Even a small capacity single-drive device with a slow robotic disc mover can eliminate the need for an operator to hover off a recorder to change discs every fifteen or twenty minutes. A top-of-the-line multi-drive system with integrated disc printer and sophisticated software can provide high-speed, flexible disc publishing capabilities to an entire staff. Linking staff members to this publishing option through the Internet or an intranet is now well within reach.

This chapter describes each of these three classes of optical disc duplicators, along with some examples of applications and related technologies. We discuss some of the business issues raised by the capabilities they afford. Finally, we look beyond CD-R to DVD duplication, and examine the possibilities for this new high-density disc format from the perspective of duplication technologies.

Duplication Terminology

In most technical fields, specialized vocabularies or jargons develop that use unfamiliar terms or use common words in uncommon ways. Compact Disc duplication is no exception.

Replication is the term used in the CD industry to describe the process of manufacturing discs from raw materials and pre-encoded data.

Duplication, in this context, describes the process of publishing a number of identical discs by recording on pre-manufactured blank media.

Disc autoloaders are systems that combine robotic disc handling with recorders and sometimes disc printers (and sometimes just printers) for unattended disc publishing.

Disc printers are devices that use one of a variety of methods for printing a design on a disc's label surface. These printers are frequently found integrated into disc autoloader systems, especially at the higher end of the market.

Disc analyzers are CD-ROM readers with special circuitry and data collection capabilities, designed for testing CD or CD-R media in a variety of ways.

Throughput is a term used in production planning that means "widgets per hour" produced in a manufacturing process.

Cycle time refers to the time required from start to finish for one production unit.

Blank media and **printable media** are not always the same thing. Some recordable (blank) CD-R discs are not printable by some disc labeling devices, particularly those that use inkjet print engines, which require a special surface. The best starting point when choosing media on the basis of appearance is to follow the printer manufacturer's suggestions. Most disc label printer makers test a variety of media to see how well it works with their printers, and are happy to share their findings with customers. Recorder manufacturers also usually have certain brands of media they recommend for optimal compatibility on their machines. Finding media that works well on both sides (printing and recording) can be a real challenge at times.

Some Basic Concepts

Before looking at some of the actual equipment and processes involved in CD duplication, the next sections examine some of the basic concepts underlying the technology.

Recordable Media

Recordable media is available in many formats and physical types. Not all are suitable for use in disc duplication.

CD duplicators are designed to use CD-R media rather than CD-ReWritable (CD-RW). CD-RW discs have the following handicaps:

- CD-RW discs are generally considered too expensive for publishing

- Recording to CD-RW media is slow (currently, 4x is the maximum recording speed for this type of media)

- CD-RW discs are not suitable for distribution since many older CD players or CD-ROM readers cannot read them

Production Efficiency

A duplicator's effective production speed is a combination of several factors: the number of recorders used, their recording speed, the efficiency of the robotics in autoloaders or the operator's hand-loading the recorders, and the time required to set up the system for a production run. Another factor that might be encountered is the speed of a disc printer, particularly if it is integrated into an autoloader system. At the time of this writing, most duplicators are available with 8x recorders, which can record a full 650MB (74 minutes) in about 11 or 12 minutes. Yet another aspect is whether the systems can work in asynchronous mode, or if they require all the drives to be loaded before starting to record. An asynchronous system with fewer drives can sometimes have higher throughput than a synchronous system with two or three times as many recorders.

System Configurations

Many duplicator systems can be reconfigured with faster recorders as they become available or affordable. However, since most duplicators use special software drivers to control their systems, the consumer is dependent on the manufacturer to perform the integration. It is not advisable to simply pull out a slower recorder and replace it without first consulting the manufacturer to be certain the new model drive will work in their system.

Additionally, some duplicator autoloaders are designed to be custom-configured or expandable to match consumer requirements. For example, more recorders, faster recorders, disc printers and other devices, such as disc analyzers, can be included out of the box. These kinds of items are sometimes available as after-market add-ons.

Uses for Disc Duplication

The range of possible uses for disc duplication has steadily expanded as enterprising individuals and organizations continue to find innovative ways to exploit the technology.

Short-Run Disc Production

This is the application for which CD-R duplication was originally designed. Sometimes a requirement exists to produce a significant number of discs in a short time, but less than the 1000 or more copies many replicators require as a minimum order for mass production. A "short run" suitable for duplication may be as few as one or as many as 1000 discs or more.

On-Demand Disc Publishing

For a new start-up disc publisher, or a frugal established one, disc duplication is a welcome alternative to replicating and warehousing thousands of discs. Disc duplication can also be a viable approach for disc titles that have small—but perhaps long-term—markets. A master CD can be kept on hand for each title, and duplicated as the need arises. In my publishing shop, we archive the disc image file and artwork for labels and packaging on a single disc when the data space permits, for storage efficiency and version control. That way, all the necessary files are certain to be together whenever a new run is required. When stocks run low, we pull out the archive disc, copy the production disc image to a new duplicating master (or hard disk) and make as many new discs as we need for immediate use, without sinking resources in to copies that would require warehousing and could even be wasted if any changes might be made to the contents before they are shipped.

Workgroup Disc Recording Applications

Some companies have discovered that certain duplicators make excellent workgroup publishing devices, as well as (or even instead of) being an effective means of producing relatively large numbers of discs. Generally, autoloader systems that have networking capabilities work best in this role.

High Security Publishing

One of the disadvantages of CD publishing for many years was that content material, which might contain highly sensitive information, had to be turned over to another company for production. Sometimes this is simply not acceptable, regardless of how trustworthy a replication company might be.

In one incident (which has been verified by a trusted source) an entire replication facility was taken over by an unnamed US government agency for a week, after agents were trained by the replicator's staff to use the equipment. Not everyone with sensitive material can afford the millions of dollars that exercise must have cost, but with CD duplicators, this kind of heavy-handed and expensive approach is no longer necessary. Disc duplicators make it possible to control the entire production process in-house for a much smaller cost, and with far less trouble and risk.

Time-sensitive Disc Production

Using mass replication for discs that must be produced with a very short turnaround time can be extremely expensive, assuming a replicator can be found who is able and willing to fit a fast-turn, low-volume job into their schedule. Having a production run done in a week or less can double or even triple its costs when compared to a more leisurely two- or three-week job. A large-capacity CD-R duplicator with autoloader can produce up to 1000 discs faster and probably cheaper than a replicator, even though each disc requires more time to encode with duplication than replication. Replicators' schedules, the necessity of delivering source material and finished products, and time required to create a glass master, mold stamper, and screens for label printing all might make out-sourced replication a more time-consuming process than in-house duplication.

Workflow Issues to Consider

When making a decision about whether to use duplication or replication, or an autoloader rather than a disc copier or tower, one of the primary items to consider is how many discs can be produced in a specific time. This includes more than just recording time; one must also take into account the time required to load the discs into the recorders, and how long it takes to label and package them. But first you must figure on the time required to produce the initial master, whether that is recorded on a CD to be copied, or mastered to an image file on a hard disc.

Data Image Mastering Time

Even after the disc contents have been created, they have to be *mastered* to put them in the right format for encoding on a disc. CD files, no matter what data format, must be assembled into a single file, since the data on a disc is laid down in one long, continuous spiral. When the planned disc content consists of many small files in several layers of subdirectories, mastering can take quite a long time, depending on the clock speed of the computer and the efficiency of the mastering software and computer operating system. This is because each file and directory must be opened and its size calculated for indexing and copying. Many small files require more processing time than a few large ones, and directory structure also affects the speed with which this operation can proceed. Naturally, if a mastered disc is already available for copying, this part of the time estimate equation can be disregarded, but if mastering is required, be sure to include that in your time requirement estimates.

Also, if working in a network environment, it is important to have the image master created on a file system local to the recorder, to prevent buffer underruns that could result if network slowdowns or interruptions occur. CD recording requires a constant stream of data to the recorder, and if that is not available, the recording stops with an error and the disc is ruined.

Recording Throughput Time Requirements

How long it takes to record a set of discs depends on how much data is to be recorded, the speed of the recorders, plus handling—moving the blank media into the recorder and removing the completed disc. If disc transfer is done by hand rather than with an autoloader, it can add several minutes to each batch. Even with an autoloader, the efficiency varies among systems. The best way to ascertain the time required is to keep track of the time over several sessions, and take an average, not forgetting the time needed to unpack and stack blank media on input spindles if using an autoloader, and the time required to unload and package the finished discs.

Autoloaders have vastly different efficiencies, depending on whether they use a synchronous or an asynchronous recording strategy. Some high-end auto-loaders are available which allow asynchronous recording and disc handling. You can either record several different titles at the same time, or you can have a single title proceeding through several different stages of production in a highly efficient manner. A potential complication in asynchronous duplication is that an open printer or recorder drawer might interfere with disc handling. The robotics control software must be designed to accommodate all the operations that are occurring at any given time.

The asynchronous strategy can also support the loading of drives and initiation of recording while additional drives are being loaded, or other operations—such as discs being moved to a printer—are being performed. With many type of duplicators, when you start a batch of discs, it loads all of the drives and then starts the recording operation. When it is finished recording the current batch, the equipment will unload the discs, load a fresh batch of media, and then resume recording. Typically, a delay of two or three minutes between batches may occur. With a large number of batches and a large number of discs, this can introduce significant delay in the duplication process.

The formula for calculating the delay is:

*delay=(print + handling time) * (number of drives)*

This delay could be as much as 8 to 12 minutes, or even 24 minutes on an 8- or 12-drive system with a slow printer.

Asynchronous duplication strategies, by their nature, require high-end SCSI host adapters and high-speed processors to support the operations. These features translate into more expensive equipment than similarly equipped synchronous devices. But, if time is more important to your application than the additional cost involved, you should consider selecting a duplicator that supports the asynchronous approach.

Disc Labeling Time

The time required for disc labeling is naturally dependent on the method of labeling used. A disc printer used with an autoloader duplicator may be employed in-line or offline, for instance, or paper labels may be applied to completed discs all at once in a batch, or as each disc comes out of the recorder.

Packaging

Equipment for automatically packaging CDs does exist, but it is generally too expensive to be considered for the small volumes to be handled in a duplication operation. Time must be allowed for this final step in the production process, but luckily it usually doesn't take long, and some of it can be completed on the first discs duplicated while others are still being recorded and labeled.

Choosing Recordable Media for Duplication

Some key factors affect the suitability of media for use with duplicators, as explained in the following sections.

Labeling Issues: To Print or Not to Print

Printing directly on a disc results in an attractive label that can be customized for each disc if desired. Issues that should be considered include cost of printing equipment; cost of printable media versus prelabeled or paper-labeled media; availability of software and trained operators to generate label artwork; and questions of how the printing process might affect disc data life expectancy.

Note: For discs that are being prepared for archival storage, ISO 18925-1999, the international standard for Disc Storage, recommends that no label at all be printed or affixed to the disc surface. This ensures that adhesives or solvents in the ink will not negatively affect the data during long periods of storage. Of course, this standard is written primarily for archivists, so if your discs are intended for more temporary purposes (up to 5 or 10 years usefulness), it is not necessary to follow it strictly. Also, some paper labels and adhesives are acid-free, so they may not have a negative impact on disc life expectancy.

Pre-screened Labels

Some duplication users choose to have blank media silk-screened with their artwork before recording. Screened labels have the highest-resolution images, but this practice does mean that many discs will have the same design, unless an area is left in the artwork for overprinting with a disc printer. Some shops have experimented with including a field of printable ink in a pre-screened design for this purpose, but experimentation with different ink formulas is necessary to find one that works with the disc printer you use, and there is the matter of lining up the existing artwork with overprinted text or graphics as well. It is not impossible, but does require specialized equipment or a great deal of manual manipulation to achieve acceptable results.

Disc Printer Options

Several types of devices are available now for printing directly on the surface of CDs. All of them can be used as computer peripherals, and several can also be integrated into an autoloader system with or without disc recorders, for automated production. Desktop disc printers fall into two general categories, depending on the kind of printer technology they use: wax transfer and inkjet.

Wax Transfer

Until recently, wax transfer disc printing was only available from Rimage, and only in monochrome, but now at least one other company has introduced a similar printer, and both they and Rimage have printers capable of multi-color printing. Advantages of this method of printing are that the resulting image is smear-proof, and they can look as sharp as screened images for the monochrome variety. The polychrome printing is generally not as high-resolution as inkjet printers are capable of, but the manufacturers are working to improve that. Physical limitations of the wax transfer process prevent much higher resolution for this type of printer, though. Also, the polychrome printers currently have a combined resolution of only 300 dpi, or 100 dpi per RGB color.

Inkjet

Inkjet disc printers are less expensive and provide more colorful output than wax transfer printers, but they require media with a special printable surface, and the resulting label can smear if rubbed with a wet finger, some worse than others. Also, some media and ink combinations don't work well together, so experimentation with different brands may be necessary to obtain desired results.

Paper Labels

Some printer manufacturers, such as Trace Digital, recommend using recordable media that has a blank paper label already affixed for disc printing. One reason for this is the inconsistencies encountered in media surfaces, which can vary significantly from disc manufacturer to manufacturer, resulting in difficulties when attempting to print high-quality images to a disc. By using a uniform paper surface when printing, the results can be controlled much more consistently. A paper surface, since it is more absorbent and less resistant to the applied ink, tends to provide superior image quality when used with inkjet printers.

Media Packaging

Blank media is delivered from the manufacturer in several different kinds of packaging. Some of these methods of packaging are more suitable for mass duplication tasks than others.

Jewel Boxes

The most expensive way to receive blank media is in plastic boxes like those used for commercially packaged CDs from a retailer. Not only are the boxes more expensive, but before the discs can be recorded, they must be unwrapped and removed from the boxes, which is a time-consuming manual chore. However, these packages can be reused for delivering or storing recorded discs, so it is not all waste.

Spindles

Some media manufacturers sell blank recordable media "bare on a spindle." These spindles are made of metal or plastic, and hold anywhere from 50 to 250 discs. Purchasing media in bulk on a spindle is usually economical. Spindles are also convenient for loading discs on an autoloader in preparation for duplication.

Bee-hives

A "bee-hive" is a small spindle package with a rigid plastic cover that is held on with a screw-on nut. These packages got their name because they look a little bit like the flattened cones used to cultivate honey bees. Bee-hives usually hold 50 or 100 blank discs, and are a moderately inexpensive way to buy and store bulk media.

Shrink-wrapped

Sometimes blank media is simply wrapped in heavy plastic and shipped with no other packaging. While this is an inexpensive way to receive media, it does present some hazards since it is possible for the discs to shift and scratch one another during shipment and handling. Spindles or bee-hives provide more protection from such hazards, and are easier to store.

Duplicator Classifications

CD duplicators can be classified into the following groups:

- *Disc copiers* are the most basic units; they are generally standalone devices designed to simply make multiple copies from a single prerecorded disc.

- *Tower duplicators* combine several manually loaded recorders in a single cabinet, allowing multiple copies to be produced very quickly.

- *Automated duplicators* employ robotics to move media from spindles to recorders and, sometimes, to printers. These units provide the most publishing versatility, but they are also the most expensive devices.

Duplicators could be categorized as either using embedded controllers or open systems. In general, embedded controls allow great simplicity of use, but using open systems gives manufacturers the chance to include many extensions to a system's capabilities, such as the ability to record to different formats or to add printers or disc analyzers. This is the usual trade-off between ease of use or powerful features.

Any of the classifications we use in this chapter could be subdivided into embedded controls or open systems. If open systems are chosen, standard printing, premastering, and so on, can be added, but the operator then must be a more sophisticated (and higher paid) computer user, not simply a copy machine attendant.

Examples of our three original classifications of duplicators appear in the following sections.

Chapter 8

Disc Copiers

The copier class of duplicator is usually found in standalone configurations. No external computer is required for standalone copiers, and operation is usually very simple, sometimes with only a "start" button. Anyone in an organization can be quickly trained to operate this type of equipment, which makes it very practical for casual, on-demand duplication.

Most disc copiers, like other classes of disc duplicators, are advertised as being able to successfully copy almost any compact disc format. Frequently, disc copiers do not have mastering capabilities. This means that a separate recorder must be available to create original content discs, since only pre-encoded discs can be duplicated.

Since they usually do not offer many sophisticated features, can copy only a few discs at a time, and require hand-loading, disc copiers are usually at the low end of the duplicator price scale.

Used to duplicate relatively small numbers of discs at a time, these devices are sometimes described as being an "appliance" since one simply pops in blank media and a disc to be copied, pushes a button, and when the process is finished, out comes the completed product. It's just like using a breakfast toaster, except that there are no crumbs.

For small businesses and workgroups who occasionally run multiple copies of a CD-ROM, musicians who want to produce audio CD demo discs for distribution, or even hobbyists who need inexpensive, easy-to-use disc copying capabilities, these machines can be exactly right.

Examples of the Disc Copier Class

There are a large number of manufacturers and integrators who offer devices in this class.

Alea Systems One of the first companies to see the potential for desktop disc copying was Alea Systems of Italy. Now based in California as well as in Milan, Alea was showing a version of their single disc copiers as early as 1993.

Model 121 A very simple but effective design from Alea Systems is called the Model 121. This external standalone copier is nothing more than a box with two drives—a CD-ROM reader, and a CD-Recorder—with a "start" button and an activity indicator. Put a disc to be copied in the reader and a blank disc in the recorder, push "Start" and a few moments later your disc is copied.

Interactive Media Corp. KanguruCD Duplicator This unit provides both standalone operation and PC-connected CD recording with models ranging from a 4x parallel-port interface design to a 6x SCSI-interface design that copies 1 to 3. Automatic format detection allows you to simply drop a disc into one tray, push the start button, and remove your copy in 10 to 20 minutes. With the high-end model, up to 12 discs per hour can be duplicated. For more details, visit the site: *www.kanguru.com/cddup.html*

Tower Duplicators

Fitting in between copiers that can record only one or two discs at a time and autoloaders that handle hundreds or thousands in a day are the devices we call *tower duplicators*. A tower is essentially a case or rack that holds several recorders, and is usually driven by a microcomputer.

Depending on the configuration and associated software, towers might work more like copiers, and only blindly copy pre-recorded discs, or they might offer mastering capabilities as well.

Tower duplicators are well-suited for companies or individuals who have a need to periodically produce a number of discs in a short time, but not on a constant basis. This class of duplicators is less expensive than automated systems, since no robotics are involved in recording or printing discs. These functions must be handled by manual loading, which makes the tower duplicator less than ideal for serious production environments. However, because some tower cabinets contain up to 30 recorders, the duplication throughput can be significantly greater than low-end disc copiers.

Most of the tower configurations do not include sophisticated software that would allow control over individual CD recorders; instead, the software is generally tailored for simply generating multiple copies from a disc image that may be stored on a host computer and then recorded to disc during the duplication process, using an embedded controller.

Examples of Tower Duplicators

Tower duplicators are available from several manufacturers. A few examples of units follow.

CDCyclone The CDCyclone mounts up to 30 drives in a single cabinet, providing the highest volume recording capacity available in this class at the time of this writing. The unit consists of a standalone enclosure that contains the

Chapter 8

selected number of CD recorders and a dedicated host computer that drives the duplication process.

Figure 8 - 1 **CDCyclone T-30**

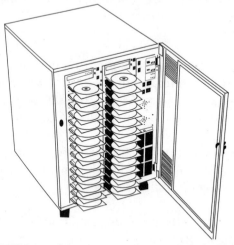

NSM NSM produces towers in several sizes, from the small peripheral unit shown in Figure 8 - 2 to larger standalone units featuring their own built-in host computer and up to five CD recorders.

Figure 8 - 2 **NSM Towers**

Automated Duplicators

Automated duplicators extend the capabilities of the optical recording medium significantly, offering the equivalent of a printing press designed for compact discs. These units are optimized for unattended production and workgroup applications and sometimes include networking interfaces so they can be accessed in the same manner as a network printer (including the ability to queue jobs, as necessary).

Although some tower duplicators also include network access, they still must be attended—someone must load and remove media as the recording process takes place. In comparison, automated duplicators include robotics to cycle the blank media to the recorders, as needed, and then to unload media to an output spindle. Some units also automate the process of printing the media, either through an integrated inkjet printer or a printer using a wax-transfer process.

Another characteristic of this type of duplicator is the relatively sophisticated software that is used to control the mechanisms involved in the robotics and to synchronize the recording process. The proprietary software controlling these operations may or may not have premastering capabilities. If premastering is not included in the process, a disc image must be first created using a suitable premastering program. The simplest units in this class may consist of nothing more than a disc copier with an autoloader connected to it. Some of the high-end systems provide full premastering in combination with autoloaders that can handle up to 400 discs at a time.

Many different unique devices exist in this class, each of which serves a different segment of the market, with multiple configurations available and widely varying capabilities. Since these kinds of units can be fairly expensive, one of the most common applications is sharing a single automated duplicator shared across a network with many individual users. Queuing software is used to mediate in cases where there are multiple requests for a recorder; jobs can be prioritized and serviced in the same manner as network print queuing is performed.

Not all automated duplicators offer full networking capabilities. Microtech and Rimage are examples of manufacturers who have carefully integrated network access into their equipment. Microtech units contain their own controlling computer. Both companies software operates under Windows NT. However, while using similar concepts, the systems from these two firms are quite different from each other. In fact, one of

Chapter 8

the most striking facts about the autoloader class in general is the diversity of approaches used by the various manufacturers to provide solutions for similar requirements.

Examples of the Autoloader Class

Each of the following units employs a single robotic arm that can handle one disc at a time. The emphasis is on the simplicity of the robotic mechanism; by avoiding complexity, the reliability of the units is improved and the costs can be constrained to a reasonable level.

MediaFORM

The diskette drive in the MediaFORM 3706 lets the operator insert a diskette with label information for the integrated disc printer. An LCD panel on the lower right side of the cabinet provides status indications while operations are being performed. The stack of drives contained in the cabinet are aligned so that the drawers extend within the reach of the central robotic arm. From the input stack on the left, discs are loaded into individual drives and then when the recording has been completed, the discs can be fed one at a time to the printer unit (shown as the cabinet on the far left side of the unit). When printing is complete, each disc is retrieved by the arm, which rotates and drops it on the output spindle shown on the right.

The robot involved in this process has two degrees of freedom: it rotates and it goes up and down. No lateral or angular movements are used to transport the discs. Once again, simplicity is engineered into these units to keep costs down and to eliminate potential problems that could arise from overly complex systems.

Figure 8 - 3   MediaFORM 3706

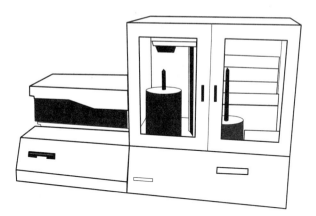

Microtech ImageAutomator

The robotic arm on the Microtech unit has one degrees of freedom: rotation. Picking up and dropping discs is accomplished by a second robot housed within the rotating arm. Discs on the input spindle are lifted within the range of the rotating arm by an elevator. As you can see in Figure 8 - 4, discs are picked up from the input spindle and then rotated into position to one of the drawers in the CD recorder stations. Recorded discs are extracted from the drawers and deposited in the printer, shown to the rear of the unit. When printing is done, the completed discs are extracted, rotated into position, and dropped on the output spindle.

The ImageAutomator is designed to be adaptable to different levels of activity and different operations through its configurable structure. The central platform with the robotic arm rotates through up to six stations that can be configured with different devices, as required. An organization can purchase and configure an appropriate number of up to 12 recorders, 2 Primera inkjet printers, or 2 I/O stations for its operations. Through software settings, the robotics can be programmed to service each of the individual stations during duplication operations. Thus, the Microtech system is the most configurable platform of its kind for disc duplication.

The ImageAutomator includes an image aligner device (shown in more detail in Figure 8 - 6) that can be configured to adjust the image printing to conform to pre-screened discs that have a blank area for overprinting. The optional image aligner utilizes a digital camera that adapts the image to be printed to the orientation of the disc that is deposited into the printer tray. This particular approach works very well with paper labels that have been pre-printed with an image, as well as previously silk-screened media. The degree of precision for this technique is also somewhat better than units that require that the media to be rotated into a particular position before printing can begin.

Figure 8 - 4   **Microtech ImageAutomator**

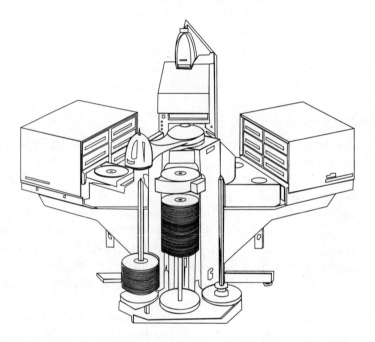

Cedar

The Cedar autoloading duplicator, distributed in the United States by MicroBoards, Inc., employs an integrated printer, the Primera Signature III, into the compact cabinet. A soft brush attached to the robotic arm is used to clean the media's recording surface before depositing it in the recorder. Dust-laden media can cause significant problems during recording, so, if the brush itself is kept clean, this feature is a welcome addition to the duplicating process.

The robotic mechanism on this unit is exceptionally simple with one degree of freedom. The arm moves up and down to deposit or extract discs from the trays that extend from the recorders or the printer. The stacks of input and output discs also rests in an extendable tray, which

Automated Duplicators

slides to one position when blank media is being selected and then moves to another position to accept recorded, printed discs.

Figure 8 - 5   **Cedar CD-R Desktop Publisher**

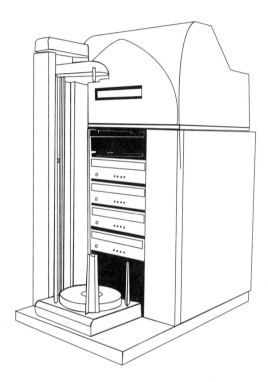

Rimage Protogé

The Rimage Protogé is very popular for workgroup applications, since it can be controlled over a Novell NetWare network. It also has a smaller footprint than many other autoloaders, allowing it to be deployed in situations where the physical space available is a limitation. This Rimage unit has an attractive design with smoked Plexiglas doors and it is sometimes found in a prominent position in the reception area of a corporation, where custom discs containing product information or presentations can be generated on demand for visitors and then handed out by the receptionist. While not really designed for heavy-duty production, this unit provides reliable, mid-level capacity duplication operations with the added benefit of networking.

Chapter 8

The printer used with this duplicator aligns print operations according to an alignment mark that must be present on the blank media. The disc is physically rotated in the printer tray until it is properly aligned for the image to be printed.

Trace Digital

Trace Digital has optimized their units for in-line printing capabilities (or off-line capabilities—they offer some autoloader units that don't even have installed recorders).

Printing discs presents a number of different challenges, which have been addressed by Trace Digital. Several different printers from OEM Epson are used in Trace Digital models. This autoloader disc printer incorporates a staging area. After a disc is printed, it is moved to a holding area to allow the ink to thoroughly dry. After the next disc is printed, the disc in the holding area is deposited on the output stack. This reduces the chance of wet ink contaminating the media surface of the adjacent disc when stacked.

Trace Digital also recommends the use of pre-labeled media—media that includes a glossy, inkjet paper label that accepts a high-resolution image more effectively than coated disc surfaces.

Disc Printers

The devices mentioned here are by no means all the products available in this class, but they are representative of those currently available in the marketplace.

Inkjet Disc Printers

Craig Associates International (CAI) ColorScribe This company pioneered the concept of desktop disc labeling in 1993, not long after the first desktop CD-Recorders were introduced in the US. CAI currently offers two models, the ColorScribe 6000, a manual-feed device, and the ColorScribe 9000, which is a printer integrated with an automated disc handler with 100-disc capacity.

Primera Signature III The latest model from Primera, the Signature III, offers 1200 dpi resolution (in both directions), advertised as the highest available in this market. This drawer-loaded printer is integrated into several duplicator autoloaders, including those from Cedar and Microtech Systems. It uses a proprietary printer engine and ships with software for both Windows and Macintosh operating systems.

Trace Digital Trace Digital sells several autoloading disc printers, both integrated with recording autoloaders, and a standalone autoloading printer, the Power-Printer II. This company is also the parent of Affex, who make manually loaded disc printers. All Trace printers, including the Affex models, are of the inkjet type, based on the high-resolution four-color (CYMK) Epson printer engines.

Affex HardCopy and CDArtist-Plus These manually loaded inkjet printers (based on high-resolution Epson print engines) use a plaque with indentations to hold discs as they pass through the printer. A unique feature of this line is that they are able to accommodate the popular "Business Card" shaped CDs by using special plaques designed for these 18 to 50MB recordable media. Affex and its parent company, Trace Digital, offer media prelabeled with glossy, inkjet-friendly (and acid-free) paper labels that are particularly well-suited for these printers.

Wax Transfer (Ribbon-Based) Printers

Rimage PerfectWriter & Prism The first wax-transfer disc printers were introduced by Rimage early in the life of the CD-R industry. This technology offers monochrome or color printing that has several advantages and some disadvantages. The advantages include the fact that monochrome images when properly designed can almost rival silk-screening for density and sharpness. The wax is also smudge-proof—a distinct advantage over most inkjet disc printers.

Disadvantages include the higher cost of these printers and their relatively low resolution, since they have a physical limitation of 300 dpi, even for the three-pass color version (the Prism) which has a combined dpi of 300 over all three bands of color.

Nonetheless, the monochrome PerfectWriter has long been considered the high-end disc printer against which others are measured. Several duplicator manufacturers integrate this printer with their autoloaders, and Rimage itself is among those who do this.

Chapter 8

Guidelines for Media Handling with Duplicators

Even though many CD duplicators are designed to be as simple as a copying machine, they are still precision instruments and the media integrity can be compromised if subjected to mishandling or abuse. The following guidelines should be considered whenever performing disc duplication to ensure the best results.

- Good quality bulk media is essential to successful recording operations. If you are producing large volumes of recorded discs, you don't want to produce a lot of faulty discs. Make sure that you choose high-quality media that is designed to accommodate the speed of your recorders. Media designed to work successfully in a 4x recorder may not work at all in an 8x or 12x recorder. Disc manufacturers formulate the dye and engineer the media to critical tolerances to ensure successful recording at higher speeds. Ensure that your media is up to the task.

- High humidity environments can cause discs to stick together on the spindle. If the duplication operation is to run unattended, the operator loading the spindles should ensure that the discs separate freely (fluttering the discs slightly while dropping them onto the spindle can help ensure that they are not sticking together). A relative humidity of approximately 45% is ideal for duplicator operation.

- Very dry environments can also cause problems—static electrical charges can prevent discs from separating freely as they are uploaded from spindles. If the humidity level is significantly below 45%, a humidifier placed in the room where the duplicator is being used can help avoid this problem.

- A clean environment favors duplication operations, since dust or dirt can foul the recording activities. While not requiring cleanroom conditions, CD duplication will be the most successful if the air in the room is free from smoke or dust and the duplicator mechanisms—including recorder trays, printer bins, and robotic assemblies—are kept as clean as possible. If using compressed air to clean out any of the cavities in a drive or printer, make sure that the product does not contain propellants or other chemicals that might leave a residue on discs or drive mechanisms. Pure compressed air is best for this application.

- Since simpler CD duplicators might be operated by personnel who are not skilled in computer operation, make sure that adequate training is provided before turning someone loose to duplicate discs. Media handling techniques should be taught along with other fundamentals that apply to CD duplication.

- Handle media only by the edges and always use a delicate touch. Rough handling can result in data integrity problems. Fingerprints or debris on the media's recording surface can impede data recording.

- If you are printing on the media, make sure that the print surface provided is compatible with your printer.

- Recordable media is sensitive to heat, humidity, and light exposure. Avoid extremes in each of these areas to ensure the best results. Treat your blank media as if it is in an archival environment (cool, dry, and dark) and you'll eliminate many potential problems.

- Provide plenty of table space and spare empty spindles for materials staging.

- Use a cart to move media between work areas.

- Don't carry a stack of media without a spindle or something to prevent the discs from shifting and possibly scratching each other.

- Be sure to remove the protective blanks from the top and bottom of new spindles of media.

Issues Involved in Printing on CDs

Most desktop printers were originally designed to print on paper, which is thin, flexible and absorbent. CDs are relatively thick (compared to paper), rigid and have a non-absorbent surface.

To modify an existing printer so it can print on CDs requires manufacturers to design special carriages, and sometimes to alter the cases of printers made by other companies. (Several disc printers are based on models from Canon and Epson.)

Printers that use water-based ink (inkjet type printers) require media with a special surface. These surfaces have been developed by many media manufacturers, using a variety of base colors and different absorbency and appearance characteristics.

Chapter 8

Wax-transfer printers can be used successfully with most "printable" or plain-lacquer coated "shiny surface" media, or media with a scratch-resistant surface.

Finding exactly the right media for your particular label design and a type that works well with your printer can be a challenge, and is largely a matter of trial and error. Until media manufacturers and distributors can be convinced to package their media products with more informative labeling, even finding the same product more than once may be difficult. Experiment and shop around are the best suggestions I have for users of these products, and let the makers know we want media to be informatively and intelligently (and consistently) labeled, so we know what we are buying.

Pre-screened CD-Rs

One method for applying a label to a disc is to pre-screen the blank recordable media with an image and then use an alignment mechanism to overprint additional title information on the surface. Some manufacturers, including Microtech and Rimage, offer this approach for producing a look which is very close to a commercially replicated and silkscreened CD.

Different alignment techniques are used to marry the overprinted image with the pre-screened image. The Rimage unit uses a mark on the media that is used to control the alignment. The Microtech unit uses a digital camera to capture the image, which can then be manually aligned using the software provided with the equipment. Once trained in this manner, the overprinting is handled automatically as a normal part of the printing process.

Using Pre-labeled Media

Sometimes it makes sense to record on media that already has a design on it, adding only identifying information to a high quality silk-screened label. This is possible with a system that "sees" the label orientation and lines up overprintable text or graphics to be added at recording time. At the time of this writing, two such systems were available, one from Microtech Systems, and the other from Rimage. Both use the Rimage wax-transfer printers, because it is not practical to overprint on UV-cured silk-screened labels with inkjet printers since the ink cannot be absorbed by such a surface.

Issues Involved in Printing on CDs

Print-alignment Devices

Microtech Systems ImageAligner Used in combination with either a Rimage printer (Prism or PerfectWriter), this device can be added as a station on several of Microtech's modular duplicators. It consists of a digital camera, which "sees" the existing label on a disc in the printer's tray, and software that overlays this image with the graphic or text to be overprinted, then lines them up according to the way it was trained to do during the setup for that production run. The image training can be stored for reuse later, saving the time required for training if production is done in several batches.

Figure 8 - 6   **Microtech ImageAligner**

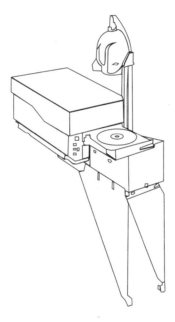

Rimage AutoPrinter

The Rimage AutoPrinter uses a sensor that detects a mark that must be previously inscribed in the inner hub of preprinted discs. This allows the software to instruct the positioning mechanism to rotate the disc into the correct orientation for the image to be overprinted. The AutoPrinter's disc handler is based on the same robotics and rotating disc bin configuration as their popular Protegé disc duplicator, but without the recorders. Using a standalone disc printer autoloader in conjunction with an

Chapter 8

autoloading disc recorder eliminates the bottleneck frequently encountered when the printer's cycle time for the number of discs to be handled is slower than the recording autoloader.

Figure 8 - 7 **Rimage AutoPrinter**

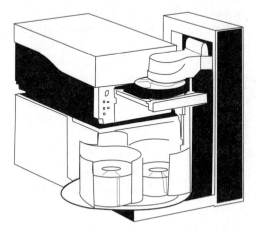

Paper Labels

While the use of paper labels has been deprecated by archivists and media testing & data storage standards groups, for discs that are not necessarily intended to be used for long-term data storage paper labels may provide a reasonable alternative to printing directly on a disc's surface. Care should be taken to ensure a label is perfectly centered on the disc, since labels that are even a tiny bit off-center may cause playback problems in high-speed drives. Several label applicators are available to make placement easier than lining them up manually. Some disc printer manufacturers even recommend using media with pre-applied paper labels to avoid problems sometimes encountered when using printable media. Paper labels have the added advantage of providing a slightly absorbent printing surface, which may display a printed image sharper and more vividly-colored than possible on printable CDs.

Packaging Duplicated Discs

Some consideration should be given to the method of packaging completed discs once they have been recorded on a duplicator. Depending on the method selected, a certain amount of time must be allocated to meet the packaging requirements, which can add significant time to the complete duplication process.

Jewel Boxes

Jewel boxes are a manually intensive means of handling the packaging of discs, but they have the advantage of adding a degree of polish to the presentation of a completed disc, particularly if a jewel box insert is produced to go along with the disc. If you choose the jewel box as the preferred method of packaging, be sure to allow sufficient time for assembling the inserts and packing the boxes at the end of the duplication process. This method also requires additional handling of the discs, which can generate scratches or other damage if not performed carefully.

Disc Sleeves and Envelopes

Disc sleeves and envelopes can be more easily integrated into a high-speed production process. A variety of materials are available, many of which can be printed using a screened or four-color printing process. The inexpensive approach, of course, is to simply use a plain sleeve insert, which should be a soft enough material so as not to abrade the disc surface when inserted or removed.

One innovative approach which has recently been introduced is a Web-based design tools that allows templates to be applied for quickly preparing a camera-ready design with just a Web browser. The design can then be printed on discs or custom sleeves, by the service provider or the customer. For more information about this technique, refer to: *www.eprint-network.com*

DVD-R or DVD-RAM for Duplication?

In the DVD world, rewritable media preceded write-once discs in the open market, just the opposite of the history of CD-R and CD-RW. However, the rationale for using write-once discs instead of rewritable discs is the same for both varieties of Compact Disc, the original CDs and DVDs. Rewritable DVD and CD media is more expensive than write-once blanks, and is not readable on all drives. If discs are being duplicated for general distribution, when the characteristics of target readers are not predictable, using write-once blanks is a safer choice than more limited-use rewritable media.

Chapter 8

When Will It Happen?

DVD-Recordable media and recorders are available, and can theoretically be integrated into duplicators, but at the time of this writing there have been no DVD-R duplication systems announced. They will become available soon, but at the moment that format is too new and not yet popular enough to be practical for duplicator manufacturers to invest in it when the CD-R duplicator market is much larger, and still growing. However, several manufacturers have told me their systems are "DVD-ready" for the time when the demand exists.

Summary

The techniques and equipment involved in CD duplication and automated disc publishing cover an enormous amount of territory. Business and corporate users, as well as independent publishers, have a broad range of choices when deciding how to best duplicate discs.

The possibilities are so wide ranging that many potential uses have yet to be discovered. Through the entire history of optical recording, the users have always come up with new potential applications that the manufacturers had never even anticipated, and each new generation of equipment and software builds on this progressive experience.

For example, initially manufacturers believed that CD duplication equipment would be primarily used for simply generating multiple copies of a pre-existing disc. New classes of equipment were designed and software was re-engineered when it was discovered that many users wanted a mechanism for producing one-offs in small quantities for distribution (for example, software companies producing multiple beta releases of a software title for testing). Inhouse training, corporate database distribution, multimedia presentations for salespersons or seminar instructors, business-to-business parts catalogs, office enterprise applications, and similar kinds of applications each present differing requirements, each generally better suited to one kind of duplicating approach rather than the alternative approaches. Prior to purchasing equipment, evaluate your requirements carefully and make sure your requirements match the capabilities of the system the you acquire. With the right combination of equipment and software, you'll have a streamlined way of creating and distributing digital content with the remarkable efficiency and economics of the compact disc, multiplied a hundredfold or more.

This chapter was written by Katherine Cochrane of The CD-Info Company. Katherine can be reached at katherine@cd-info.com.

9
Practical Applications

In this brave new digital world, virtually anything that can be converted to a binary format can be put onto a disc and distributed to computer users worldwide. For organizations and corporations that routinely distribute information to support their operations, the practical applications of optical recording offer ways to reduce costs, increase efficiency, and communicate with staff members and customers in new and creative ways.

In the early enthusiastic days following the widespread acceptance of the CD-ROM, many development companies jumped onto the multimedia bandwagon and invested hundreds of thousands of dollars, and sometimes millions, into productions ranging from interactive games with digital video clips featuring actors and expensively commissioned musical scores to multimedia tours of the great museums of the world. Traditional book publishers flirted with the medium, online database services experimented with releasing content on CD-ROM, and libraries of fonts, graphic images, music clips, and other rich media assets found their way to CD-ROM. A few companies flourished and titles such *MYST* and *From Alice to Ocean* demonstrated that multimedia CD-ROM titles could sell in the millions of copies. During this period, many companies also overinvested in title production, experienced poor sales, and disappeared from view in a year or two.

Some areas of CD-ROM title development are still very healthy and more adventurous companies are beginning to experiment with the additional storage capabilities of DVD-ROM and DVD-Video. But, in some ways, the era of the blockbuster multimedia title seems to have been replaced with a more finely focused, more practical approach to the medium. There

Chapter 9

will still be bestselling CD-ROM games that reach hundreds of thousands of customers, and CD-ROMs will remain the distribution medium of choice for the vast majority of software applications for some years to come. But, to a large degree, most companies releasing titles on CD-ROM with multimedia components are making their content as dynamic as possible using Web-enabled links. This combines the best of both worlds: the inexpensive, universally available distribution medium of the optical disc with the dynamic, readily accessible, rapidly changing face of the World Wide Web. Both have their own unique advantages and each complements the other in a number of ways.

While the era of the blockbuster titles may be over, individuals and corporations worldwide are enjoying the benefits of inexpensive optical recording for a multitude of communication tasks. Networked CD-ROM and DVD-ROM towers provide terabytes of data to intranet and Internet users. Training videos on DVD, certification programs on CD-ROM, and corporate databases on disc provide a ready method to produce and distribute key information about company processes or products in a much more dynamic and cost-effective way than can be achieved by means of printed books, manuals, or reports. The skyrocketing sales of optical media and recording equipment of all formats demonstrates the strong interest in optical recording as a vitally important organization communication tool. As DVD-ROM expands on the capabilities of the CD-ROM, new areas of enterprise and communication will certainly develop as a consequence of the rising interest in this format.

This chapter explores some of the practical methods by which recordable CDs and DVDs can solve the problems of businesses and organizations and make them more effective at their daily tasks.

Email Archiving

With the increased dependency on handling basic communication tasks with email, many email users begin receiving dozens and sometimes upwards of a hundred messages a day. If you've given out your email address in response to product registrations, special online offers, Internet purchases, or similar activities, the volume will continue to escalate as you begin receiving the electronic equivalent to junk mail.

Many email applications now include filtering options, to allow you to consolidate incoming mail into appropriate storage areas. The question is: How long do you really want to keep email on your system? If allowed to reside on the system indefinitely, you'll begin seeing a fairly substantial

Email Archiving

reduction in your available storage space, particularly if you get messages that include attachments frequently.

Optical recording offers an effective solution to this problem. You can archive email to CD-R or CD-RW discs on a periodic basis—perhaps once a year. To be able to retain and access the specific details you need to later refer to your email messages, you can first convert each message to Acrobat format and then use Adobe Acrobat's Catalog utility. Catalog lets you create a full-text index of your message contents simply by selecting a range of folders to index and then starting the catalog function, as shown in Figure 9 - 1.

Figure 9 - 1   **Acrobat Catalog options**

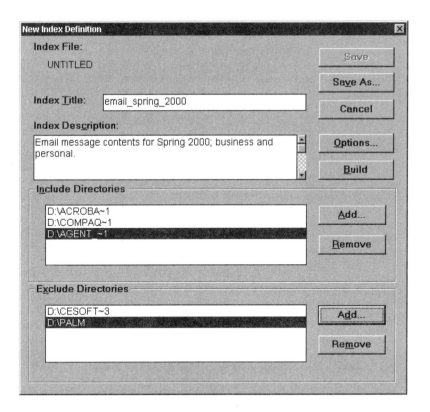

For this application, use CD-R if you plan on archiving messages for long-term storage. The write-once characteristic of the medium will prevent you from erasing and overwriting any of the data at a later time.

If your storage needs are more temporary, CD-RW lets you store almost the same amount of information, but also lets you reuse the disc if you decide that you can purge the messages at some point in the future. This extra flexibility can be valuable in production environments where the requirements and storage needs change very frequently.

The basic process could work something like this:

- Using the printer selection options for your system, select the Adobe Acrobat PDF Writer as the selected printer. PDF Writer is adequate for basic printing needs and can be treated as simply another printer on the system.

- From your email application, choose the range of messages that you want to archive.

- Start the printing process by choosing the print command.

- Some email applications will automatically assign numerical names to each printed email message sent to the PDF Writer; others may initially require your input unless you can configure them to print automatically. Obviously, it's much more convenient to let the names be assigned automatically.

- Run Acrobat Catalog on the files and folders that you want to index, making sure that the created index file is stored in a location where you can easily find it and include it when you burn a CD.

- Using your favorite recording application, copy the PDF files containing your mail messages and the index that you created to disc. You can then perform text searches throughout the entire range of indexed PDF files, locating even obscure references from past communications. For example, if you had an email conversation with a colleague about the Coen brothers' movies, you could perform a search using the terms "Coen" and "movie" and quickly locate the message among hundreds or even thousands of email files.

Exchanging Musical Project Data

The Internet has encouraged the growth of collaborative projects where programmers, developers, or musicians can contribute to the creative effort from around the world. While the Internet works well for small scale data exchange, it still falls short in bandwidth for distributing uncompressed raw audio files, which can occupy 10MB per minute of sound. For complex multitrack projects, which might easily have source files exceeding 300MB for a single three-minute song, the Web is not practical for exchanging the necessary volume of data.

For the cost of a first class stamp (or a bit more for international mailings), musicians can take advantage of the portability of optical recording to exchange both the source audio files and mix information. Vegas Pro, a remarkably capable multitrack editor from Sonic Foundry, uses a process known as non-destructive editing, which allows source files to be processed and mixed without altering the original file content.

This feature opens up a number of creative possibilities. Musicians can progressively build a project file that includes all of the multitrack elements to be included in the final song. Once all of the members of the project are working from the same set of source files, project files can be exchanged to illustrate the effects of different mixes and combinations of tracks on the overall sound.

Vegas Pro projects can be saved in two different formats:

- **Vegas Pro Project**: Contains the data that defines a single project, but does not contain the actual media files that are used within the project. This is the most portable way to distribute project information, assuming that all project members have access to the original media files. This project file includes the track effects, envelope information, bus assignments, output properties, and all of the key data associated with a particular project.

- **Vegas Pro Project with external media**: Includes both the project definitions, as described in the previous paragraph, as well as the actual media files associated with the project. All of the elements of the project are copied into the same folder. The folder can easily be burned to CD for distribution to everyone who will be collaborating on the content.

Chapter 9

Once all of the collaborators have the necessary media files stored locally (in equivalent volumes and folders), ongoing mixes and edits can be exchanged be sending the Vegas Pro Project file (without the external media). This feature allows rapid exchange of the sound processing effects, track edits, and so on that will shape the sound of the project.

Figure 9 - 2   **Vegas Pro editing window**

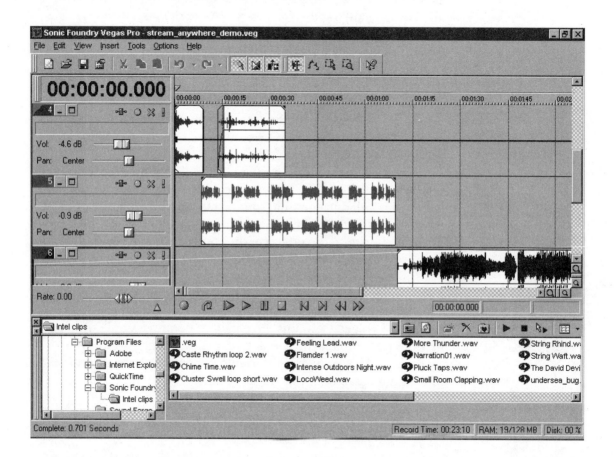

When Vegas Pro projects have been finalized, the program renders the audio information into any of several different formats:

- Advanced Streaming Format: Microsoft's standard format for Web streaming
- Wave: the most common audio file format used under Windows

- Audio Interchange File Format: the most common audio file format used under the MacOS
- RealSystem G2: the streaming audio format designed and supported by RealNetworks
- Windows Media Audio: an audio-only format that delivers content in streaming and downloadable versions
- Vegas Pro Project: the master file that contains the definitions of the audio handling for each individual track within a multitrack project, as well as the overall project data

This flexible rendering system makes it easy for musicians to produce a final CD-quality version of a mix and then use the same project file to generate streaming audio content for posting on their Web site or the Web site of their record label or distributor.

Distributing Cost-Effective Digital Press Kits

The conventional press kit from a high-tech company or most mainstream businesses consists of a glossy, four-color folder packed with more glossy 8.5x11-inch data sheets, white papers, and any other marketing literature that will fit in the package. While impressive in a shiny, showboat sort of way, this kind of press kit is definitely expensive—both in terms of the four-color printing, fancy stocks, and mailing costs, as well as in terms of the environmental impacts of distributing short lifespan materials using difficult to recycle, tree-based paper stock (the glossy coating complicates the recycling process).

A more benign way to deal with rapid distribution of full color marketing materials is via the CD-ROM or DVD-ROM. The natural luminosity of a computer's monitor gives color presentations the vibrant, saturated effects we enjoy in projected or back-lighted color slides. With a bit of imagination, static images can be recast as dynamic animations or even digital video. Sound—including voice-overs and background music—can be included to add more texture and emotion to an informational piece. The full range of interactive effects can be pulled into play to deliver product information or corporate background data more compactly and in more compelling terms than simply dumping a pile of data sheets onto a journalist's desk.

Of course, by making a CD-ROM or DVD-ROM Web-enabled, the audience can immediately respond to the material by following the provided links, collecting additional information or more current details about a product or services.

Recycling techniques are being developed for optical discs and several facilities already exist that accept optical discs. Discs can be stripped of their aluminum or gold layer, melted down to the raw polycarbonate, and turned into automobile dashboard components, football helmets, or office storage containers. If the discs are shipped in a recyclable fiber sleeve instead of a plastic jewel case, these sleeves can easily be reused or recycled along with other paper waste. New sources for the fiber for sleeves, such as industrial hemp and kenaf fibers, offer an environmentally sound alternative to paper-based disc storage systems. For businesses that are actively reshaping their processes to better support sustainability over the long term, the optical disc makes a tiny footprint in comparison to the enormous waste generated by a conventional press kit.

Lightweight Digital Video Exchange

Digital video tools have come of age and they are revolutionizing the way that visual information is communicated. While the VideoCD format originated several years ago, and achieved some success, primarily in Europe and Japan, as a medium for provided video content to consumers, the capacity was limited to shorter films and video projects. The quality of the MPEG-1 video content was roughly equivalent to VHS. The DVD-Video format, on the other hand, provides higher quality video output (using MPEG-2 file formats) and longer playback times, making it ideal not only for consumer applications, but for business training and corporate presentations, as well.

For producing presentations for playback on DVD players, a DVD-R unit is desirable, but some products also support creation of DVD content on recordable-CD players. For example, DVDit! from Sonic Solutions (*www.dvdit.com*) offers an option to output the MPEG-2 files to CD-R. You can also output to DVD-R, if you have access to the equipment, or to hard disk drives or removable media. For maximum compatibility, a DVD-R unit is the most direct choice, but also the most expensive option.

For easy exchange of digital video, you can also use the rewritable form of DVD: DVD-RAM. With a DVD-RAM and some rewritable discs, large volumes of digital video can be conveniently exchanged. While it is always possible to exchange digital video data on the actual video cartridges, the receiver then has to import that content into the computer, using a compatible DV camera and a high-speed port, such as a Fire-Wire port. An optical disc is a much more direct way to both exchange digital video information and conveniently access it from the computer.

While the DVD-RAM approach does require compatible equipment on both ends of the communication path, it removes the tedious importation process otherwise required to bring the data into the computer for editing or further work. Simply by inserting the DVD-RAM disc into a drive, the receiver has random access to the full range of content stored on the disc.

Optical discs, of course, are lightweight and easy to ship, so for anyone working with digital video content, this media has much in its favor for exchanging the extremely large files generated by digital video camcorders.

Case Study: Optical Storage for Digital Video Applications

Genesis (*www.gnsis.com*), a video production company based in Walnut Creek, California, finds recordable CDs a useful medium for archiving their completed productions. With corporate clients ranging from Adaptec to Sun, Genesis produces a variety of corporate video material, including presentations, marketing and advertising demonstrations, training materials, and so on. I talked with co-founder, Kevin Deane, about the role of optical storage in their workflow and the direction of their business.

Can you give me a capsule history of Genesis and how you got started?

My partner and I started off doing things in Los Angeles. We had some film work optioned—working on a screenplay—and soon after that we had a television show optioned. We were working with a producer and he hired us to work on this file, but, ultimately, it didn't go anywhere. It was becoming too difficult to work with the gentleman who the story was about. Finally, after beating our heads against the Hollywood wall for such a long time, we came back to the Walnut Creek area and decided to get a real business going. We learned so much about how Hollywood works that we feel we can go back there better armed next time.

Do you work in both film and video?

We don't get to work in film very much anymore. Most of our clients aren't spending that kind of money for the work that we are doing. I would love to do a lot more film work. We do everything in BetaCam,

Chapter 9

which is pretty much the broadcast standard, and everybody seems very happy with that for almost everything. There are some film projects that come along now and then; somehow those seem to end up going to ad agencies and not to the people that we typically deal with on the corporate level. We certainly have the capabilities to work in film.

How do you use recordable-CD in your work?

Primarily for distribution and for archiving. We commonly use discs for storage, since each CD-ROM is able to hold 650MB. We archive a tremendous amount of material that we create here for our clients. There is not reason for us to archive a video clip on CD, because we can just go back and pull it off of tape. On things we have created, where we have done compositing or special graphics treatments or an animation, we don't want to have to rebuild the content if the client asks us to revisit their project. So, we store it on recordable discs. It works out very handily.

How long are most of your corporate presentations?

Somewhere around 20 or 30 minutes. We often shoot someone within a corporation who is doing a presentation. Most of that kind of work gets immediately transferred to VHS right off the bat. For that kind of work, you make a BetaCam master and then a dozen VHS copies and you never hear from it again. We also do a fair amount of trade show work and marketing work.

Are you looking at all to DVD-RAM for gaining a little extra storage capacity for archiving?

We tend to avoid jumping on the bandwagon right away. I think there are still too many format issues in the DVD world. We're a relatively small company. There is my partner and I and our office manager. Also, a part-time editor, and the rest of the people we hire as freelancers. For us to go out and dump the money into something that next year turns out not to be the format of choice is a pretty scary thought. I'm a lot more interested in the DVD-ROM side of things. If we could give a client their high-end video—and all the other things that we could do with some appropriate authoring—take the whole interactive CD-ROM idea and do it DVD, where you have so much more space. You could do some pretty interesting things. Whether or not the clients would still be willing to pay for it—I still think you end up having competition from the Web.

Case Study: Optical Storage for Digital Video Applications

When you give clients previews of videos that you have done for them, what format do you use?

We use MPEG-1 video in VideoCD format. It's designed so they just have to pop it into a CD-ROM drive, click on a button, and play it back. Then they can give us a call and let us know if they like it. We're doing the same thing—depending on the length of the clip—via the Web. We upload it to our site and I just send a hyperlink to the client. They can download it and view it at any time. The reason we use both VideoCD and the Web, to give you a good example, a 26.5-minute presentation came up in MPEG at 272MB. Far too much to send over the Internet. In comparison, we can send a 30MB clip, about 3 minutes of video, to a client who has a T-1 Internet connection in just a few minutes. Because we can put the video in their hands, clients don't have to come here and edit with us, necessarily, until we get right down to the end and sometimes not at all. Depending on the complexity of the project, with some clients that we have already worked with, we have an understanding of what they are looking for and we have done maybe our fifth or sixth video project with them. You don't have to actually have them come up and edit unless they feel like they want to get out of the office. Walnut Creek, being the place it is, people do come here and say, "Great—I'm going to go shopping. I'll see you guys in about an hour."

Any other practical daily uses that you have for recordable CD?

We will occasionally burn our music cuts onto CD. My partner sometimes does original music for our clients, if they don't want to get something out of the library. We always prefer that they do original music. We burn the songs to CD when we are done.

What tools do you use for music production?

We use Sonic Foundry ACID. Vegas Pro. We also use Sound Forge. Acoustic Mirror. I think we have almost all the ACID loops.

For more information about Genesis, contact:

Kevin Deane
Genesis

2223 North Main Street
Walnut Creek, CA 94596
925-926-1344
www.gnsis.com

Chapter 9

Other Uses

The applications described in this chapter aren't intended to be a comprehensive listing of the ways that optical discs can be used in business, but merely to suggest some ideas that might help trigger thoughts on other ways you can take advantage of the large storage capacity and universal playback of CD-ROMs and DVD-ROMs. Other possible uses include:

- **Business-to-Business product or service catalogs**: companies and organizations that engage in business-to-business sales of products and services of sometimes measure their transactions in hundreds, rather than the hundreds of thousands that are more common in the retail trades. Printing four-color catalogs in low volumes to support this kind of business is sometimes not cost effective. In comparison, an online catalog describing products or services can easily be constructed in electronic format—such as Adobe Acrobat or HTML—and delivered on CD-ROM or DVD-ROM to potential customers. Optical discs can be run off as needed in small batches using CD duplication equipment. Add a disc printer to add some final polish to the CD appearance and you have a convenient delivery mechanism for communicating product information with customers. A Web-enabled order form can also easily be incorporated into this approach.

- **Software Beta Testing**: Software development groups frequently go through several beta test cycles while producing a final polished version of a software product. For those doing the beta testing, the most convenient way to receive the beta software is on CD-ROM. An installer utility can be used to make the software installation process similar to the final product, allowing each beta release to identify the full range of problems that might be encountered by the end user.

- **Storage of digital images**: Digital cameras abound in the business world, capturing every type of image from standard product shots to more fanciful advertising and marketing imagery. CD-R and CD-RW offer the perfect storage medium for high-resolution digital images. Brøderbund's PrintShop Multimedia Organizer can be used to quickly identify stored photographs on disc, as well as other rich media assets. Since many photo processors are also using recordable CDs to scan and return images from processed film to consumers, you can also maintain a library of images captured with film-based cameras.

- **Backup of departmental or organizational files**: Improved backup applications that compress files on the file and can store backup data on sequential optical discs make it possible to use CD-R for many routine backup operations. Corporations and organizations that want random access to files that have been backed up in a format that is more convenient than tape can rely on CD-R for many different types of applications. Backup utilities that work with tower recorders are obviously more convenient, since no one needs to be present to load blank discs.

Summary

Optical disc recording applications fit into an extremely wide range of daily operations involving large volumes of data. The different uses suggested in this chapter may inspire you to devise other applications for the versatile optical disc.

10

Simple Authoring Techniques

Creating a title to publish on optical disc doesn't have to involve thousands of dollars of development tools and a building full of staff members to carry out. Excellent, useful CD-ROMs and DVD-ROMs can be produced on the desktop using commonly available tools, such as HTML editors, Adobe Acrobat, your favorite word processor, and similar inexpensive applications. This chapter examines some of the techniques available to those individuals and companies that want to get started publishing to optical media in the simplest manner possible.

Publishing to Disc with Adobe Acrobat

Adobe Acrobat files have the extension .PDF—which stands for Portable Document Format. This suffix says it all. The portability of Acrobat files comes from a single common format for documents with readers that can be installed on all the major platforms: Windows 95/98/NT; MacOS, and UNIX/LINUX. Acrobat files have become a common means for distributing any kind of formatted document, from the hundreds of tax forms issued by the Internal Revenue Service to a current prospectus for a stock offering on E*trade.

Part of the appeal of Acrobat is the ease with which documents designed for electronic publishing can be created. The technique relies on printer drivers that convert the output from your application of choice—whether Microsoft Word, Adobe Framemaker, or Corel Draw—and change it into a PDF file. The resulting PDF file contains all of the elements that were present in the original source file, include fonts, page layout, embedded graphic images, line art, and, in some applications, HTML links.

Acrobat files can be structured for very high-resolution output, suitable for submission to print houses as document masters. They can also be compressed using several different utilities that trim the graphics and reduce the overall file size for more suitable distribution across the Internet. When Acrobat files are distributed on CD-ROM, Adobe permits the distribution of the Acrobat Reader installer for each appropriate platform, making it simple to construct cross-platform CD-ROMs or DVD-ROMs. If someone doesn't have the current version of the Acrobat Reader installed, the installation process takes only a few minutes.

Adobe Acrobat includes some fundamental multimedia capabilities, as well. Straight text documents can be enhanced with animations, sound, and video elements. Embedded URLs can point readers to additional information on the Web. The form processing capabilities of Acrobat are also a very interesting feature. Using forms laid out on a page graphically and using JavaScript, many different types of user data can be collected and processed. The processing can include submitting actual product orders to a secure site on the World Wide Web. A downloadable catalog can be enabled as a purchasing tool with the addition of an embedded interactive order form that connects to a properly configured Web site. Potential customers who might be reluctant to make a purchase on the Internet because they feel pressured while actively online can browse an Acrobat catalog offline and complete a purchase at their convenience.

PDF Writer or Distiller

Most versions of Acrobat include two forms of the print driver that creates the PDF file. A basic driver known as the PDF Writer performs quick and effective conversions of documents to PDF format from any active application. PDF Writer does the best job on documents that are primarily text and do not contain numerous high-resolution graphics.

The next step up in overall PDF quality involves two separate processes, first converting a document to a PostScript file and then distilling the PostScript file to PDF format using Acrobat Distiller. These two steps are integrated into a single operation in some Adobe applications that are closely coupled with Acrobat—such as Adobe Framemaker and Pagemaker. The quality of the resulting output file when using Distiller is substantially higher than when using the simpler PDF Writer, and this is the recommended technique to use when producing most PDF documents for use on CD-ROM or DVD-ROM. Unless you need to produce a single compact PDF file for both Web and disc distribution, Distiller will provide more finely rendered output and superior graphics. The trade-off is that you will generate larger PDF files.

If you are not working from an Adobe page layout program (such as FrameMaker or PageMaker) when generating a PostScript file for distilling, first select a PostScript printer from the list of printers available for your computer. It doesn't matter if the printer is one that you actually have connected to your computer. Choose the option to print to a file and select a destination for your print file. Ideally, the name that you select for the print file should include a .PS extension so that it will be easy to identify for distilling. Once the print file has been created, start Acrobat Distiller and make any appropriate settings for the file output. Acrobat 4 provides a simple selection process that can tailor the distilling operation for screen viewing, print applications, or press. Each has a different level of compression, ranging from screen viewing (producing the most compact files) to the press option (generating the very large files necessary for commercial printing). Select the file from its storage location and start the distiller.

Indexing a Document Set

If you have a word processor, Adobe Acrobat, and a CD-R burner, you can produce handsome documents in electronically published form that can easily be distributed on CD-ROM. If you produce large collections of documents about one topic or a number of related topics, you can make it easier for your audience to navigate the information by using Acrobat Catalog to create a full text index for your document set. With a properly created index, boolean search strings can be used to quickly search the full range of contents for your document set, even if it consists of thousands of pages of information. Using Catalog for this purpose is a simple matter of defining the files and folders to include in the indexing process and starting the operation from within the program. The index file should be included on disc along with the PDF files that are included in the search.

As shown in Figure 10 - 1, Acrobat Catalog offers numerous options for controlling the indexing process, including the ability to determine the group size that applies to CD-ROM-based indexes.

Figure 10 - 1   Acrobat Catalog preferences

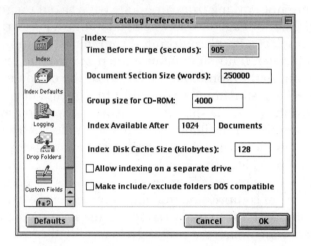

Post-Production Processing in Acrobat

Once you have created a PDF document or a document set, Acrobat includes a number of features that allow you to enhance the documents for electronic publishing. Some of the tools you have available include:

- **Forms creation**: you can create elaborate, interactive forms within Adobe Acrobat and overlay them on existing pages of a PDF document. These forms can include drop-down menu selections, validation of form entries through JavaScript, option button choices, calculation of shipping amounts and product costs, and similar features. Many of the options can be created through a menu-driven process, but if you want to get into some more custom areas, such as advanced forms validation techniques, you may need to know some JavaScript to accomplish the necessary tasks.

- **Audio or video elements**: following the creation of a text document in PDF form, you can embed buttons or other selection options that trigger the playback of audio files or video images within the Acrobat document. If you leave areas within your page layout for these items to be inserted, the final page will look more polished and professional.

- **Hyperlinks**: PDF documents that include URLs referencing Web pages can extend the effectiveness of your communication. The

Acrobat document can serve as a core reference that also includes external resources that add to the overall value of the content. Including pointers to information that tends to be dynamic, such as product price lists or dates for company tradeshows, can make your Acrobat document much more timely as well. An Acrobat document intended for installation on a CD-ROM can even link to other Acrobat documents that are posted on the Web. Links are easy to create from within the program. You can also create hyperlinks between topics on locally available Acrobat documents, which is a nice way to supplement the search abilities of Acrobat Catalog.

- **Custom document creation**: Acrobat also has the ability to splice pages together and extract sections from other Acrobat documents. You can easily construct a custom document from several different Acrobat documents using this feature. For example, if you had a number of different electronic brochures, each highlighting a specialized service or product, you could create a single Acrobat document custom-tailored to a customer's needs or requirements using the extraction and insertion commands. Ideally, all these Acrobat pages should be based on the same form factor—you don't want to try combining 5 x 7-inch pages with 8.5 x 11-inch pages.

Automated Acrobat Production

If you work from an Adobe page-layout program, such as FrameMaker, you gain several benefits during the creation of Acrobat documents. FrameMaker features are carefully integrated with the capabilities of Acrobat so you can generate a number of important links automatically simply by choosing to save the FrameMaker document as a PDF file. Instead of having to tediously add the table of contents links and the index links, FrameMaker produces these during the PDF creation. Someone browsing through the index of your electronic document can click on any of the index entries and jump immediately to the corresponding Acrobat page. You also gain the benefit of a nested set of bookmarks, built out of the table of contents in your FrameMaker document, that can be displayed in the bookmarks pane in the Acrobat Reader window. A hierarchical set of bookmarks is a very easy way to navigate through a complex Acrobat document and this can be very time-consuming to create manually. Having FrameMaker automatically generate the table of contents data and index links can save many hours of production time.

Chapter 10

Figure 10 - 2   **Adobe Acrobat bookmarks pane**

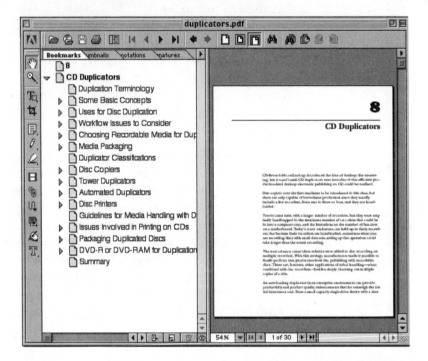

The trend in document production programs is toward single-source applications. By adopting single-source principles, you can create one set of document files and then convert them to other formats, as required. For example, a single FrameMaker document can be quickly converted to an Acrobat document, an HTML document, or a file designed for portable viewing with FrameMaker's own reader application.

Acrobat offers the enterprising electronic publisher many effective solutions to inexpensive, high-speed document production. The cost-of-entry is low and the results look very professional. Since the Acrobat Reader is free and easily available, either through direct download from Adobe's site (www.adobe.com) or via an installer that can be bundled with the rest of the files on your CD-ROM, your audience can access your content with minimal fuss.

Inmedia Slides and Sound

One of the most accessible forms of multimedia is the slide show. Even someone without a great deal of computer training can quickly master the basics of this form of presentation. The typical steps include:

- Collecting a number of images for use in the presentation, including photographic slides, illustrations and artwork, scannable images or objects, or computer-generated images or visual aids

- Obtaining any other needed assets, such as royalty free music clips or pre-built animations

- Putting the images to be used in sequence and choosing the transitions from image to image

- Adding a music bed, placing background music wherever appropriate in the show

- Adding any additional voiceover narration to complete the presentation

- Fine tuning the whole work for overall tempo and effectiveness

Once the files have been prepared in this manner, they can be transferred to CD-ROM, along with the playback engine that will present the material. This kind of presentation can be prepared in a high-end package, such as Macromedia Director, but you may not be willing to invest several hundred dollars if your requirements for this type of tool are only occasional.

One tool that makes this process very simple is Inmedia's *Slides and Sound*; available for approximately $50, Slides and Sound has a very well organized user interface and allows even novices to become quickly productive. The program generates an executable file (for use on Windows-based systems) that can access the full range of media assets included in a presentation. Presentations are made more interesting by a very wide selection of transitions from slide-to-slide, and support for a variety of audio formats, allowing voiceover commentary and background music to be easily added to a project.

For a project of this sort, content can include both originally produced digital assets, such as images captured through a digital camera or artwork imported from a scanner, as well as images or sounds from clip libraries. If you selectively choose your content from some of the higher quality stock image libraries, you can come up with some extremely striking visual effects and top-notch presentations. For screen-ready viewing (as opposed to print-ready applications), many stock image houses offer downloadable JPEG files that can inexpensively be incorporated into a project. For example, The Stock Market (*www.stockmarketphoto.com*) offers both downloadable images as well as extremely high resolution visuals, available on CD-ROM, for print or broadcast projects. Choosing judiciously from among professional-level content of this sort, you can quickly assemble a powerful collection of assets that will enhance your communication efforts considerably.

A running slide show, as opposed to a typical interactive multimedia project, often takes a linear path through the content from beginning to end. This, of course, can be an advantage if you have a particular message to convey to your audience. The communication flow and playback is determined while you author the project and the viewer can't deviate from the path.

The disadvantage, of course, is that the playback for a tool of this sort is limited to the Windows platform. If your goal is to reach a wider audience, including MacOS and UNIX users, try one of the other tools discussed in this chapter, such as Adobe Acrobat or Blue Sky Software RoboHelp HTML 2000.

Blue Sky Software RoboHelp HTML 2000

Single-source documentation was discussed earlier in the section about Adobe Acrobat publishing. RoboHelp HTML 2000 is another product designed to allow you to create one set of source files that can then be outputted to other formats. For example, RoboHelp HTML 2000 lets you create:

- **WebHelp**: a cross-platform three-pane help environment that can be used on the Web or on removable media, such as CD-ROM or DVD-ROM. WebHelp is based on Java and is a good choice for playback from optical media.

- **HTML Help**: the form of help incorporated in Windows 2000. Designed for playback on Windows 95/98/NT/2000 systems, HTML Help can encompass both application help, as well as standalone systems designed to explain a concept or teach someone about a product.

- **Printed documentation**: source files created in RoboHelp can be quickly turned into Microsoft Word documents with full indexes and automatically generated tables of contents.

RoboHelp has the advantage of being a very robust production environment. Having been through several releases, the program has been polished and improved over the years and it offers many features for complex document production. It also includes a number of tools for debugging very complex hyperlinked projects, including utilities for identifying faulty links and outdated topics.

Tapping into the Java Virtual Machine

The Java-based version of RoboHelp's output, termed WebHelp by Blue Sky, provides the greatest cross-platform compatibility for projects distributed on CD-ROM. WebHelp projects transmit equally well across the Internet as they do on optical media; several accessory files must be included along with the project file to ensure Java playback on the appropriate virtual machine. These files are generated automatically by RoboHelp during the compiling process and can be uploaded to the Web or stored in a directory on a CD-ROM, depending on your chosen method of delivery.

RoboHelp Development Environment

The development environment, shown in Figure 10 - 3, mirrors the appearance of the final output, whether you are porting content to WebHelp or HTML help. The pane of the left side of the application window represents a collapsible tree of the topics included in the project. Content can be restructured and reorganized on this tree as required. The project author can also navigate through individual topics through selections made in this pane.

The pane at the bottom of the application window serves different functions at different times. It can display the status of the compiling process when you are preparing a help file for release.

The high-end version of the RoboHelp package, RoboHelp Office 2000, includes an additional output format: JavaHelp. JavaHelp, developed by Sun Microsystems, is a delivery system that works under the Java virtual machine. As such, it can be used and viewed anywhere there is a compatible playback environment, either through the Java capabilities of a browser or the built-in Java virtual machine capabilities within an operating system (whether Windows, UNIX, or MacOS). This output format, because of its cross-platform nature, can be adapted to CD-ROM delivery modes very easily.

Figure 10 - 3   Main application window in RoboHelp HTML

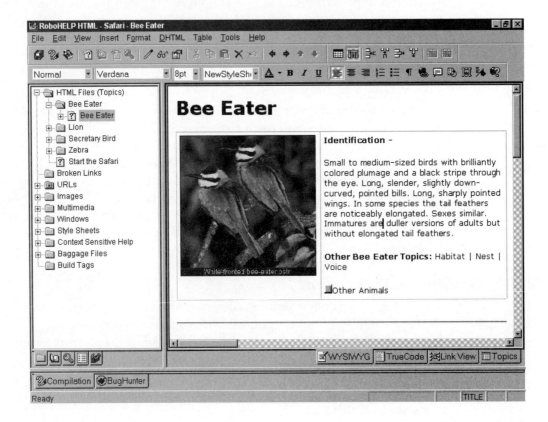

To simplify content creation for novice authors, RoboHelp includes an abundance of wizards to direct someone through the proper steps to carry out a task. For example, the indexing wizard, shown in Figure 10 - 4, guides someone through an automated process of generating a full index for a project based on keywords, selected terms, or other factors.

RoboHelp also includes tools for image manipulation, such as image resizing and bit-depth redefinition. Hotspot creation tools let an image be modified to include regions that can be hyperlinked to specific topics or to pages on the Web. RoboHelp also features a graphics locator utility that can search your local and network drives and display previews of any available graphic images.

Figure 10 - 4   RoboHelp Index Wizard

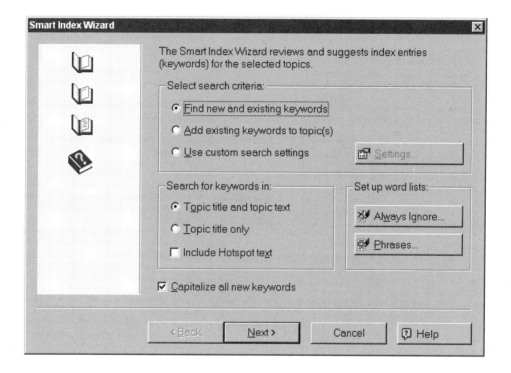

RoboHelp is not the most inexpensive package for authoring in the market. Just the standalone RoboHelp HTML package can cost in the range of $250 - $300. The full RoboHelp Office 2000 typically retails in the $600 - $700 range for new purchasers (those who are not upgrading from a previous edition). The advantage of this package, however, is that it is a full-featured and extremely versatile communication tool, equally capable of producing comprehensive help systems for the Web as for either CD-ROM or DVD-ROM. As an investment for the serious developer, it offers some compelling benefits.

Developing Content with an HTML Editor

One of the easiest ways to create interactive multimedia for a CD-ROM or DVD-ROM project is by simply using your favorite HMTL editor. Whether you enjoy working in Microsoft Front Page, Macromedia Dreamweaver, or a simple text editor, HTML offers some very clear advantages as an authoring medium.

Chapter 10

- Tools for creating and debugging HTML code are highly refined and very easy to work with, including visual editing tools (such as Dreamweaver) that let you preview the HTML formatting as you develop it. These tools are also typically very inexpensive.

- Properly designed HTML projects can be viewed on any platform as long as a Web browser is available to the audience.

- Multimedia elements, such as sound and animation, can be easily integrated into a project using either plug-ins or the native capabilities of the latest browsers. Dynamic HTML also offers some interesting functionality that can make an HTML project exhibit characteristics that are usually associated with standard software applications.

- The learning curve for authoring content in this format is certainly less for anyone who has some rudimentary experience in putting up Web pages, which, of course, most developers do.

- HTML content can quickly be repurposed for posting on the Web.

The biggest consideration when developing content in HTML involves naming conventions and differences between absolute and relative links that occur within the HTML pages. Tools such as Dreamweaver offer control over the naming of the links, allowing you to test link continuity locally, and then restructure the links and names as necessary for porting to the target destination. This valuable feature can help avoid many last-minute headaches testing a CD-ROM one-off that contains HTML content.

HTML on CD-ROM has become a popular vehicle for many companies to take advantage of the best features of inexpensive optical media, while still working in a familiar development environment. The following example, Case Study: Blackhawk Down, relates the experiences of a major metropolitan newspaper when converting a successful series to HTML format for release on CD-ROM.

Case Study: Blackhawk Down

The Philadelphia Inquirer gave weight to the term in-depth journalism when they produced a multi-part series on their Web site and then ported the content to CD-ROM to reach an even greater audience. The widely heralded project, *Blackhawk Down*, documents the United States ill-fated involvement in Somalia and chronicles the misadventures of the troops and individuals caught up in a military operation gone bad.

I talked with John McQuiggan at the Philadelphia Inquirer, who offered his perspective on the creation of the CD-ROM and the associated production issues.

Was there much work involved going from the Web site to CD-ROM?

We had to make a lot of conversions from ASP pages over to HTML. It took a lot of experimentation with the images to get them right for the CD-ROM. The video also had to be converted to a different format, which the folks at KR Video did. So, there was more experimentation going on there. Part of the reason that it was difficult to do all this: the decision to put the project onto CD-ROM came up out of the blue after the Web site was already up. Somebody said, "Hey, wouldn't it be great to put this on CD-ROM and offer it with the reprint?" I said, "No, no, no. It wouldn't." [Laughs] There is a helluva lotta work that would be involved in doing this. We decided to do it anyway. U.S. Interactive was the group the really pulled it off. They basically just told us what had to be changed.

So, some work was done inhouse and some was done at U.S. Interactive?

Yes, U.S. Interactive was the company that did the development of the CD-ROM for us. Inhouse, we needed to convert the formats of a lot of the text files, the video, and the images, so they would be in the proper form for the CD-ROM.

It looks like the CD-ROM design is very close to your Web site appearance. Is that deliberate?

We considered making it a CD-ROM presentation, entirely reformatting it, and presenting it the way we would have if we initially built it as a CD. The cost and the time involved were prohibitive. The newspaper was especially keen to jump on it very quickly because of popularity of the series in the newspaper. We just didn't have a whole lot of time to turn it around. Consequently, we decided to put it on the CD-ROM as a Web

Chapter 10

site. There was a fair amount of discussion as to whether that was a wise thing to do and if people had Internet access, why would we need to put it on a CD-ROM for them. They could just click on a site on the Web. Ultimately that is the direction that we went in.

In retrospect, do you feel it was a wise decision? With CD-ROM, you have the advantage of instantaneous access.

It was a selling point for what we called the Collector's Edition, that was offered for sale after the series ran. Some people ordered it just because it was available on CD-ROM and they didn't have to fool around with the Web to read the site. So, it was nice. Word of it has spread far and wide, we've heard from some military colleges that have included the CD-ROM as part of their course offerings. (I don't know the names of them off-hand, but professors from various military schools have called up and purchased multiple copies so that they could show them to their classes.)

We ran out recently and we had to reorder more CD-ROMs. We now have versions of Blackhawk Down as a video product, a book, an audio book, and the CD-ROM version.

When the book came out, all of a sudden there was another surge of orders and many inquiries about the CD-ROM, so we decided to have more made.

Was the PBS documentary release a one-hour special?

Yes, it was a one-hour documentary. There are pieces of that on the CD-ROM. I believe that we put them together using QuickTime.

That was another issue that we ran into when producing the CD-ROM. We had to get various licensing and permissions. I kind of got a crash course in getting a license for QuickTime and we put a browser on the CD-ROM, too, just in case people were expecting it to open up as a CD-ROM production.

For the QuickTime licensing, what was your experience with Apple?

It wasn't difficult. It was a matter of tracking down the information on their site and then walking through the process and filling out some forms. And then submitting the forms to them. Apple then mailed something back to us to sign. I signed and returned it. It wasn't that difficult.

It was royalty free?

Right. There were no royalties involved.

How would you describe your role in this process, John?

I was the coordinator between our staff here and the U.S. Interactive people and KR Video and the Inquirer newsroom. The focal point where all the people intersected.

From your perspective, do you have advice for developers who are working on projects like this, where many things have to be coordinated in a fairly short period of time?

In retrospect, I wish we had anticipated the need for the CD-ROM before we put the Web site up. We could have saved ourselves a whole bunch of work.

What would you have done differently?

The formatting of the HTML pages—we would have done it in a way that we wouldn't have to do a conversion. There were some pieces of the site that required rechecking of links, because file extensions were changing and formats of audio and video were changing. So, it was a big project—almost 30 installments. And, it became a little bit of a nightmare to go back through it and check all the links and make sure that everything was still working. We have pretty good software for link checking, but we kept stumbling over some things that we had missed. It made it quite an adventure.

Any idea how many individual elements are on that CD-ROM?

It's hard to say. I would say it's in the hundreds. I'm not sure it would reach the thousands. Hundreds of HTML pages, and audio clips, and QuickTime.

Do you have a sense whether people are still visiting the core Web site?

As of about two weeks ago, we ran a report against our log files. It is still getting about 3000 page views per day. That does as well as some sites that we update every day. It's amazing, the staying power of this thing.

Chapter 10

Has the CD-ROM distribution met expectations?

Frankly, we haven't found a way to extrapolate and come up with a new way to market what we do through CDs. In this case, I think, it was the content rather than the format that sold it. There was just so much interest in this topic.

So, word of mouth has been a pretty powerful promotional tool?

Yeah. Plus the fact that the series was syndicated after we ran it. We could see new surges on the site as it hit each new city. It would run in Seattle and we would get a surge of hits. And it would run in another paper in Georgia and we would get a surge of hits. It worked its way across the country.

For more information on Blackhawk Down, refer to the site: *home.phillynews.com/packages/Somalia*

Summary

The price of admission to the world of CD-ROM authoring is relatively low, as can be seen by the various approaches outlined in this chapter. You can get started and actually create quite sophisticated projects with tools as simple as a plain vanilla word processor or HTML editor. The quality of the content and your imaginative approach to presenting it can go a long way towards distinguishing your work from your business competitors or other developers.

11

Audio Recording for Music Enthusiasts

We've moved past the point where any computer user can produce their own custom audio CDs on the desktop for under a dollar a disc. Optical recording offers even more opportunities to produce, store, and record music in digital formats. For example, new tools and techniques now make it possible to store several hours of MP3 music on a disc and play it back through a component-style audio unit. Digital sound processing tools coupled with optical recording equipment make it possible to go beyond the restrictions of stereo and create six-channel audio using surround sound technologies, including Dolby Digital. As processor speed increases bring computers to performance levels equivalent to the supercomputers of a few years back, equipping a home studio with incredibly powerful hardware is within the reach of almost anyone.

You can create your own custom music discs, restore your favorite albums from the 40's, 50's, or 60's using digital sound processing, create your own music using a variety of powerful and far-reaching compositional tools, or master an original audio CD of your own music for replication and independent marketing.

Just the few topics mentioned easily represent another book's worth of material, so this chapter will provide some examples of the capabilities of optical recording tools in this area and preview a number of options that are available to you. Two case studies and an expert's view section present topics of interest to many sound engineers and musicians: how to create an Enhanced CD, how to create a Web-enabled CD, and how to process sound for DVD applications.

Chapter 11

Vinyl Restoration

Goal: To transfer the audio tracks from an old jazz recording to audio CD

Tools: Sonic Foundry Sound Forge 4.5; Sonic Foundry Vinyl Restoration plug-in, Asimware HotBurn

A lot of the good music recorded onto vinyl records has never been remastered and released on audio CD. Many excellent works from the swing era, or early blues masterpieces, or even little-known folk records from the 60's can be found inexpensively in used record stores, tossed out unceremoniously by their owners when audio CDs took over the world in the 80's. If you have a collection of vinyl records and would like to convert some of your favorites to CD format before the clicks and scratches get totally out of control, the following procedure will guide you through the process. This technique may also prove useful if you're working professionally in the audio industry and need a streamlined method for retrieving music from vinyl for a historical project, play, film adaptation, or multimedia project.

For a test case, I chose an old copy of a record by The Dave Brubeck Quartet originally released in the early 1960's, *Jazz Impressions of Japan*, for which I've been unable to find in an equivalent updated CD version. This well-played album is littered with clicks, scratches, and that familiar background crackle that characterizes most phonograph records of early vintage. The goal will be to eliminate as much of the intruding noise as possible using plug-ins to Sound Forge 4.5 and then to burn an audio CD from the cleaned up tracks using HotBurn.

To bring the audio into your computer so that you can work with it in digital format, you need the highest quality Analog-to-Digital conversion that you can manage. This might be any of the following options:

- An internal sound board with RCA connectors for connecting your component sound system to your computer, or built-in circuitry in your computer that accomplishes the same task. Many low-end sound cards aren't up to the task of high quality A-to-D conversions, so if you're sound board isn't high caliber, you might want to consider upgrading.

- An external device, such as a DAT recorder, that includes digital outputs. The recorder can perform the A-to-D conversion and you can then port the signal in digital format into your computer using a sound board that supports direct digital transfer

For this example, I chose the second approach, connecting the output lines from my stereo component system to a Mackie 1202 mixer to a Panasonic SV3700 Digital Audio Tape deck. The SV3700 has much better A-to-D conversion than my native computer sound board, so it's a better component to use in the chain.

Figure 11 - 1 **Components for Recording**

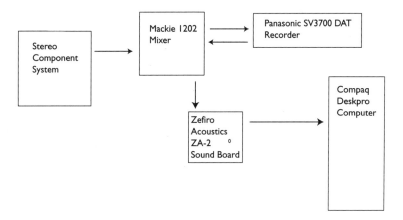

Tinker with the signal levels, the output on your stereo amplifier and the input of your recording device, until you get consistent levels, as close as you can get to 0 dB without clipping. Keep in mind that unlike analog clipping, clipping in the digital realm produces distinctly obnoxious artifacts that are difficult to remove. If necessary, set a lower input level; you can always cleanly boost the digital signal later if you need to obtain a higher signal for creating your audio CD.

Recording to DAT also gives you an good medium to work with as you bring the digital audio into Sound Forge. Ideally, once you have the sound recorded in digital format (PCM), you want to keep it in the digital realm. Additional D-to-A and A-to-D conversions will lower your overall sound quality.

In my case, I used a Zefiro Acoustics ZA-2 sound board to input the recording from the DAT and bring the content into Sound Forge, monitoring the levels, as shown in Figure 11 - 2. To perform the vinyl restoration, it's easiest if you bring in the audio one song at a time, even if you've recording an entire side of the album uninterrupted. Sound editing generally is easier to perform if you're working with audio files that are between 2 and 6 minutes. You'll also want the songs in individual files in preparation for burning the audio CD.

Figure 11 - 2 **Recording to Sound Forge**

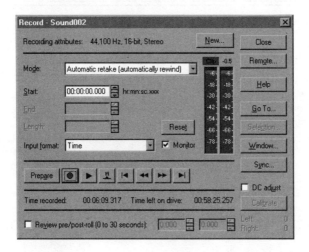

Performing the Restoration

Once you have brought the first song into Sound Forge, take a first look at the graphic signal representation, as shown in Figure 11 - 3. Ideally, you want a good strong signal to work with—if the levels are a bit too low initially, use the Normalization option in Sound Forge (or other sound processing program) to bring the level up to a consistent setting, making sure that you avoid clipping. Use the same level' (you can save your normalization settings) for each of the audio files that you import for processing.

The Vinyl Restoration tool is a separate package available from Sonic Foundry. If you don't already have it, you need to purchase and install it to perform the next step in the process.

Figure 11 - 3   **Audio Signal in Sound Forge**

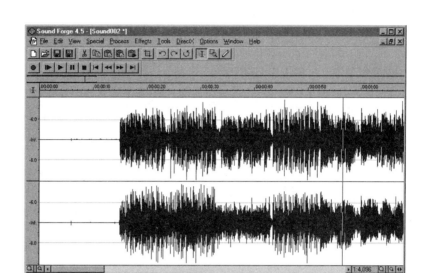

The Vinyl Restoration option, once the plug-in has been installed, is accessed through the DirectX menu. As you can see from the screen in Figure 11 - 4, you can use the slide bars to very precisely control the restoration settings, from the click removal amount to the frequencies affected to the noise floor. Under the Name drop down list, you can also select General Restoration. Sonic Foundry's presets do a respectable job for many different types of recordings and you may want to start with this selection and test the results.

To get a sense of the sound processing at work, you can highlight a section of your audio file and then use the Preview button to listen to the restoration as performed on the selected portion of the file. Once you've tinkered with the sliders and chosen the best settings, use the Selection button to choose All Sample Data and then click OK on the Vinyl Restoration dialog to start the process. By default, Sound Forge initially creates an Undo file, so that if you're not happy with the results, you can select Undo from the Edit menu and return to the previous file.

Once you've performed the Vinyl Restoration, you can use other Sound Forge options to further refine and enhance the sound quality. The Noise Reduction plug-in helps remove any other stray disturbances in the audio file. You can also use equalization, normalization, or other process-

Chapter 11

ing effects to improve the overall sound. Use effects judiciously, as you can seriously alter the character of the original sound file by going overboard on the effects processing.

Figure 11 - 4 **Vinyl Restoration Options**

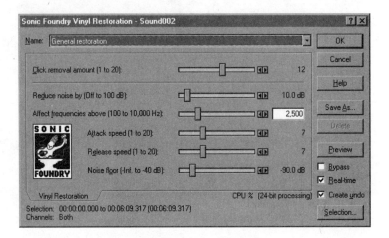

Once you have the file in good shape, trim any extra blanks areas from the beginning and end of the file and save it as a standard .WAV file in PCM format. Audio files for CD use should always be prepared as 16-bit stereo files sampled at 44.1K bps. Don't be tempted to try to change the sample rate to get more songs on a disc—it won't work.

Repeat this process for each of the songs on your album. You'll find it's fairly time-consuming and laborious—probably a task best suited for your favorite albums and custom collections, rather than something to do for every record in your collection. Save all of the restored audio files in a single folder. The next step is to produce an audio CD from them.

Producing an Audio CD

Almost every CD recorder package includes a program for producing audio CDs. Some are of the drag-and-drop variety, where you simply drag files into a creation window until you've filled the capacity of the disc to be burned and then click the record button. Other professional-caliber applications, such as Sonic Foundry CD Architect, give you more precise control over low-level aspects of the recording process, such as the gaps between tracks and the contents of the P and Q subcodes. For this example, we're not planning on producing a master for replication purposes, so a simple audio CD application will suffice. Sony's HotBurn provides an easy, failsafe path to audio CD creation.

The first screen in HotBurn offers a selection of Wizards, Layout Managers, and some Other Options, as shown in Figure 11 - 5. HotBurn was developed by Asimware Innovations, Inc. and is distributed under license by Sony Electronics, Inc.

Figure 11 - 5 **Sony HotBurn Introductory Screen**

For this example, we'll choose the option from the Layout Manager, **Create an Audio CD**. The window that appears represents the track listing, which, of course, is initially empty, that will be used to design the contents of the audio CD. You can either use drag-and-drop techniques to pull WAV files into the listing window or use the **Add Audio Files** command from the **Track** menu to build the disc contents listing.

HotBurn deliberately keeps the options simple when you are creating an audio CD. In the simplest situation, you can order the files, add blank space between individual tracks, and choose **Track-at-Once** or **Disc-at-Once** for the recording operation. If you select the **Disc-at-Once** option, HotBurn gives you more precise control over the track positioning and index marks. This features allows you do a reasonably good job of preparing an audio CD for replication, if necessary. For many home users, however, the **Track-at-Once** mode gives you more flexibility in building the

Chapter 11

CDs and also allows you to take advantage of CD-RW discs for storing music.

Figure 11 - 6   **Track Listing**

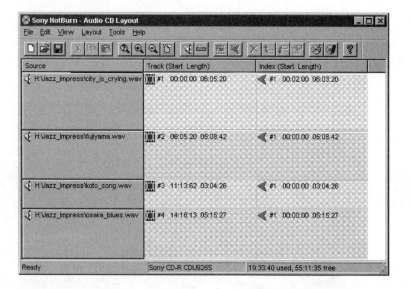

HotBurn is fairly flexible when it comes to the source files containing the audio content. It converts the audio files on-the-fly to the standard CD-DA format (otherwise known as PCM format) required for creating the audio CD. Files can be resampled, if required, and byte reordering is performed for those file formats that vary from the CD-DA norm. The options are shown in Figure 11 - 7.

For the best results, however, you should perform conversions and any digital sound processing on the audio files before you reach this step, using any standard sound processing application. Sound Forge XP 4.5 is an inexpensive yet versatile application for this purpose.

Figure 11 - 7 Specifying the Audio Features for Conversion

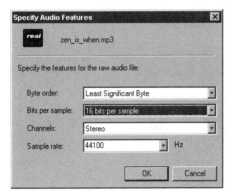

If you use Track-at-Once recording, HotBurn presents a simpler set of choices, as shown in Figure 11 - 8. You still have the option to insert blank spaces between individual tracks, but you can no longer control the precise index mark locations, as with Disc-at-Once recording.

Figure 11 - 8 Audio Layout for Track-at-Once

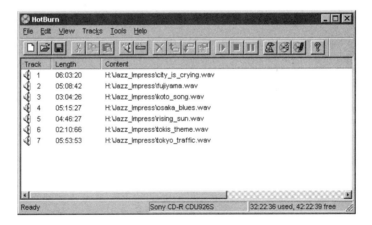

When you have organized the audio layout to your preferences, click the Write this CD button. HotBurn displays a tabbed panel that lets you set any final options, such as setting a slower recording speed (if your system does not do well at higher speeds) or deciding whether to close the session when done, to let the disc be readable in CD players. As the record-

Chapter 11

ing operation takes place, HotBurn indicates the progress in a dialog box, shown in Figure 11 - 9.

Figure 11 - 9   **Progress of the Write Operation**

When the write operation is complete, you have a bona fide audio CD ready for playback in a computer system CD-ROM drive or in a conventional CD player. You can skip between tracks, just as with a commercial CD, and playback times for the individual tracks will be indicated.

HotBurn offers a wizard for creating an audio CD, shown in Figure 11 - 10 if you feel you need a bit of extra help with the process, but most people should find the audio layout approach very easy to handle—even if they don't have prior experience making CDs.

Figure 11 - 10   **Create an Audio CD Wizard**

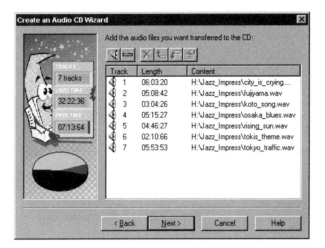

Case Study: Creating an Enhanced CD

The process of creating an enhanced CD takes imagination, fortitude, and a seriously warped sense of humor, as is evident in this interview with the co-founder of Right Angle, Inc. (*www.rightangleinc.com*), Kelly Meeks. When the content is on the outer edges of contemporary humor, sometimes you have to pull out all the stops to keep in step.

What can you tell me about the Jerky Boys comic that appears on this CD?

Johnny Brennan, the guy who does most of the voices, is just insane. I don't know how else to describe it. I've never seen anyone as fast on their feet as he is doing this stuff. It's amazing.

So these are real characters from the audio portion?

Oh yeah, everyone that you see there. You've got Kissel, Pico, Jack Tors, Frank Rizzo who is the tough guy, and poor Sol Rosenberg, who is the neurotic guy who everything bad happens to. And Tarbash. These are the personas that these guys take on when they get on the phone and start crank calling.

I guess it's not necessarily the type of program I'd like to send to my mother.

It could ensure my place in eternity, I'm afraid.

Chapter 11

We were big fans of the Jerky Boys, in general, and when we were given an opportunity to be paid to work with them, we said, "Hey, we've got to do this."

It looks like you've done an enormous amount of work in Director to put this all together.

Yeah, in about four weeks, by the way.

In four weeks? I'm astounded.

That's with a crew of three people.

You did most of the programming in Director?

Any code that was written for that was done by me.

It looks like you've been pushing the capabilities of Director a fair amount with some of this stuff.

People think of Director as, at best, a non-linear presentation tool. I think it is infinitely more powerful than that. I just finished a clip-art program that Director uses to manage a couple thousand pieces of clip-art, cross-platform. It's a very powerful development system. It certainly has some limitations—it's no C++ or anything along those lines. But, if you're talking about creating a real multimedia experience, there isn't a heck of a lot you can't do with Director.

While writing an earlier book on CD-ROM technology, I interviewed a lot of game producers. Many of them would use Director for prototyping the game and doing some of the early renditions and then would jump off and either develop their own game engine or find some other tool to support the high-speed things they wanted to do.

We usually try first to do things in Director. In the event that we can't, we're certainly capable of using other development systems.

Were you working sixty and seventy hour weeks to build this thing in four weeks?

Yeah, it was crazy. First of all, what I say in the event that this actually gets ink, we normally don't make our living by taking on insane deadlines. But this was an opportunity for us to work with the Jerky Boys. Again, like I said, we're big fans. It was something we really wanted to do. I don't believe that Mercury had previously done an Enhanced CD. There was a learning curve on their end. When all of that was said and done, we spent a lot of time with Johnny in our studios in front of a microphone doing the sound work for us. Once we got the art approved, it was up to me to crank out the program.

Are the other components, such as the music and sound effects, from libraries of royalty-free material?

Yeah. Johnny is pretty good at doing his own sound effects. We didn't have to draw too heavily on third-party libraries, but we do have access to several inhouse royalty-free generic sound-bed type libraries of sound effects. Those get used pretty heavily.

I found the whole production very professional throughout. The sound adds a lot to the overall effect—people sometimes overlook the audio parts of a production.

To me, multimedia means sound. In some respects, I think sound is one of the easier things to do right. I mean, it is a lot more cut and dried. We've got art people in house; actually, our art director Eric Talbot did the vast majority of the art for the project. For the sound stuff, though, having Johnny to do the voiceovers was great. It demonstrates that the project really came from him, which is what we were trying to get through.

How much space did you have on the CD-ROM for the interactive portion?

We were told from the onset that the budget for the audio portion in the first session was about 55 minutes, which meant that we had about 150MB to work with.

That's not a huge amount for a project of this scale. From the beginning, you were really having to budget every Megabyte of space on the disc.

You better believe it. I think per partition, the Mac and PC sides are about 43 to 45 Megabytes, respectively. We did have some additional room to play with, but, quite frankly, I think we were fighting the clock as much as anything. Heaven forbid if we had six months to work on something—I think we'd really go to hell at that point.

You'd be totally insane by the end of the project.

Yeah, we'd be doing the Thorazine shuffle or something. That's something that we're actually very interested in doing with them—to get a crack at a real time frame and to do something that is totally off the wall.

There still doesn't seem to be mass market acceptance of the whole Enhanced CD format. Do you think you're up against a giant obstacle getting people to buy into this notion of merging audio and multimedia on a CD?

Initially, yes. I think there is a barrier at a number of different levels, one of which is the technology. I think there are a fair number of people out there who don't have multisession-capable drives. Between firmware updates and software driver updates, there is a lot you can do to fix that. Most companies do have some upgrade path to deal with that sort of thing.

The other thing is the issue of retailing. The music industry in general is still cutting it's teeth on what they can or cannot do with Enhanced CD. When we were first talking to the Jerky Boys about doing this, I went out and bought about every Enhanced CD I could get my hands on. Without getting specific, quite frankly I thought that most of them were pretty lame. Invariably, what you got was digital video of the act or something.

What we wanted to do was something totally different—as opposed to a lot of QuickTime video and lyrics (which really didn't apply to the Jerky Boys anyhow), we said, "Hey, let's do some character development here." They're interested in taking the Jerky Boys into animation and doing all these other things, which I think would be great, but we thought this would be good interim step to give people a chance to experience the Jerky Boys. We didn't think the digital video or the traditional ways of doing Enhanced CDs was the way to go. We took a totally different approach and said, "Hey, let's create an environment. Let's create a Jerky World here and let people cruise around in it."

Any research on what the platform barriers are to playing Enhanced CD?

Multisession capability is really the only barrier. I don't know if you're familiar with all the different modes of hybrid-based CDs there are. You've got Mixed Mode, which has the Track 1 problem. You can write the audio portion after the interactive portion, but that literally gets written as Track 1 on the audio side. So, you go to play that on your stereo and you get all sorts of porpoise sounds and damage your speakers. That to me is a non-solution. At some point, even I would forget and play Track 1 and who needs that?

Case Study: Creating an Enhanced CD

We really felt that the Enhanced CD is the most viable mechanism for doing it, even though there might be a goodly number of people out there that don't have multisession drives. Because, it works totally in the audio portion of it. It works totally in the interactive portion of it as long as you've met that one hardware requirement. It just seemed like the way to go.

Did you do your own inhouse prototyping using a CD recorder?

We did. Then we let Disc Makers do the final mastering. I had never mastered a commercially released Enhanced CD before and I figured at this point, what a great way to get Disc Makers involved. At this point, I can't see any reason to master an Enhanced CD myself when I can have someone more experienced do it.

How did you supply the material to Disc Makers?

We gave them a CD-R master of the audio and a CD-R master of the interactive portion.

So one was formatted to Red Book standards and the other was a basic Yellow Book CD-ROM?

Exactly. And the two together were a Blue Book Enhanced CD.

Tell me more about your company.

Right Angle as a core company exists as three people. Myself, I kind of head up the technical division, if you will. Kevin Russell, who is my partner. He and I own Right Angle. Kevin handles most of the graphic design, layout, 3D work, sound editing. He does a tremendous amount of work. Eric Talbot is our art director and our Photoshop guy. He does all the pure, physical artwork.

Do you want to give the 30-second capsule description of how the company got started?

Kevin and I were working for the individual who started the Teenage Mutant Ninja Turtles. A guy by the name of Kevin Eastman. We were running a publishing company for him. We pretty much thought that we had done everything there that we could at that point. We were dealing with comic-based properties at that point. We had just finished up the Crow

Chapter 11

movie deal with Brandon Lee—some of those types of things and we decided to capitalize on that success and go off on our own.

I had worked with and for Apple for about 8 years and had a strong technical background and Kevin has always been the marketing and creative brain of this company. We decided to strike off on our own and see what type of trouble we could get ourselves into. And it didn't take long...

Were you working out of the Cupertino office of Apple?

No, at the time I was doing it, they actually had offices in Marlboro, MA. When I moved to Massachusetts to work for Kevin Eastman in the publishing company, that's how I hooked up with Kevin Russell and one thing led to another, and now I guess we're both going to hell... with this Jerky Boys thing.

Are you worried about your reputation suffering with people out there saying, "these guys are totally over the edge..."

I'm sure there is plenty of that, but personally I am proud of the association. I mean that. They're not out there hurting anyone. There are having good, clean fun, for the most part. I think you don't have to look around much to find a whole lot of worse things out there. Plus, I just think they are hysterical. I think Johnny and Kamal are both literally comedic geniuses. We are proud of the association.

Have you done other Enhanced CDs?

We recently finished the Enhanced CD for the Conan O'Brien show, as well, which is also going to released through Mercury. It is pre-dominately a digital video based thing—some of the funnier video clips off of Conan's show. The audio portion of the CD is going to be musical acts that have been on the show. We've got some pretty outrageous stuff on that, as well.

Sounds like you're seriously committed to working in the Enhanced CD format.

I'll tell you why. To me it is really intriguing. It is a great way to get into different markets, I think. If it is done properly, it can be more than the sum of its parts.

Many of the companies I talked to when I wrote my first book on CD-ROM technology have vanished.

Case Study: Creating an Enhanced CD

We see Enhanced CDs as a way to bring new life to the medium. The Jerky Boys conceivably will sell a million copies. It is pretty rare for a multimedia to sell like that, unless you're MYST. Those are numbers that you don't get in the software industry. We're committed to further exploring those areas and seeing what we can do to get more and more into that market.

It's true. You get 50,000 sales in the CD-ROM industry and you're looking pretty good.

You're popping the champagne corks.

Is it fair to say that you're going to try to balance your work between Web presentations and optical disc media.

Yeah, we're hoping that it will be a 50-50 type relationship.

And you see the two technologies as being complementary?

I do. Again, this is where Director is at its strength, I think. With the ability to take something you've done and without complete pain and suffering, get that thing live on the Web. It's pretty cool. Your ability to do that as an Internet-aware CD is even more powerful. We're definitely working to develop those types of situations more and more.

The Web in many respects still doesn't deliver on the true multimedia promise that everyone thinks it should. But, if you've got a CD that you can put the media-intensive assets on (the digital video and things of that nature) and use the Web to update it in a fashion that makes sense to do—like text and simple things—it is a pretty powerful combination.

Not too many people seem to be talking advantage of linking from disc to the Web.

It scares a lot of people from a development standpoint. There are a fair amount of things you can't control. You load this thing up and you double-click on something, if their TCP-IP connection is screwy or something like that, then you are presumably not getting assets that you need and then what do you do? If you use the ones that are stored locally, it gets pretty fat pretty fast.

That's true. There are a lot of variables thrown into the mix.

In custom situations, where you're doing specific work for a specific client, you can tailor that interaction a bit more finely, but for the gross consuming public, it's still a tough nut to crack, I think.

Any words of advice for developers who may be considering getting involved in Enhanced CD?

Yeah. To me, the best way to do it is with your Director-type presentation. If you're using Director in the development end I think it is a pretty straight-forward process. The key is to make sure that you're managing assets very closely, depending on the audio portion of things, you might be asked to cram an awful lot into a small data area. You just have to really have a good handle on the techniques involved in doing that. It's not rocket science. It really isn't. It's quite doable. The tough part is just making things fit. It's probably smaller than what you're used to working with.

Most music albums run about 35 or 40 minutes?

No, you're probably talking on average 40 to 45, I would think. Which gives you quite a bit of room to work with, but again where so many people use digital video in an Enhanced CD environment, digital video is big. You take a few minutes of video and Cinepack that stuff down and you're talking about multiple Megabytes of size. It's real easy if you want to work with digital video in an Enhanced CD environment to chew that space up quickly.

The key is knowing your audio budget going into it and planning to use less than what you think.

Are you involved in any digital video projects?

Well, we did the Conan O'Brien. That, to us, was kind of fun because the video clips themselves were pretty hysterical. You've got Don Ho singing Beastie Boys songs. You've got the Zappa Brothers jamming with John Tesh. The kind of stuff that you belly laugh over. We did all the work on that and it came off pretty well.

We like to work on fun projects. We come from an environment that's been tied in with a lot of entertainment activity—the Teenage Mutant Ninja Turtles, comic books. We committed ourselves to doing this multimedia stuff, but really working on having a lot of fun in the process. We've been very fortunate in that the majority of the stuff that we've done has not been boring. We've done some kid's projects. We've done a Tattoo clip-art CD. That's fun stuff. And certainly the Jerky Boys would qualify at the top of that list.

Was the animation done using onion-skinning in Director?

Yeah, a lot of was done in a pure, cel-based fashion. Literally doing overlays from a paper sampling. Scanning that stuff in, computer-coloring them, and then using Director to onion-skin the things out.

Are there any tools out there that do decent cel-based animation without having to resort to paper?

I think if you get at the really high end that there are. And, certainly, you've got you're traditional 3D packages out there which can be used for doing stuff pretty effectively. I guess, for us, we've found that the best way to get the best end result is to start things off by hand, get it into the system and then really use the computer for what it is really good at. And that's manipulating things after the fact.

Case Study: Making a Web-Enabled CD

This case study continues where the last one left off. Right Angle Inc. took advantage of the scheduled re-release of the Jerky Boys CD to update the content and add Web-enabled components. This case study describes how they did it and also offers some perspective on the market trends for multimedia on optical disc versus the Web.

Maybe we can talk about what you have done to the earlier CD to make it Web enabled.

One of the things we did to make it Web enabled: We made the Web portion of it dynamic. We used a Cold Fusion driven back end that then feeds into the Enhanced CD, if the person viewing the content has online capabilities.

Was this project completed fairly recently?

This was done last year [1999]. Basically, what happened is this. The label, which is Mercury records, felt compelled to charge more for the enhanced version of the CD. The industry standard is to say: "we're including this additional CD content for free." We released a limited edition in limited quantities—a situation that appeared to present barriers to proper distribution. I know that the Jerky Boys, as an act, were unhappy about it, so they decided to re-release it. I was insistent about having an opportunity to freshen the project. Some time had expired and the bar had been raised as to what is cool and what is not cool on the Web. By adding the whole Web-aware component, we felt we had a way to make the project more evergreen.

Does the extra content look seamless while you are online playing the CD? Will it actually pull in new content automatically from the Web?

It will pull in new content and the primary attraction to the online component was contest driven. When you get to the main screen you have the opportunity to play what in essence is a needle-drop type of contest. You're delivered a sound clip via the Internet and you try to guess what track and what album it is from. The results are entered into a database-driven Web page on the Jerky Boys site. They were pulling winners out of it weekly or monthly. We wanted to have some incentive built into it, in addition to the ability to deliver changing assets without having to reburn a disc.

Is the new Enhanced CD on the market now?

It is out there. Sales are about a year old at this point, so it is certainly not setting any sales records at the moment. I believe it went gold—which is half-a-million copies.

The Jerky Boy themselves are a real interesting machine. From their discography, they sell somewhere in the vicinity of 5,000 to 10,000 units a week. It really is quite phenomenal how well prank phone calls can pay. We'd like to think that we had a little to do with the sales of the most recent effort.

Can you get into the mechanics of how you pulled all these components together and did the Web-enabled part?

Sure. As always, whenever you are talking Enhanced CDs, the primary consideration from the start of the project is real estate. By that I mean how much room you are going to have to work with on the CD-ROM after the audio portion is complete. The difficulty there is that you and the artist have some latitude in determining how many tracks are going to be included. We usually have to set a limit that might even be a false limit on the high side of the audio, just to ensure that we are not going to run out of room if they add a bonus track.

Case Study: Making a Web-Enabled CD

On the plus side, with the Web enabling, you gain another whole extension to the content.

Sure—if you want to shuffle things in and out via the Web, you can do that. The real trick from a programming standpoint is ensuring that if they do not have Web access, you are still able to apply a consistent look and feel to the project. If they have no component to get to those assets, you have to make sure that they have other assets that will work. Try to design as much flexibility into those hard-coded assets on the CD as you can to make sure that it still works for the long-term. Mechanically, this approach added two or three additional layers of consideration—you have to deal with the ramifications of someone not having Web access.

Did you Director again for the design tool?

We're still using Director. At the time of this last project, Director 7 was relatively new. I think that we stuck with Director 6.5 to ensure that everything would work. Director 7 introduced some fairly radical changes. Everything tested fine, but it added considerable size to the Projector file. Since I wasn't using any features in Director that required version 7, I went with the smaller footprint.

Did you use that Apple Media Tool for doing the Enhanced CD part of it?

Yes, we used AMT for mastering the enhanced portion.

From what I can see, there aren't a lot of tools out there for doing low-level work on Enhanced CDs.

Exactly. I would have to say, unfortunately, that is true. There really isn't anything else.

It's not even supported by Apple any more, is it?

One of the key people involved in developing it for Apple is still trying to support it as a tool for generating the QuAC files. It is still a little bit of voodoo to make these things work correctly.

Is it safe to say that most of the people making Enhanced CDs now are relying on this early tool?

Chapter 11

Yup. To the best of my knowledge, I haven't seen anything else that will do it. Pretty much, everyone I know who is doing that sort of stuff is dusting off the AMT 2.0.

That is kind of incredible. Aren't there still a fair number of Enhanced CDs being released?

Sure. There certainly are.

At the same time, it seems as though there a lot of new companies getting into CD and DVD premastering.

Indeed. So, there is definitely money to be made there. There is a certain Hollywood appeal to it—it's a high-profile kind of project. We do not aggressively pursue those kinds of opportunities. We have found that the ones that come to us are the best ones. In the meantime, we pay the bills by doing a lot of Web-site and straight multimedia work.

Where did ColdFusion come into play? Are you using it to deliver the dynamic data?

Exactly. What we wanted to do is to make sure that the content portion of the Web-enhanced CD was random, but random in a way that would allow the Jerky Boy management to log onto their site and via a username and password, handle the contest-related information without having to know HTML. By using ColdFusion, which is from Allaire Corporation in Cambridge, we were able to set this up. ColdFusion supplies an ODBC-compliant level interface from the Web to the database technology. You can create a database of information that is then used to dynamically generate the Web content.

ColdFusion will meld the graphics and text and tags together to build the Web pages?

You got it. ColdFusion is an interesting animal in and of itself.

It really is. We have developed around a proprietary shopping cart system using it. We have entire Web sites for clients now that are entirely, utterly database driven. So, they can log onto their site and change the entire contents of their Website without having to know a lick of HTML. This is done by using browser-based forms to get data in and out of the database, which in turn drives their Web site.

Doesn't it create an enormous amount of maintenance for whoever has to load the content on a regular basis?

It is a challenge. The customer has to understand that in taking this approach, there is a certain commitment they are making. We've had customers pay us a relatively substantial amount of money to develop a completely database-driven Web site for them and then they realize the full scope of the work that has been placed back on them.

Invariably, once they get in and find they're not completely overwhelmed by the process, they realize they have a tremendous tool at their disposal. By using ColdFusion to drive the contest portion of the site, it was very very simple for the Jerky Boys to get in there and upload an Shockwave audio file, a short clip from one of the takes, and indicate what CD and what track it is from. Also, we were able to capture the incoming information from the site visitors and to write that to a database. Contest winners could then be selected from those who made the correct choices.

Are there other parts of the CD that are Web enabled?

Yes, There are some game components to the CD—the audio for which varies if you have Web access. One of the games on the CD is: Shoot Jack Tors in the ass with a potato gun.

I remember that classic game.

A classic, classic game. If you had Web access, you would get a whole symphony of different reactions that would vary, based on streaming audio, when you shoot Jack. Little things like that that the Jerky Boys felt fine-tuned it.

The difficult part in dealing with people who are as creative as they are, they don't ever want to stop. With a CD—both in an audio format and in an interactive format—you've got to stop at some point. By using the Web to provide new access to those types of things, it let them feel as though they were more in control and helped them feel more creative.

Do you have tips and guidelines for other developers using similar kinds of components?

We had to think long and hard with the crushing deadlines that were presented to us. The key is to think about what you need to do a year after this project is done if you want to shut it off. How do you deal with a bad connection, for instance? And how is that going to impact the content that should be streaming and isn't. We got to the point where we were trying to evaluate the user experience by unplugging their Internet connection halfway through it. Those are the types of tests that absolutely have to be considered when doing this type of Web-enabled project. You have to find a way to control the quality of the experience when that does happen—because, believe us, it is going to...

Given the timeframe we had available, we did a whole lot of testing. For example, here is a system with no Internet connection at all. How is that going to work? How is going to work with an AOL-based connection? At the time, this certainly changed every rule. If you were dialed up through AOL, that added a whole other heap of application-related issues. We had to do a lot of testing on AOL-based systems to ensure that we were judiciously using memory in a way that wouldn't cause too much overhead.

Do you normally use outside test services?

We use a couple of quality checking test services. Not from a focus group standpoint. But just from a virus checking standpoint, and ensuring that directory content is complete and clean. We require that the record company pay for that. Although, this issue is something that we pretty consistently have to argue.

Really?

At some point they become accountants. You can finally get them to understand that we are doing this testing to protect them. It is usually not that expensive, either. Based on the kind of detail that you want and what kind of reporting, it might cost $1000 to have them really hammer the system. They will also test the installation routine on a number of different platforms, which can be difficult to do inhouse.

By the time all the audio tracks were filled up, what did you have left on this one for the multimedia?

We had the entire presentation in cross-platform form within about 30MB. I believe there were 48 minutes of audio on there. There were a couple of additions at the very last minute. It was good that we budgeted the space in the fashion that we did. We definitely would have had more room to work with, but we felt comfortable with that limit. Considering the premastering and indexing and everything else required in the interactive portion, we made sure that there would be no space conflicts at all.

Do you get much feedback about the interactive material?

Tons of fan mail—bearing in mind that we also do the Jerky Boys Web site, as well. People love it. They love the Jerky Boys, anyhow. We were fortunate to have a good act that let us do that kind of outrageous material.

Do you find that your clients are always requesting a Web-enabled component when they have you do a CD-ROM for them?

We are finding that people are shying away from doing a lot of CD-ROM-based work at this point. As the bandwidth starts to increase, the propensity of DSL and cable modems, you have the opportunity to deliver pretty cool Shockwave content more quickly through the Web.

We do a fair amount of sales corporation presentation CD-ROMs that we like to make Web-aware, because it gives them the opportunity to create the project once and then bring in that fresh content when they need it. It becoming more difficult to find a compelling reason to do just the CD-ROM portion of it.

Do you think that situation changes with DVD-ROM becoming more popular?

I think it definitely changes things. Certainly, the opportunities of what you can do on DVD are almost staggering—the amount of data you can cram onto one of these things and the video quality!

Do you have people asking you to do DVD work yet?

Yes and no. They seem to be interested in it and they want to know if it is something that they need to do. More often than not, it doesn't take too much of a question and answer to discover that they can probably do 90% of what they want to do on the Web, period.

A lot of the game developers seem to be flirting with DVD, but there have only been a handful of high-profile releases, so far.

You have something of a standards issue from the onset—not a huge one, but from a developer's standpoint, it is a tremendous risk if you put a lot of effort into a format that could be obviated. The mastering equipment is more expensive. It is a whole new level of investment to get into that end of it. I think there are an awful lot of people who want to test the waters with a big toe, but that is about it.

We have done well as a development media house dealing with CD-R based mastering issues. Even though there are many inexpensive CD burners and software, it is still hard to avoid making a lot of coasters if you don't have some kind of understanding about how these things really work. But, it is getting increasingly more difficult to get people to bite on the CD portion of things.

It seems like the games market is still pretty healthy.

The game thing to me is always real interesting. It is arguably one of the most delivery intensive things that you can do. It is just going to be an awfully long time before there is enough bandwidth out there to deliver that kind of action via the Web. I think CD-ROMs and DVD-ROMs will be kind of immune from that kind of debate. With some of the compression improvements for video, it is an intriguing time to do video designed for the Web. Cable modems are starting to come of age, even out in the boondocks (where I live). I still can't get that kind of service, but my business partner who lives four miles down the road has it. The speed at which he can surf the Web with that kind of setup is ridiculous—he is pulling in 300K a second on his cable modem.

Any new markets springing up for multimedia?

We have found more Web-based opportunities, but we're also doing a lot of interactive, online content that is also database-driven. Interactive components get fed information that is being generated dynamically from the Web site. It is gratifying, because it gives you a level of control that you always want to have when doing this type of work. You can really make it work down to the smallest detail. That kind of content is becoming more and more popular.

Mechanics of Building an Enhanced CD

Although many premastering applications provide a means for burning an Enhanced CD, most of them provide a minimal approach to the process. Many developers are still relying on a copy of a tool originally made by Apple and then discontinued, the Apple Interactive Media Toolkit (AIMT). This stalwart product enjoys continued life through support from one of the original developers.

Mechanics of Building an Enhanced CD

Disc Makers, who frequently prepare Enhanced CDs for clients, list the following steps as fundamental to the creation process.

Note: If the audio material is supplied on a DAT, the first step is create an audio CD that includes an index and ISRC codes. Any additional post-production work, such as equalization and normalization, should be applied prior to burning the audio CD.

1. Using the CD-Copy program from Astarte, extract the tracks from the audio CD as a single image file (rather than individual tracks). Ensure that the audio settings for CD-Copy are set for capturing ISRC codes, writing a CD-DA file, and reading index values.

2. Using Adaptec's JAM program, select the CD-DA image file created in the first step. Check to ensure that there are no missing ISRC codes. Also check the index values and the size of the spaces between individual songs. Using the multi-session record option, record the session to a blank CD-R, but do not close the disc.

3. Place the CD-R on which you have created the session into the system's CD-ROM drive. Open the Apple Interactive Media Toolkit program and choose the **New File** command from the menu. AIMT identifies the track data on the disc. Check to ensure that all the fields that appear in red contain data and select **Make QuAC File** from the menu. A .CDQ file is then created.

4. Remove the CD from the CD-ROM drive. Using Adaptec's Toast, select **Enhanced CD** from the format menu. From the utilities menu, select **Create Temporary Partition**. Based on the storage space required for the data portion of the CD, size and create the partition.

5. Drag the .CDQ file created in Step 3 to the temporary partition. Select the Data option from the Toast window and select the temporary partition from the volumes available. When prompted "Use QuAC file?", answer "Yes."

6. Open the temporary partition and identify the folder called "CDEX-TRA" and the QuAC file. Copy files that are intended for shared use and all Macintosh-only files into the temporary partition.

7. Open the CDEXTRA folder and copy all files intended for Windows-only use into the folder. If you have created an AUTORUN.INF file for autostarting the CD under Windows, overwrite the file that appears in this folder.

8. Create an alias of the shared files residing in the temporary partition. Drag the alias files or the equivalent folder representing all the aliases into the CDEXTRA folder. Rename each of the files or folders that were copied to remove the alias suffix.

9. Organize the contents of the temporary partition for viewing by the Macintosh user. The structure you create will be maintained when the CD-ROM is viewed.

Close the windows and click **Data** in the Toast window. Select the temporary partition and ensure that the **Optimize for Speed** option is active. Choose the **Write to Disc** option to start recording.

With this organization, the files contained in the temporary partition will be accessible to the Macintosh users. Windows users will see the CDEXTRA folder at the top level and additional files and folders contained in the CDEXTRA folder.

Expert's View: Working with Surround Sound Audio

One of the characteristics of DVD-Video and the DVD-Audio format is the use of multichannel sound, characterized by the mysterious sounding 5.1 descriptor. To clear up some of the confusion over surround-sound and audio processing for DVD, I talked with author and sound engineer Rudy Trubitt (*www.trubitt.com*).

What's the difference in delivery systems between film and TV sound versus music-only playback systems?

One big difference between sound for pictures and for Hi-Fi has been the number of speaker channels. Since its introduction stereo has become the music delivery method of choice—although we did flirt with Quadraphonic sound back in the 70's. On the other hand, Film has flirted with more than two-speaker sound for a long time. The original *Fantasia* was designed to be played back in a theater set up with nearly 30 different speakers and channels. But, there were only a handful of theaters that could play the thing the way Disney intended. More recently, around the time of the first *Star Wars* film, it became common to release films in multi-channel sound. One reason for the additional speakers is that if someone is sitting off to one side of the theater or another, it's important that the apparent source of the dialog is anchored to the location of the actor, as seen from the listener's perspective. If you're sitting off to the right, the actor is somewhere off to your left, yet if you were just listening to regular stereo, you would hear the sound coming from directly in front of you. This can cause a feeling of disconnect, as the visual location of the actor doesn't match the audible source of their dialog. So, a center channel speaker, in addition to the left and right, means you can hear the actor's voice coming from a point matching their on-screen location, more or less.

Then, to get effects that envelope the audience, they started adding loudspeakers along the sides and the rear of the theater. That gives you the sense of environmental sounds outside the picture, bringing you in more to the environment of the film. There are several different ways to deliver this multi-channel sound. The current scheme for this is discrete multi-channel sound, such as Dolby Digital, but let's not get into that just yet. Until recently, here's how surround sound usually worked in cinemas and home theater systems: Our audio program would be carried as a plain-old stereo signal and that was processed to get four semi-independent channels. This is the way that most earlier cinema systems worked, and this is also true for the bulk of home theater systems out there.

You take a regular stereo signal and by processing it in the analog domain (or digital domain), you can extract information that is common to both left and right and turn that into the signal that will be fed to the center channel speaker. Then, you can look at information that is different between the left and right, and throw that into the surrounds. Now what we have done is take a two-channel audio signal, with left and right, and because some of that information was correlated, essentially panned to the center, and some was decorrelated, which is often the case with ambient reverberant sounds, we've turned it into four actual separate signals. You've got left, center, right, and a single signal that goes to the surrounds, even though that might be going to more than one actual loudspeaker.

The term "matrixing" is used to describe this process—looking for the things that are common and looking for the things that are different and then extracting those elements and making them the center and the surround signals. In the home today you can find that scheme implemented on equipment that is labeled with the Dolby Pro Logic logo.

Is that intended only for systems that have four speakers?

Tapes or broadcasts encoded with Dolby Surround are compatible with two, four or five-speaker systems. If the viewer has a plain old mono or stereo television, they don't have to do anything, just sit back and listen. If there's a home theater receiver involved, a few options come up. If you have only two speakers, turn the Pro Logic circuitry on your receiver "off," and you'll hear a regular stereo signal through your better-sounding left and right Hi-fi speakers. If you have four speakers, (left, right and two rear surrounds), turn Dolby Pro Logic "on," and select the "Phantom" center channel mode. Finally, if you have five speakers (adding a center) turn off the "phantom" channel and you'll get separate left-center-right-surround.

This is totally different from Surround Sound?

No, in fact it is, indeed, Surround Sound. You've got something coming out into the rear surround speakers to create an enveloping sense of space, and a center channel to help focus the viewer's attention on the on-screen dialog. Generally, dialog in film and TV mixes are panned to the center. As a result, when you play back through Dolby Pro Logic, the voices will end up coming out of the center-channel speaker. And, if music has been panned with a lot of elements to the left or the right, then you will hear those coming out of the other two speakers. It really helps you stay centered on the dialog, which is often the most important thing to keep the story moving along.

Again, an important benefit of Dolby Surround/Pro Logic is that it is a regular stereo signal, so that it can be carried by VHS movie rentals, and it can be broadcast. What you will often see on the media will be a little logo that says, Dolby Surround. Sometimes on network television shows you will see that little Dolby Surround "where available" logo. That is telling you is that this stereo signal has been mixed to provide surround sound if listened to over a system that is capable of decoding that information. It's a little confusing, but "Dolby Pro Logic" and "Dolby Surround" are two halves of the same process: Dolby Pro Logic identifies equipment capable of decoding media (VHS tapes) or broadcast or cable signals encoded with "Dolby Surround."

Are there other techniques that can replicate a multiple-loudspeaker experience from just two speakers?

There are systems that attempt to do that. For many years, there has been a term (that I consider an unfortunate one) called 3D sound. The theory is that you can take a pair of loudspeakers and by processing the sound a certain way, create the illusion of a completely enveloping three-dimensional sound field like you would hear out in the world, as opposed to listening to point-source loudspeakers in your living room.

Is there any easy way to explain how that works?

There is an easy way to explain how it works—it doesn't. I'm being a little harsh, but I've never heard sounds coming from behind me while listening to a pair of speakers that were physically in front of me. Now, there is a simple and highly effective technique called *binaural audio*. That is where you take a mannequin head—an anatomically correct dummy head—and stick microphones in its ear canals. Then, one listens to the resulting recordings over headphones. That works. You get extremely natural, lifelike-sounding audio.

Binaural sound is simple and cheap and it has been used for years. The only problem is that a lot of people don't listen over headphones. You want something that maintains some of that image over loudspeakers. There are schemes that attempt to recreate that multi-speaker experience, but from just two speakers. The main problem is that they are coming out of two loudspeakers and those sounds are bouncing around the room in the same way sounds bounce around the room when they come from loudspeakers (as opposed to actually emanating from the positions that you are trying to pretend that they are coming from). Those systems tend to work best only when you are sitting exactly in the "sweet spot," directly centered between the two loudspeakers. As soon as you begin moving off to one side or another, the illusion begins to collapse.

Now, I don't mean to imply that these "3D" audio systems do nothing. What I can hear quite readily from the better examples of this technology is that particular sounds emanate from a point that is beyond the arc that is extending from your left speaker to your right speaker. So, you are hearing something, for example, that sounds to be further "left" than the physical position of the left loudspeaker. Digital signal processing is getting better and faster and cheaper; I haven't heard the most recent incarnations of these techniques, but I have yet to hear anything that was as convincing to me as actually having a loudspeaker off to the side or behind me. I have never heard something like it was behind me from a pair of speakers in front of me.

How is a sense of space and position created in game audio?

Many modern games create quite believable spatial orientation. For example, if you're running down a road and you hear crickets chirping off to your right and you turn your character around 180 degrees, the crickets will be chirping on the left. A lot of that is done just by means of clever panning. With a lot of these games, we will use a lot of these 3D algorithms to try to push sound out beyond the range of the loudspeakers. And, you've got the visual cue of the game and your mental idea of where you are and what you are looking at—how you are oriented—to help reinforce the cues that they are providing soundwise.

What is cool (and difficult) about the game environment is that they are computing this processing in real time based on user input, as opposed to working with a film soundtrack where you always know that when the tornado hits the grain silo, the roof of the grain silo is going to come bouncing across the road behind our heroes truck and fly off the screen to the left (we just saw *Twister* again last night). I have a Playstation in the living room and I enjoy games that use the same kinds of effects.

How does surround sound technology fit into the DVD realm?

Unlike "matrixed" surround sound, the DVD scheme actually stores the sound channels as discrete entities—instead of trying to mush everything into stereo and then extract it later. When we talk about 5.1 sound, we are talking about an independent left, center, right, rear-left, rear-right (there's your 5) and the point-1 is the subwoofer, also called low-frequency effects or LFE channel. With it, we can just divert the low-end rumble of Godzilla's stomping feet into the subwoofer while the other speakers are playing the sounds of the whizzing jets and people screaming and running.

The advantage of 5.1 sound over the stereo matrixed surround is several fold. The first is that the rear channels on a Pro Logic system are of limited bandwidth. They are not getting a full-range signal. They are only good from 100Hz to 6 or 8KHz, comparable to AM radio. It turns out that they work surprisingly well even though they are not full-on Hi-fi. It really not optimal for things like the sound of wind, grass, and crickets- you're really not going to get the bright, airy, realistic rustling leaves sound out of something that is only good up to 8 or 10KHz.

The other problem is that because you are matrixing, the image tends to jump around a little bit. When someone starts talking the center channel, information that is playing out of the surrounds might get ducked—the level basically drops down when a person starts talking and when they stop talking, it pops back up again. This is especially obvious on broadcast television, where lots of dynamic range compression is applied to the audio, but you can hear it on VHS movies, too. Also, you can't have separate sounds in the rear left versus the rear right, because the same signal is going to both of those speakers. With a discrete 5.1 soundtrack, all channels have the same frequency response and the image stability is vastly improved, because you are not matrixing. Now you're not worrying about a loud sound in the center channel pulling in everything else inward, spatially speaking. So, it is a much more stable multi-channel sound image that is created.

If you are watching a DVD movie of *Bridges of Madison County* and you've got Clint Eastwood and Meryl Streep out sitting in some field somewhere talking, if you are watching it on a TV broadcast, when they speak, subtly but perceptively you will notice that the chirping crickets in the background tend to get quieter during the instance while they are actually speaking-the cricket level is riding up and down. With a DVD version of the same film, the spatial position of the crickets is going to be rock solid. They aren't going to be affected by what is happening in the center channel.

Sounds like to accomplish this, you need a whole new set of audio tools.

That is basically true. You would have to follow the multi-channel film sound model, rather than the stereo, or two-channel music production model. Yes, you do need some different tools and different aesthetic judgements to make. We'll talk a little bit about that but before we do, let me finish the thought on Dolby Digital.

We've mixed our program to five discrete channels plus we've got this low-frequency effects channel. We store them on some medium and now we want to stick them on the DVD. While, the DVD doesn't have enough bandwidth to carry if we use the compact disc data rate, 16-bit, 44.1KHz, that requires 10MB per minute for stereo. Now, let's make life easy and say we are talking about 6 channels, even though that is not quite true, that would mean three times that, now we have gone to 30MB per minute—you just can't get the data off fast enough to do video and all that sound. So, you need a way to compress the data (we're talking about digital data compression as opposed to audio signal processing compression) to reduce the data rate of all this audio. That is what Dolby Digital does. It is a data compression codec in the same way that RealAudio is a codec for compressing mono or stereo sound, so that you can download it over a modem. Dolby Digital makes it possible to get all of those channels of discrete sound onto a DVD and then off again within the constraints of how much bandwidth the DVD player has—how much data can you get off a disc at one time.

Actually, Dolby Digital was originally designed to get digital audio onto a piece of movie film for cinema distribution, but the same resource-limitation issues apply to DVD as well. Note that there are other competing codecs, for example DTS-Digital Theater Sound and a Sony system called SDDS (Dynamic Digital Sound). When you go to movie theaters you may see a Dolby Digital trailer at the beginning or maybe it will say DTS, maybe it will say SDDS. But in the home theater world, it is pretty much Dolby Digital.

Is Dolby Digital the successor to Dolby AC3?

They are one and the same. Originally, Dolby began calling this process by their internal code name—AC3, short for "audio codec number 3." Later they changed their mind and said, "Wait a minute. We don't want you to use AC3 as a term to reference this process. Now we're calling it Dolby Digital." But by then the cat was out of the bag—you will see equipment that has a connector on the back that will say AC3 and you will hear people talk about AC3. I avoid saying "AC3" out of deference to Dolby's marketing department, but AC3 and Dolby Digital are one and the same.

Are there differences between the approach you are describing for DVD-Video and the still evolving DVD-Audio format?

The production process, up to the encoding will be the same. But there will be different audio codecs that offer higher sound quality than Dolby Digital for audio-only DVDs. Unfortunately, DVD-Audio is still on hold awaiting the release of players. A well-publicized hack of the DVD encryption scheme has introduced a delay in the introduction of the players. Maybe they want to revisit the encryption spec before they start shipping products. It's very unfortunate, because so many DVD players have been sold, that a lot of people will be confused or annoyed when new audio-only DVDs come out that won't play on the DVD players they already own.

Does any of this relate to game development?

There are systems that are using Dolby Digital for sound playback. The problem is that it is very computationally intensive to be converting those discrete channels of audio to the Dolby Digital signal. There aren't any games that I am aware of that in real time, as you turn your head in the game, pan the sound around in a Dolby Digital signal. They are using it more for pre-rendered ambiences (that doesn't change based on the perspective of the player in the game). But then those "pre-rendered" sounds have to be mixed with the real-time game audio.

It's very messy right now, but there is talk about soon being able to create the Dolby Digital data stream "on-the-fly," which is really, really what we need to have happen in order for games and DVD-video to share the same playback system in the living room. Microsoft is talking about on-the-fly Dolby Digital for their forthcoming X-Box game console, but that's not due until fall 2001. Maybe someone else will come up with an implementation before that. In the mean time, there are other specifications, like Creative Labs Environmental Audio, which are better positioned for real-time 3D positioning of an object that is interactively moving around. But this requires its own set of four speakers, it won't be able to use the 5.1 speaker system you're using for DVD-Video, unless you're prepared to re-wire the living room back and forth.

Is it extremely expensive or difficult to set up a home studio to be able to produce multichannel sound?

It is definitely doable. I'm working on a short film project now where I will be implementing at least the Pro Logic level of surround. What you need in your monitoring environment for starters: you need a 5.1 loudspeaker set up where you are mixing. Instead of mixing to stereo, you are mixing to five or six channels. Let's say that you have an 8-track multi-track tape recorder—like an Alesis ADAT or a Tascam DA-88. Your mixing console would have to be set up to mix to multiple outputs instead of just two.

You could do this with any 8-bus mixing console. It creates a lot of problems in figuring out how to pan across bus pair. If you are looking for a product that simplifies this, there is a tool called MTrax made by Mini-Tonka software. They have a system that is very nicely set up for doing multi-channel panning. Steinberg's Nuendo audio editing program offers similar features, and there are others coming on-line every month.

What is the next step in the process?

Then you need to actually encode the audio to do the data-compression step we talked about earlier. You have your multiple channel audio stored either to a multitrack tape recorder or you have mixed to a bunch of disc files. So, you have six files on your hard drive that represent the left, the center, the right, the rear-left, the rear-right, and the subwoofer channel.

You need to encode that as if you were going to take a track from a CD and turn it into an MP3 file.

You need to perform this data compression operation-that is something that comes from Dolby and is licensed by Dolby. You can either buy it as a hardware box or a software version. For Windows, the Sonic Foundry folks are licensees of Dolby. You can get a Dolby Digital encoder software package from them for somewhere around $1000. You give it your six sound files and it creates a Dolby Digital file that you can then use as part of your DVD production process. Astarte has a version of the Dolby Digital encoder that runs on the Mac.

The single Dolby Digital file contains all six tracks?

Correct.

Summary

It's hard to do justice to the importance of digital audio recording to optical discs in the span of one chapter. This whole area has been one of the reasons that CD-R technology took hold in the imaginations of music lovers around the world. Inexpensive CD burners turn a desktop computer into a recording machine capable of delivering audio discs with the pristine finish of any commercially pressed disc.

The advent of 5.1 audio delivery, with its close correlation to DVD-Video and the emerging DVD-Audio format, will catapult digital audio sound quality into new levels of richness and purity. The best part of all of this is that you can work with the processes involved on your desktop computer and copy finished projects to optical disc.

12

Interactive Music Design

One of the most effective ways to enhance the contents of a CD-ROM or DVD-ROM title is through music. With a variety of music presentation formats, ranging from the compact shorthand of MIDI to Dolby 5.1 Surround Sound, developers must make critical choices when budgeting space for music, as well as finding the best means to integrate the music within a given title.

This entire chapter is devoted to a single Expert's View. George Sanger, aka The Fat Man, has made a career out of the skillful application of music to CD-ROM games and other interactive media titles. From his base of operations in Austin, Texas, George brings together other talented musicians and composers—the heart and soul of Team Fat—and crafts memorable musical themes that frequently accompany the best-selling CD-ROMs in the gaming market. His yearly Project BBQ attracts the key figures in interactive music—both from the technology side of the industry and the musician's side. If you're contemplating adding music to a project that you're developing, the following interview should give you the appropriate perspective to proceed in the right direction.

Expert's View: George Sanger Talks about Interactive Music

Do you work exclusively in MIDI or do you also do full audio tracks?

Absolutely—full audio. My lips hurt—I was playing trumpet and french horn all last night. It's killing me. Yeah, any blessed thing that will make noise and will sound a little less like a piece of plastic and a little bit more like a heart beating is a good thing. We at Team Fat will leap all over that like boots on a rattlesnake.

But you also do strictly MIDI soundtracks as well, don't you?

Oh, sure. I did the first General MIDI sound track to a game ever.

How do you feel about some of the expanded formats, such as MIDI-XG?

General MIDI got off to a pretty shaky start. I started doing General MIDI when there was only one device that supported it—the Roland Sound Canvas. I made a big deal about it in the manual for *The Seventh Guest*. *The Seventh Guest* went out and sold a million and a half copies when the previous top-selling multimedia product had only sold 20,000 copies. So, all of a sudden in swell foop [chuckles theatrically] —God, I love that—multimedia stopped being the "0-billion dollar industry"; it started being an actual viable business. And, General MIDI became The Thing. So, these General MIDI instruments started coming out. Basically, they did not play back the music as predicted. There was no spec—there was nothing in the General MIDI spec to say how loud one instrument sound should be compared to another. So, instrument sounds would just get lost or would come blaring out suddenly. It was a real mess.

So, I started a campaign and a service to fix that. I never really made any money on it, but I think that between us (FAT Labs) and the eventual acceptance of what we were doing by the MIDI Manufacturers Association, we got it where General MIDI now is pretty much a workable standard.

That required a lot of effort. And, it paid back well in terms of saving people from having to listen to terribly distorted music.

I thought I saw your name attached to approval of a sound board that had XG capabilities.

If there was a quote from me—I love those Yamaha guys and I love those Yamaha instruments—it was probably something like this: The product was balanced very well for General MIDI and that when you play back General MIDI on that instrument or you write it on that instrument, you're working within a very compatible system.

I won't associate my approval with MIDI-XG, although I am really enjoying my XG stuff, but I don't believe that we've got a set of standards now

Expert's View: George Sanger Talks about Interactive Music

that really works. By expanding that out, you're only cluttering the horizon.

I think I've also seen quotes from you on finding ways to increase the expressiveness of MIDI music, which of course is one thing that XG does, but it doesn't help if no one out there can hear the extra nuances that you put into the music.

Absolutely. The battle for computer game music over the last ten years has been all because of a misunderstanding—that people believe that features have a place in music for computers, rather than compatibility.

I will no longer consider endorsing anything that puts features over compatibility. It's all about standards and sticking to them. And, that is one thing that Red Book audio has going for it. With CD-ROMs as opposed to MIDI over the Internet have the advantage that they're more than likely to play. More likely to work. All of our MIDI stuff, I always felt like, "Well, how would the Beatles feel if there was a sticker on their albums that said CAUTION: MAY NOT PLAY BACK CORRECTLY?

You mentioned that you are doing pure audio for the CyberStrike title, is that right?

CyberStrike 2. That's right.

Do you worry about integrating the audio with the game when you're doing the music?

Currently, the biggest consideration in doing the music for CyberStrike is disc space. I believe that we've got the integration and the flow and the mechanics of it all worked out, or certainly we haven't seen anything that was going to stop us from doing it well. Really, the biggest issue right now is that there is only room for about 15 minutes music on each of the two discs. That's got to be distributed in a way that alleviates boredom rather than accentuates it. That's the challenge in that case.

We are opting to stick with 15 minutes of Red Book rather than 30 minutes of 22Kbps music. That's the way that this project is going. There is always that option of reading a WAV file in—streaming a WAV file. That's always a possibility if the programmers are feeling up to it.

A lot times I have to sound out the mood of the programmers and see if they're in the mood to be harassed by some musician. What they feel like they want to go out and kick ass on and what they don't. A producer like Ron Gilbert—and there aren't many producers like Ron Gilbert who run

Chapter 12

Humongous and Cave Dog (they did all the Putt Putt games and they did Total Annihilation)—will go out and program stuff from the ground up just to make something that plays well. He'll program a sound system regardless of inconvenience. He'll make a real commitment to an audio system. They'll do that at Origin Systems, too, and they'll do that at EA from time to time. And at LucasArts. But they're aren't that many other companies you will take a programmer and say, "OK, dude, do it. Make us something that will get the best possible audio out of this."

A lot of people will ask about the capabilities of John Miles's system, which is called (I think) the Miles Sound System or AIL, audio integrative library, but that is a standard system that most game developers use. Or, they'll ask about the capabilities of Direct Music, which is a Microsoft development. They're kind of curious and they may investigate what they can do easily. But, that is the kind of thing I deal with. I look at how into that people want to get. I have to be a little bit flexible about that.

So, you've got to adapt your approach to the music to how the programmers are going to treat the audio?

Yeah, I have to adapt how much music goes in to the level of importance or resource allocation that the programmers and developers want to put into it.

Frankly, as much respect as music gets, it still is not on most people's radar. A person who shoots a movie says, "Well, it's going to be 30% of the movie experience. I'll give it (I don't know what their formula is) 4% of the budget. 0.5% of the budget (or whatever it is).

And then they take that formula—it's been done a million times—and they go to the union guys who know how to do it right and they make their movie.

Someone who is making an interactive product has to do everything that the movie guy has to do. Picture a pie chart of this. You take the whole pie chart of making a movie, with it's little wedge which is music, and squash that into a wedge—it's maybe a quarter of the pie chart of the game developer. Think of how many cutting edges a game developer has to be at. I mean, even if you've ever been at any cutting edge—which you have—you know that it's exhausting to be even at just one.

They have to be at 4 or 5 before they even get to the music cutting edge. They have to take care of programmers. They have to market innova-

Expert's View: George Sanger Talks about Interactive Music

tively. They have to come up with a whole way of telling a story interactively that usually has to be innovative. It's like they have to create a whole new genre every game. Because a game has pressure on it to be a different work of art from the last game. So, music will get squished off of their resource allocation radar. However, their vision of the game usually entails some expectation that the music should feel as high quality as movie music for 40 hours of game play.

There is this sense you have to have a "just do the damn music" budget stretching out over 40 hours of movie quality experience. Which is really the challenge of our business. We're trying to make that impossible thing happen. Some people have done it, I think. It's an amazing thing to think about.

Is there anyone else out there similar to you specializing in interactive music?

Yeah. There is Tommy Tallerico, to whom I refer to as my "rich brother." He is quite the showman. He is a fun fellow and seems to be doing pretty well with the business. He's got a show and what not. Drives himself a Ferrari.

Tommy Tallerico—he's a character and he's a good man. He's been doing this for a long time. There are a lot of other people who have dabbled in this who carry themselves as if they're waiting for another film or TV gig to come along.

There are folks who have been in this part of this business and have aligned themselves more with one company. So there are some cats that I really respect. The team at LucasArts. The team at Brøderbund. The team at WestWood. Kessmai. There are some real cats out there. And, I know I'm missing a lot of them.

Team Fat is the original out-of-house source for music for games.

Are you at the point now that game companies will regularly seeks you out and you don't have to scramble for jobs?

It is absolutely a struggle. And that's without batting an eye, still saying, "I'm the biggest name in music for this." And that's without batting an eye saying that my prices are reasonable.

I've known a lot of exceptional musicians with plenty of talent who are doing things other than music to get by.

With reputations, too. It's a juicy irony that I can be the top person in my career and be struggling that way. I believe it is rooted to the nature of the business and I am far from complaining. In fact, I'm going to take a trip in three days to Seattle and Portland to see about buying a Rolls. So, I can't complain. It's a used Rolls. It's a cheap Rolls, but in so doing I am forfeiting the right to bitch at all about my beautiful, adventurous life.

Merely being the top guy at your gig, or even having a reputation, I mean—what if there are 8000 guys who are actually better than me. You can't judge music that way. It's like judging talking.

It's a gift from God. It allows us to express something that is in us. That is in everybody. We can express it to each other in our own individual way, and people recognize what the other person is saying. That's like this little miracle. But, to expect to make money off of that is just a wild expectation. For some reason, I have managed to do it. And not only that, I've managed to succeed for many years and I've also managed to completely love my life and my business all that time.

The fact you can make a living playing music and you don't have to go to an office or a factory is something.

Yeah, I'm sort of in an office, but I can burn incense if I want to and I don't have to wear shoes. I can even hang bullhorns off my equipment rack.

You know what I like these days. We've done what I call "music-only CDs." We laughingly refer to it as our innovation in interactive. The music-only CD is 100% compatible with all the equipment out there. When you press Play, it always plays.

So, you have three audio CDs now that are available through your Web site?

That's right. They were all done at Disc Makers. According to my record company (as in "I had a meeting with my record company today, but he got a flat tire"), it was a very positive experience. There was like one little flaw in each—just in there to let the evil spirits out—but, on the whole, it was extremely smooth. We were really pleased.

Can you tell me about the CDs? Are there excerpts from your game releases, or are those tied up in copyrights?

Expert's View: George Sanger Talks about Interactive Music

A lot of them are. It required too many phone calls to get permission to use the titles of games on some of these. The first album—*SURF.COM*—is actually the sound track from a game that stiffed pretty badly. It is a 3DO game, the follow-up to the game Twisted. It's a TV game show that takes over a communist country called Bezernia. Team Fat is the American cowboy surf group that plays as the pit band for this game show. In which, you can win valuable cash and prizes and the winner gets the trip of his dreams to America. The sticker on the *SURF.COM* box is: "Texas surf music for a communist game show." And, that's exactly what it is.

But there's no real reference to the game. The game is called *Zhadnost, the People's Party* and I understand that it is quite a wonderful game. Both people who bought it really enjoyed it.

I'll have to look for that.

You can dig it up. You can rent it. You can probably buy it for about two bucks. You'll need to buy a 3DO player.

The album plays like a great surf party blast. It's very surrealistic. The games producer gave us a license to be surreal. And we made a lot of awful vacuum-tube like noises. It's very enjoyable. One of our goals in that was to completely mar the glossy finish of computer products. As John Perry Barlow says, the thing that is missing from a computer experience is the smell of cow shit. And, we put the smell of cow shit back in there, by god.

That's the first album. The second album is called *Seven Eleven* and it is the compiled sound tracks from *The Seventh Guest* and *The Eleventh Hour.*

So you did get releases for those?

Yup, everything is on the up and up. It was just so much hassle to try to even... this was the same thing with *Zhadnost, the People's Party*. We got a better more artistic product this way, but it just required 20 phone calls to get anybody to call back. Because they're really not thinking about music. So, we got all the permission to use the music. The one thing that was left to negotiate was to allow us to use the name of the game and they just couldn't get it together. We worked on it for three years. People were too busy getting hired and fired and wandering around through the building.

Chapter 12

And it just really wasn't a priority. They love me, they respect me. But, they're just not thinking in those terms. And the same thing with *The Seventh Guest* and *The Eleventh Hour*. We got all that negotiated and I just couldn't get clear with the legal department to use the names of the games.

But it is the sound track and it's a beautiful piece, too. It is all the best out of it. It flows real nicely. The first half is all the tunes with lyrics. There are a couple of funny tunes in the beginning, including, Mr. Death. And then it kind of winds down through a couple of the go-behind cut scenes and then it rambles on at this slow pace and plays all the important game play songs—really nice recordings of them. It hits pretty hard with a bunch of attention-getting stuff and then allows itself to drop back to the background. It's real good.

The third one is all of our funny music. It's called *Flabby Rode*. That's got all the music where we would finish writing all the stuff that people would ask us to do for a game and then we'd say, "OK, we fulfilled the contract, right?" And they would say, "Right." And we'd say, "OK, now we're going to get started. Just let us do something that you're not expecting."

Cause you know, they're saying, "Well, it's a war game so we want a lot of military drums and French horns and horn calls." We kind of established that as the right genre when we did the *Wing Commander* music. *Wing Commander* 1 and 2 kind of laid down the standard for that. And that's, of course, just us aping John Williams, but it got people aping us aping John Williams. And I don't know who John Williams is aping. It's somebody.

It's just a daisy chain of apes. There's your pull quote.

We got a chance to agent a theme song for *King of the Hill*. Before that TV show came out. My called me up and said, "Hey, you've got a week. Can you come up with a song?" I said, "A week! I'll have it for you by Monday!" We did five songs in about three days. And those all made it onto *Flabby Rode*. We made the short list for *King of the Hill*. But, at the last minute something else came up and you know how it is in show biz. But, those all made it on to the album.

There is one instrumental that I wrote for a game called *Savage Empire*, which is an Origin game, an Ultima game. And it is the love song that I had done in MIDI for the woman Aiella, the cave girl. But, when I wrote it, I wrote lyrics to it to help me write the melody.

Expert's View: George Sanger Talks about Interactive Music

So, on *Flabby Rode* you'll hear the lyrics for the first time ever. And it goes like this [sings]:

Aiella, soft as a flower's underwear.
Loud as a clap of thunderware.
The eagle's dare and grizzly bear. I hear you calling,
Toumbia...

There it is.

Is it the Aiella from Clan of the Cave Bear?

Oh, did someone steal the Aiella from Clan of the Cave Bear? So typical. Let's say yes... or no... It's one of those.

Do you sing on the album?

Oh, yes.

Do other band members do vocals as well?

Oh, yes. Yes. I would play it for you over the phone if it would do your interview any good.

Are you hoping the audio CD stuff becomes self-sustaining? How are you marketing it?

Music and Computers was handling all three albums with full page ads.

They just went out of business, didn't they?

Yeah. It only hurts when it hurts. Kind of like a vasectomy. I'm hoping to set up at some point a similar situation with a PC magazine or a game fan magazine. I have not been pursuing that very actively because I'm busy [laughs]. I've got shit to do, man. Like, I've got a daily game of *Star Craft* with Kevin down the hall.

Is it true that you get bored with the games pretty quickly?

Oh, you must have read that thing in *Wired?* That was quite off track.

So the article was off the mark?

Chapter 12

I think the best way to put it. is: Thomas Dolby came by and visited for a couple of days (boy, was he great company!) and he looked at me and said, "You know, I read that article in *Wired*. They weren't very kind, were they?" And that's all you can really say.

I just feel sorry for that guy (the writer). He lived with us for three days, and he just did not see the love. There was love all around him He did not see the love. So, he was thinking that people were glaring at him.

Can you imagine being in one of the most wonderful.... I mean, these guys were friends. We opened up to him. It was all there. He just didn't get it. Which means, boy, what must his life be like? His whole thing was about me eating these greasy hamburgers and stuff, which is cool, but when he first walked in, the very first thing that happened was: my daughter, who was about three at the time, started doing yoga poses. "Tree pose. Daddy, this is tree pose." It was the first time she had done that. He just stood there watching her do that. It just kind of went in one ear and out the other.

That really hurt to see a misconception like that. We love playing games. We love playing games. What he picked up on and took as truth: The Fat Man knows the games are stupid. He just picked up on that same, slightly cynical attitude that you have been exposed to: that basically people just don't get it or this is just another clone game. People aren't using their imagination. But we love games. And I've gotten into this because I love playing games. I'd rather do that than push apples any day. It's games! It's playing!

Other things that I've heard that I'm wondering about now if they're true is that you don't have any desire to do sound tracks or background music for TV shows or movies.

I'd love to do that. And we'd do a hell of a job. However, I'm 100% focusing my marketing on games for the simple reason that it's what we do. As healthy as this business is, we'll be that healthy. We're the Kleenex of music for games. We own music for games in people's minds. I can never be the biggest name in movies, because that's already gotten.

So, marketing wise, this gives us a focus and an emphasis and I'm very comfortable with it. We're squeaking by, sometimes better than others, but it's what we do. It's a firm decision on my part to emphasize that.

I'll tell you what. A movie gig would hit the spot every now and then.

Expert's View: George Sanger Talks about Interactive Music

Sounds like it would be an interesting change of pace and a chance to stretch your musical muscles.

We won't get any more new directions than we've found. The business that we've been in, our musical emphasis has been stretched so far in so many different directions on such small budgets for so many hours. And still maintaining this excellent high number of audience—you know, people hearing us. Maintaining a high level of quality throughout each project. Fast deadlines. Low budget. We're lean and mean and bad to the bone.

If you put us on a film project, you're gonna get... I mean, every game that we've done we've tried to come at...like, OK, let's make this feel so weird that it's like a Kubrick thing. We're going to stick our *Blue Danube* over the spaceship.

And for the last ten years, we've gotten, "Huh. Huh, what do you mean, just do the music?" And we've said, "No, no, no. This will be cool. This will be cool. Please let us do it. We'll do it. It will be great." And everytime we've been allowed to do that, we've completely kicked ass. Because of that kind of thing, we've been held responsible for [laughs] five or six revolutions in our business.

You've obviously made an impact on the MIDI situation and changing the perception of music in relation to games.

If you were making a movie and you wanted to get someone in there who had a different attitude, which seems to be the thing in making movies, it seems that every once in a while Hollywood kind of purges. "We want something different and wild." Basically, we've been over here in training, while doing our business. It's like we're lumberjacks and someone else is out there looking for a baseball team. We'd be hilarious baseball players for a couple of seasons. We've been swinging that ax, baby. And, I bet we'd hit a home run or two.

Music for games. That's what we do. That's what we love. That's what it will say on my tombstone.

Just for fun, let's say you're addressing a roomful of game developers at a seminar in Monterey, California. What advice would you give them?

Basically, the thing to keep in mind, whoever you're using to do music for a game, is that the more that you can allow the artist to do art, the more

Chapter 12

you can structure the relationship so that the artist is doing art and the scientists are doing science, and there is very little overlap between the two.... That will maximize your results. The biggest drawbacks in the business have been when tools were employed that made the artists jump through a lot of hoops to get the integration of the music into the game. Or, when the programmers were asked to evaluate music that they didn't know about. Or, to determine where music was to go in the game when they didn't feel capable of doing that.

A sensitivity to what is appropriate helps. I would also say, you can't do this with everybody, but if you're working with someone who has a rep, you can do very well. One of the best ways to really kick ass, is to let that person do their thing. Give them some small amount of guidance and don't worry too hard if you're not working yourself to the bone coming up with a perfect exact spec for them. It's different with engineering. It's different with programming and game design. You need a hard spec for that. But a certain amount of looseness and floppiness given to the musicians might give you a remarkable payoff. Take a risk. It might not be perfect, but you don't want to be the same as everybody else. Is that good advice?

Not bad. I'm curious about the use of tools. In the video and film business there are lots of tools to sync and play back music along with the onscreen action. Is there anything similar for integrating music with game sequences?

It is a missing art. The tool is missing. There are two or three of them that aren't accessible to anybody for one reason or another. I hope to have a meeting in about three days when I'm in Portland (picking up the Rolls), I'm going to skip over to Seattle to talk to some folks about building such a tool. I call the art music integration. I'd like to get a team together and create an integrator. LucasArts has a really good one called IMuse. But it is a proprietary system and they only use it inhouse.

There is a system that kind of does it in Direct Music. I've only met a couple of people who have managed to conquer it. For a musician to use something like that, it has to be "click and drag." You have to be able to do it when you're drunk [laughs]...

I think you've hit on one of the other problems, so many of these companies—like LucasArts—are using proprietary game engines they've developed. For this approach to work, it seems you'd need a tool to tie in to a mainstream game engine. Other than Macromedia Director, is there anything like that in the development world?

Expert's View: George Sanger Talks about Interactive Music

I'm no expert, but I do play one on TV. I do have a conference every year with all the experts and turn them loose on the issues. It's called Project BBQ. We have got 380 acres of God's country, an open bar the whole time, jam sessions all night long, and we work our asses off for three days brainstorming. And amazing stuff comes out of this.

At the first one, there was a meeting around the jacuzzi with the top guys in interactive music who came up with a spec list of what a standard interactive music language would have to have. They reported it to the BBQ group. They spun off what we call a rogue group—that's part the beauty of the BBQ, if the muse comes and sits on your shoulder and says, "You don't need to be prioritizing this stuff with these guys—go form a rogue group." then you're supposed to. They had a rogue group on that and came back and reported to the BBQ group. And someone from Apple stood up and said, "Excuse me, but didn't you just describe the functionality of QuickTime." And for the last two years, people have been sitting around with that realization and not acting. I personally believe that this thing could be built on top of QuickTime.

It's interesting what Thomas Dolby is doing with the Beatnik engine. It's pretty much controllable through JavaScript, which is pretty universal—at least on the Web—but in terms of adaptability for CD-ROM, it may not be perfect.

On one hand, if you look at the art of interactive music, you might be tempted to believe something... I have observed a tendency that every piece of interactive differs fundamentally from every other piece of interactive music, in such a way that building the tool defines the art. However, on the other hand, there are certain basic things that everybody does and that are not being done easily. So, if you just draw the line and say, "This functionality will be included in version 1. It will be based on top of QuickTime, so that other functionality is available later. I want to be able to drag and drop things into a pool from which you will choose tunes randomly, either cutting between them on a bar line or cross fading them until the game conditions change, at which time you'll draw from a another pool.

Simple stuff. Do I go from this song to this song as a cross fade or a straight cut? That easy. Build a little menu that you can control the interactivity from. You can eventually replace that menu with a game. So, it is a missing element in our field. I'm going to have some meetings. I hope someone builds it. It might turn out to be me.

I know there are a couple people who seem to be working on it.

Chapter 12

Any parting words to the game developer community?

I'd say that if there is a spirit or a muse that is attempting to help you make your game, it will not be able to help you make that game unless you get out of the way. Another way to put that is: Sometimes it is not so important who you are as who you aren't.

That sounds very Zen like.

Zen is very practical shit.

Two more quick questions. Did you listen to the Ventures growing up in San Diego?

Yeah, I grew up in San Diego. Boy, you are the homework guy. Have we ever hung out together?

I just read a couple articles from the Net where it sounded like you were interested in surf music back then. The Ventures got me interested in electric guitar, because they were the first group that really seemed to take advantage of the instrument.

Cool. Do you remember the Astronauts?

No. Were they a West coast band?

No. They were a Colorado surf band. They had a hit called *Baja* that sold 7-million copies. The leader of that band was my band director in high school. Starting in junior high, because I was drafted out of the junior high band to be in the high school band because we didn't have enough marchers.

So, I was six years under the tutelage of Robert G. Demmon, who is this charismatic, cowboy meets the music man. He taught me everything a person needs to know about showmanship and cowboy outfits [laughs].

So that is where the cowboy origins come from?

Well, I don't know. It certainly resonated with me when I moved to Texas. I had to have my hat. Bob never wore hats. The spirit of being a surf band from Texas sure felt like returning to my roots. The first guitar picking I ever did was on Bob's Jazzmaster and this big old Fender Jazz bass. They had four big blond guitars right in front of the drummer. Four guitars and the drummer. It was like a guitar army—all these matched blonde

Fenders with the matching tuxedoes. They had that great surf guitar look...And a big sound.

That's all stuff that Bob taught me. When I learned to play guitar, that's what I learned. So, that surf music turned into my roots, oddly enough.

So, you've gone on to other instruments. You mentioned playing French horn.

Yeah, playing French horn and trumpet lately. That all felt like going back to Bob Demmon and the roots, because he taught me trumpet and French horn. They had a couple of surf tunes that Bob played trumpet and French horn on. And it's that kind of spirit—a little bad French horn is a helluva lot better than no French horn. If you're going to make a mistake, make a big one. All of these little sayings, rules of thumb about music—I've two hundred of those on file in mind from Bob Demmon. It's kind of got this strange, Colorado surf music roots. Yeah, I'm into surf music.

Parting question. Who has better Mexican food: California or Texas?

State law requires that I answer Texas. I'd get shot for anything else. Thank god it's the truth!

Note: Portions of this interview originally appeared in *CD-ROM Access*, the quarterly publication of Disc Makers and are reprinted here by permission. A free subscription can be obtained by visiting: *www.discmakers.com/cdrom/*

13

Business Uses for Optical Recording

Businesses and organizations are adopting and deploying optical disc equipment and software to solve a variety of data distribution problems. Sales patterns for networked recorders, as well as software designed to provide access to optical discs over the network, have been strong in recent months. Optical discs make sense for businesses for many of the same reasons that they have proven so successful in other areas: the low cost of storage, the portability of the medium, and the simplicity with which hardware and software can be configured, deployed, and maintained.

Many of the applications discussed earlier in this book, of course, also apply to businesses. Data backup and archiving always have high visibility within businesses. Distributing databases, video, audio, and other high volume content also work very well on optical disc. The difference addressed in this chapter, however, is more the matter of scale. Businesses and large organizations typically have a much more urgent need to share data, sometimes with geographically disparate staff members, and to collaborate on projects over very great distances. To defray the cost of important business equipment, such as CD recorders, devices are often placed on the network for group access, presenting another level of complexity for ensuring error-free recording and effective utilization of the hardware.

This chapter addresses those unique needs faced by businesses when integrating optical disc storage components into their daily work processes.

Chapter 13

Networking Optical Disc Drives

The networking of CD drives dates back almost to the beginning of CD-ROM technology. Early networking approaches dealt with CD-ROM drives in much the same manner as hard disk drives, linking them through SCSI interfaces and daisy chaining. Tower units soon appeared, accessible to network users through management software, but the early versions had very little built-in intelligence and access was notoriously slow, particularly if multiple users tried to access the same discs.

Jukeboxes capable of managing 20 to 40 discs became a popular network add-on—these units became a practical means of making libraries of content available to workgroup members. The first generation of jukeboxes were primarily server connected units.

The trend in recent months has been towards optical disc towers that are network attached, rather than dependent on a file server. Slow disc access times have been overcome through intelligent use of disc caching. Some systems allow entire disc libraries to be cached, the discs removed from the drives, and the data delivered at hard disk access rates. This type of system can take advantage of the current low cost and high performance of hard disk drives, while benefitting from the data permanence of optical discs and their capability of consolidating important information and helping organize data into accessible collections.

Network-attached optical disc storage also has greatly reduced setup requirements, relieving administrators of the burden of lengthy configuration processes and difficult network maintenance chores. Some examples of network-attached products are provided at the end of this chapter.

The utility of optical disc storage on networks has been demonstrated as a practical and cost-effective means of supplying data to workgroup members for several years. New developments that have arisen in the last few months involve a technology referred to as ThinServers, using network attached storage (NAS) principles. Under this approach, CDs and DVDs become similar to appliances accessed across the network, serviced by hardware and software that minimize access and performance issues.

Thin servers typically provide the ability to cache optical disc content to hard disk, where it can be more quickly delivered to network users. This approach achieves significant performance improvements over the straight CD jukebox or non-intelligent optical disc tower approach. Current generation thin servers emphasize the ISO 9600 file system over the more recent UDF system, which generally rules out the use of CD-RW discs on the network. DVD support also generally relies on the UDF Bridge format, the hybrid file system that provides backwards compatibility to ISO 9660, so while most DVD-ROM and DVD-Video discs should be accessible, those that eschew the UDF Bridge approach will not be accessible.

Some thin servers, such as the Microtech DiscZerver and the TenXpert system, can also write to recordable CDs residing on the network. Support for a wide range of CD recorders tends to be limited, with only the most popular models supported. Only one of the current vendors, TenXpert, provides support for jukeboxes through their thin server. TenXpert also provides support for DVD-RAM recording across the network, although their approach has some limitations.

Advantages of Networking CDs and DVDs

At first glance, it might seem that there is no compelling reason for networking CDs and DVDs, given the inexpensive nature of hard disk storage. On closer examination, though, some important advantages become clear:

- Data from optical discs that is distributed over the network is much more cost effective than purchasing individual disc titles and distributing them among every workgroup member.

- Networked users can access CD and DVD collections and extended data sets that might span several discs without having to swap discs in and out of a local disc drive.

- Optical discs provide a level of version control for critical corporate information, such as policy manuals, product price lists, engineering software releases, catalogs, and so on. Releasing this kind of information through shared discs on the network overcomes the difficulty of performing company-wide updates of the same hard disk resident data.

- Information that is obtained through subscription services on disc can be inexpensively shared (if licenses permit this kind of use) among network members. For example, law databases—such as Westlaw—can cost in the range of $30,000 for a single copy. Sharing this data through networked optical discs is far less expensive than purchasing multiple copies of the database.

- Data that is frequently consolidated on optical disc, such as corporate style guides with associated fonts and logotypes, can be posted on a networked CD and made available to network users.

Networking optical discs doesn't provide the ultimate solution in every data distribution scenario, but, as you can see, it offers some key advantages in many important areas.

Cost Factors

Cost reductions in hard disk drives have made RAID technology a very inexpensive proposition for organizations that need truly large-scale storage capabilities. Current costs are around $18 per Gigabyte. In comparison, optical disc systems designed for the network require additional costs—including management software, with per-user license fees, and the additional hardware expenses (whether tower or jukebox). Equivalent costs for optical storage systems with network capabilities can be as high as $48 per Gigabyte.

Optical storage systems, however, support the concept of library access; constant, unchanging data can be attached to the network without fear of hard disk drive failures or overwriting of critical data. Organizations that rely on huge repositories of physical data, such as the United States Geological Survey and NASA, can assemble sets of compiled data surveys stored on DVD-ROM or CD-ROM, make them accessible through the network, and easily extend the contents with new data when available. The permanent nature of optical discs makes this type of application more practical than the other long-term data storage options, such as magnetic tape, which has to be transferred to new media every few years. For removable storage and data interchange, optical discs offer the ability to consolidate meaningful data into readily accessible areas on the network.

Computer Output to Laser Disc

Certain kinds of business applications by their nature require enormous amounts of physical data storage, including:

- Medical imaging technologies
- Document scanning and retrieval
- Insurance and financial record storage and access
- Digital pre-press file storage and handling

- Legal data repositories
- Engineering and design applications

Computer Output to Laser Disc (COLD) is an approach that has been designed to directly process the output of applications, writing data to optical recorders. In this specialized application area, hybrid software and hardware tools have been developed to deal with these enormous quantities of data. For a certain segment of these applications where the data integrity and archival properties of the media are paramount, Write-Once Read-Many (WORM) drives are used to store the content.

For example, Plasmon offers a line of products, including the 8000 Series, that utilize 12-inch optical media for storing 30GB of data. Other systems integrate CD or DVD recorders into the process, depending on the required storage capacity. Some systems combine several recorders in a tower configuration that allows the software to write sequentially to a disc at a time until the total capacity has been consumed. Very high-end systems employ robotic arms to continue supplying blank media to the unit after the currently loaded media capacity has been consumed.

One key characteristic of this type of application is that the disc recorder is treated as an output device, similar to a printer. Images, records, scanned documents, and so on, can be directed to the output device—a CD or DVD recorder—and the information is permanently written. This manner of workflow eliminates the premastering step that is normally a part of creating a CD or DVD, saving time for those business operations that require a steady output of data to disc.

Appendix A includes a number of pointers to companies that produce products for COLD applications.

Network Storage Solutions

Read-only optical disc systems for network use are the least expensive approach for making both CD-ROMs and DVDs available for network users. Some solutions rely on software to make optical discs systems available to workgroup members. Other products bundle CD towers and management software together and create a unit that can be attached to the network as simply as an appliance.

The following examples describe typical systems currently on the market.

Chapter 13

SciNet CD-Manager 5

SciNet, Inc. of Sunnyvale, California introduced the concept of Network-Attached Storage solutions for CD and DVD. They offer software only networking packages and complete systems designed for quick installation and ease of use.

The CD-Manager 5 package is a server-resident application that provides simultaneous support for Novell NCP and Microsoft/IBM SMB clients. One unique feature of the product is the ability to partition the server hard disk drive into a series of segments, each the size of a single CD. CD-ROM data is then cached into these segments, where it becomes available at higher hard disk access speeds. SciNet calls this approach TurboDrive technology. Unlike similar systems, the CDs do not have to remain resident in the system—the access to the TurboDrive equivalents can take place even after the CDs have been removed. The characteristics of the cached CD data remains inviolate—the read-only restriction is retained and the ISO 9660 file system is still employed.

SciNet's other claim to fame is the development of their extremely streamlined, plug-and-play approach to connecting CD/DVD storage systems to the network—a process that can be completed in just one or two minutes. Termed *Network-Attached Storage* (NAS), this approach has lowered both the costs and the setup difficulty for networking optical storage equipment.

The SciNet CD-Manager 5 is available as a standalone product, with a cost corresponding to the number of supported discs on the network, or bundled with any of the turnkey CD/DVD server units produced by SciNet. SciNet has prepared the equipment with the appliance model in mind—just plug it in and turn it on and the equipment becomes immediately available.

SciNet, Inc.

268 Santa Ana Court
Sunnyvale, CA 94086
Phone: 408 328 0160
Web: www.scinetcorp.com

MediaPath MA32+

Designed to bypass the jukebox and tower approach to CD sharing, MA32+ offers a software-only solution to networking CD-ROMs. The product retrieves the contents of several CD-ROMs, stores them in a special compressed format on a server hard disk drive, and makes them accessible to network users at hard disk drive connection speeds. Multi-user and multi-CD access is fully supported by this approach, an inexpensive alternative to the hardware-dependent solutions requiring jukeboxes and towers.

A management utility included with the software provides control over the CD titles made available on the network. The product is designed for Windows networking implementations.

MediaPath Technologies, Inc.
125 F. Gaither Street
Mount Laurel, NJ
08054
Phone: 800 357 0697
Fax: 609 222 0552
Web: *www.mediapath.com*

Examples of Network Ready Optical Disc Units

Products in this category are changing extremely rapidly as the technology matures, but the following examples should give you some idea of the capabilities of current equipment at press time.

Plasmon AutoTower

Part of the obstacle to deploying a CD recorder or DVD recorder on the network can be the network configuration issues. Host-based installations can be grueling and time-consuming, plus considerable expertise is typically required to complete the process. Debugging and troubleshooting after installation can also be very time consuming.

In comparison, a class of products has appeared in the business market based on NAS principles, essentially designed to be network ready. Installation of these kinds of units typically requires only a couple of minutes or so. Plasmon's AutoTower is an example of this approach, bundling 2, 4, or 6 DVD-RAM drives into a tower configuration that is coupled with their own NetReady server. A selection of jukebox units, ranging from 120 to 480 disc capacity completes the package. The unit contains an internal 9GB hard disk caching system.

Configuration of the Plasmon AutoTower is accomplished through a browser-based interface. File system security is enforced through settings applied in this manner. The unit has a simple control panel for basic operations and for entering the IP address during initial configuration. Status and error messages are also delivered through this panel.

The system design supports multiple write operations to any of the two to six DVD-RAM unit is installed. Not all units of this sort can handle recording operations initiated from different users without suffering significant slowdowns in recording speeds.

Given the options available for the jukebox portion this product, the maximum available capacity would be 1.2 Terabytes (with a 480-disc capacity jukebox). The DVD-RAM drives support the full range of optical formats, including CD-ROM, CD-R, CD-RW, DVD-ROM, DVD-R, and DVD-RAM. Plasmon offers their own certified media to use with this product, but independent testing has demonstrated compatibility with all commercial DVD-RAM media.

Plasmon IDE

8625 West 76th Street
Eden Prairie, MN 55344
Phone: 612 946 4100
Web: *www.plasmon.com*

Axis StorPoint CD E100

Axis Communication designed the ThinServer technology used in a variety of networking products, particularly those tailored for CD/DVD storage. This approach has been widely embraced throughout the storage industry, offering a number of advantages over host-based connections.

The StorPoint server technology created by Axis, included in the CD E100 server, has also been licensed by many tower vendors and included in their products. The thin server model provides an efficient way to attach peripherals to a network without the overhead or burden of typical network operating system constraints. Thin server products are specialized implementations, operating independently from proprietary protocols, tailored for easy installation and remote configuration through a Web browser. Since the approach is independent of the network file server, I/O requests do not require additional processing time through the file server and overall network throughput is not negatively affected.

The CD E100 is designed to give up to 1000 users access to the contents of 127 CD-ROMs or DVD-ROMs. Caching to a hard disk improves the level of performance, reducing access times to approximately 10 millisec-

Examples of Network Ready Optical Disc Units

onds compared to the 100 milliseconds or more required for a optical disc drive. Internal server operations are managed by a 100MHz RISC processor. The CD E100 server is designed to work with most of the optical disc storage devices on the market, including jukeboxes, individual drives, and disc tower units. These can be integrated into networks based on Ethernet, Fast Ethernet, or Token Ring architectures. Macintosh clients can also take advantage of the disc sharing over the network.

As with most thin server technologies, management of the CD E100 is handled through a Web-browser based interface, so that an administrator anywhere on the network can oversee and control the operation. A useful utility that AXIS includes with the server, StorPoint DISCO, retrieves details about all of the networked optical discs that are accessible and consolidates this information within a single folder for convenient access.

The Axis CD E100 is a read-only storage solution—it does not support CD-R or DVD-RAM drives. Nonetheless, the flexibility and reliability of this product make it a strong candidate for optical disc sharing in many different networked environments.

Axis Communications

100 Apollo Drive
Chelmsford, MA 01824
Phone: 800 444 2947
Fax: 978 614 2100
Web: *www.axis.com*

Quantum CD Net Universal XP Cache Server

The venerable Meridian Data Systems, one of the pioneers of CD recording, was recently acquired by Quantum. One of their strongest products, CD Net, has re-emerged as a network-ready implementation that supports CD-ROM sharing across NetWare, Windows NT, OS/2 and UNIX networks. Server units can be obtained with several different drive capacities, ranging from 3-drive mini towers to 7-drive towers with locking access doors. Rack mount units are also available, supporting up to 7 CD-ROM drives.

Employing the popular hard disk caching approach, CD Net servers deliver CD-ROM content at standard hard disk access speeds by storing the relevant files in cached regions. Optical disc contents can be cached and then removed from their CD-ROM drives, freeing space for other optical discs to be accessed.

The standard network interface for the Net CD server is Ethernet—support for both 100-BASE-TX and 10-BASE-T are automatic through an internal sensing mechanism. A 4MB Flash RAM in the server can receive software updates through FTP connections whenever new manufacturer upgrades are available.

As with other network-ready units, these servers are designed to operate right out of the box. The network file servers do not need to be configured, nor do the workstation clients that will be accessing the CD-ROM drives. CD Net servers can support multiple protocols, including simultaneous operations from different protocols, in their basic configuration. Supported protocols include TCP/IP, DHCP, NetWare IPX, NetBEUI, UDP/IP, and HTTP. Management tasks are aided by support for SNMP MIB-II status reporting, allowing seamless integration with a wide range of standard network management utilities.

Quantum

500 McCarthy Blvd.
Milpitas, CA 95035
Phone: 800 767 2537
Web: *www.quantum.com*

Increasing Storage Requirements

For large enterprise networks, the demands for increasing storage capacity rise annually and tax the available capabilities of many forms of removable storage. The 650MB of storage available on a single CD-ROM disc doesn't keep pace with growing storage requirements of network users, particularly when individual systems often have 12GB of hard disk storage or more. DVD-RAM, with 5.2GB disc cartridges now available, can typically only work for full system archiving if several drives are embedded in a tower. Increased storage capacities may be achieved with some innovative new technologies currently in the works.

The 120-millimeter optical disc format that works so successfully for CD and DVD can only be increased in storage capacity by packing the data at higher densities on the disc. One technique for accomplishing this is through the use of the violet laser, which offers a 405-nanometer wavelength, compared to the 635-650 nanometers of the red laser, allowing much tighter data patterns to be detected and read. A commercial release of the violet laser by Nichia Chemical Industries of Japan will support data densities of 27.8GB on a 120-millimeter optical disc. This capacity could be increased through the use of multiple layers. Future DVDs may incorporate this technology.

Particle Beams for Data Reading

Some manufacturers are looking beyond the laser beam to other recording methods, such as the use of the electronic particle beams. One company, Norsam, has been experimenting with the use of particle beams to write to a metallic media, achieving capacities of up to 200GB for a single 120-millimeter optical disc. This write-once technology, slated for arrival in the year 2003, is currently being envisioned as being incorporated in a tower system containing ten or more HD-ROM drives, offering a capacity of somewhere around 60 Terabytes.

Fluorescent Multi-layer Discs

Another promising approach to data storage comes from Israel, where a company called C-3D has devised a recording technique they call Fluorescent Multi-layer (FM). Instead of relying on a reflective surface for data detection, as do CD and DVD, an FM disc emits light. The fluorescent media can be stacked in layers, each layer capable of handling 14GB. The initial goal is to produce discs with a minimum of 10 layers, representing the equivalent to 1.4 Terabytes of storage on a 120-millimeter optical disc. The first planned products include an FM-ROM drive, an FM-WORM drive, and FM PC-card drives, scheduled for introduction in early 2001.

Optical Super Density Format

Even tried and true technologies, such as magneto-optical are being given new life through advanced techniques. The Optical Super Density format (OSD), devised by Maxoptix, aims to deliver 40GB capacity on disk cartridges that can be read and recorded on both sides. Data transfer rates for this format are in the 30MB per second range, clearly a promising level of performance for many different kinds of business applications where hard disks can be replaced or supplemented by removable magneto-optical media.

Between now and the time you read this book, several other promising technologies will, no doubt, have entered the consciousness of the optical recording world. Networks will continue to grow in size and scope, and optical recording technologies will continue to increase in capacity to support them.

14

Using DVD with Video in a Corporate Environment

This chapter addresses the question "what do you actually do with DVD in the real world?" and focuses on video applications in the corporate environment. The chapter is not intended to be a complete manual for production and use of video (that would take an entire book in itself), but instead we present an overview of the topics you must consider if you want to use video and DVD in the corporate environment. These topics are the pieces of the front-end processes that support the use of corporate DVD.

Applications for Corporate Video

Why would a corporation want to use DVD? There are two primary uses for DVD in the corporate environment:

- As a storage medium
- As a vehicle for using video

As a storage medium, DVD-ROM or DVD-RAM is just like CD-ROM or Zip disks. It stores files, without regard to what those files contain. Some of these files may be video clips, displayed using a player like Quick-Time's MoviePlayer; some of the files may be Macromedia Director interactive modules which play video clips. This functionality is identical to that of CD-ROM, but CD-ROM technology is limited for video use by the maximum storage size of 650MB per disc. The first-generation DVDs have a minimum storage capability of 2.7GB, more than 4 times that of a

Chapter 14

CD-ROM. Second-generation DVD disks have 4.7GB per disc, and labs have already demonstrated prototypes with capacities of 17GB per disc.

As a vehicle for using video, DVD-Video is a proprietary disc structure which can play back video stored in this format. The disc structure contains what is essentially its own navigational system of *menus* and *chapters*. We'll look at this structure in more detail later.

Video is effective for communicating, but expensive. Bringing the capability to create and maintain video applications in-house makes sense, as desktop video production can easily produce broadcast quality results with today's hardware and software. Of course, each company's situation is different, and video production will range from total production done under contract through a mix of in-house and contract production to complete in-house production. You choose which approach is the best combination for your milieu.

What can you do with video in relation to DVD? Nothing new. It's just the next-generation mode of distribution and use. The specific advantages gained will be pointed out as we discuss the various aspects.

The previous uses of video in corporations include:

- In support of speaker presentations
- In multimedia training or communications modules
- On intranet or Internet, as part of communication and information systems
- In video brochures to distribute company or product information
- As broadcast-quality clips for use in television and news programs
- In trade show kiosks and point-of-sale displays
- In video libraries for all the uses above

These uses are still applicable, so let's see how you can migrate to the use of new DVD hardware. To do so, we'll begin at the end.

Delivery Options for Corporate Video

The choices of hardware and software for production and delivery of DVD-based video depend on how you actually plan to deliver the video. Therefore, we begin at the end: the delivery mode.

There are three modes for delivery of video:

- Playback from a VCR standard VHS analog video
- Playback from a set-top box (DVD player) and a television: DVD-Video
- Playback from a computer: DVD-ROM or DVD-RAM

Figure 14 - 1 shows the configurations of hardware needed for these modes. Let's look at each in turn.

Figure 14 - 1   **Delivery modes for video**

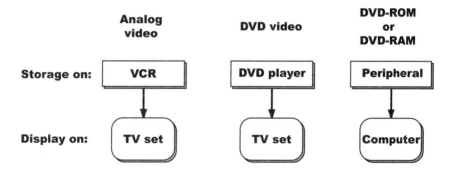

Delivery via Analog Video

This delivery mode is included because everyone is familiar with it, so it provides a baseline from which to compare the other modes.

To deliver analog video, all you need is a videocassette recorder (VCR) connected to a television set. The good news is that almost everyone uses the VHS format. The bad news is that different countries use differing, and incompatible, television standards. The main broadcast television standards are NTSC (used in North America), PAL (used in Europe) and SECAM (used in scattered places like France and Korea). This means that you can take a videocassette anywhere in the world that shares the same television standard, and be certain of being able to play it back. On a domestic level, you can take a promo of your company or products and present it anywhere in the same country. Or you can "transcode" (i.e. translate) your videocassette content into another format and play it back in a country which uses a different standard.

The production process for analog video To produce the videotape in the first place requires the traditional analog production process. This is illustrated in Figure 14 - 2.

Figure 14 - 2   Process for producing analog video

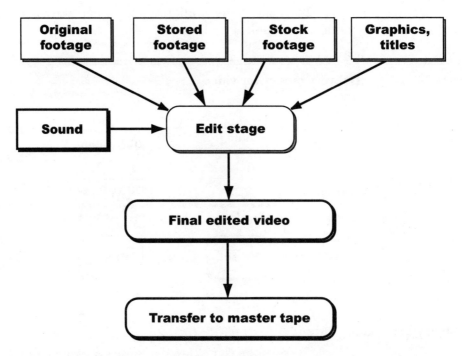

Video is produced by editing together various sources of images and sound. Since this chapter is principally about DVD and video images, we only consider explicitly the image part of video. The "sound" box in Figure 14 - 2 is actually composed of sub-components such as field-recorded sound, studio narration, music and sound effects.

The various sources of images are shown in the figure. Original footage is, of course footage shot for the current project. Stored footage is previously-filmed material, and retrieved from source or back-up tapes. Stock footage refers to clips that are purchased from video supply houses when, for example, you need a shot of downtown Bangkok and you can't afford to send someone to get the shot. Graphics and titles for traditional analog sources are usually produced by dedicated character or graphics generators.

The final edited video is transferred to the storage and playback medium, a videocassette.

A Phased Approach Traditional analog video is well known and widespread. Most corporate boardrooms and conference rooms have a VCR and television available, and therefore you can show off your production with well-understood equipment. It's also easy to show the video at trade shows and kiosks.

As a transition to the DVD milieu, you can still use analog video (i.e. VCRs and televisions) as a delivery medium, but produce the video using digital techniques. The advantages of producing video using digital techniques are indicated below. Using this phased approach allows you and your company to move to all-digital production and delivery in a step-by-step fashion: first you go all-digital for production of the video material, and, once you are comfortable and set up producing video in that way, you can add the DVD delivery described below. Finally, you can add the in-house capability to produce DVD discs.

Digital Video

Before getting into a detailed discussion of DVD and video, we need to make a short digression into digital video itself, and its various aspects and considerations. This will prepare you for the choices you need to make for DVD hardware and software.

Why Digital?

Traditional video systems are built on a video signal which is analog (a varying voltage level). The lingua franca of computers is composed of digits: 1's and 0's. Video that is converted (encoded) in the form of a computer digital file allows much more power and flexibility in its use. For example, a video clip can be sent over a phone line as a stream of 1's and 0's. This means that, say, a consultant who is producing a video for a company in another city can create a preview and squirt it over the phone lines, without bothering with couriers or snail mail. Or, video clips can appear on the World Wide Web.

Chapter 14

In addition, video in digital form can appear in interactive computer presentations, enhancing the presentation. The digital video can be stored on computer media such as Zips, CD-ROMs or DVD-ROM/RAMs, which are significantly more robust than videocassette tape.

Making video digital, therefore, allows more extended, more flexible use of the medium.

How Digital Video Works

To turn analog video into a digital file, you need a digitizing board in your computer. The board takes the input analog signal and turns it into a stream of 1's and 0's.

Figure 14 - 3 **Digitizing Video Signals**

The main consideration here is to have a system which is fast enough to capture video without missing frames. This means a capture card and an AV hard drive, the combination being able to handle 30MB per second and process it.

The stream of digits contains all the necessary information about frame rate, color, motion, and so on to capture the analog signal. Now comes the techie part: To store and use the file, you need a specific file format and you need a specific compression algorithm. Understanding these two technical terms is essential for understanding digital video.

A file format is the way of storing digital information of the digitized video on a computer storage medium. As far as we are concerned in this book, there are only three formats of significance: DV, MPEG-2 and QuickTime. (We don't consider AVI, as it is no longer supported, and it was not as powerful as QuickTime.)

DV is a high-quality, digital format that integrates well with desktop systems. There are currently three DV formats: miniDV, DVCPro, and DVCam. MiniDV is the most common, and generally is the format used by consumer cameras. DVCPro and DVCam are professional formats which are not as widely available as miniDV. The DV format is far superior to Hi8, S-VHS, and other consumer formats. DV is digital, so it does not suffer from generation loss: a copy of a DV tape is identical to the original.

MPEG-2 stands for Motion Picture Experts Group, level2. This file format has been defined by a consortium of companies, with a view to introducing a proper standard for video compression.

QuickTime is a file format created by Apple Computer. It has been a howling success, as evidenced by the facts that it is the format most used on the web, and it will be the core technology of MPEG-4. (There will be no MPEG-3. Go figure.) We'll look at an overview of the benefits and features of QuickTime after reviewing at the core item of digital video: compression.

Digital video needs huge amounts storage space. One computer screen of data is approximately 1MB in size. Since video in North America displays at 30 frames (screens) per second, that's 30MB per second of raw information. Apart from the consequent enormous storage space for raw video, computers cannot handle processing and displaying that amount of data per second. Therefore, we need a way of reducing the 30MB/second demands on our systems. The way this is done is to analyze the video frames, and look for ways of throwing away data, without apparent degradation of the image. This process is called compression. There are two ways to compress digital video:

- By throwing away data in a single frame (intra-frame compression)
- By throwing away data which differ from frame to frame (inter-frame compression

There are many ways of analyzing the video stream and throwing away data in a systematic fashion and still maintaining an acceptable image quality. Each different way uses a different algorithm (method of computation). Each way is called a codec, short for compression/decompression. Since throwing away data calls for choices and compromises in deciding what can be thrown away and what can not, different codecs are better for some types of video images than others. Some codecs require extensive calculation to compress the video, while requiring much less computation to decompress (play back) the video. This type of algorithm is called asymmetric in its compression. For example, one popular codec is Cinepak. It takes much longer to compress than real time, but plays back well in real time. Obviously, the fundamental criterion for an effective codec is to play back in real time, with no visual stuttering or dropping of frames.

The process of throwing away data in a video image introduces so-called artifacts in the image, which look like blockiness or pixilation of the image. The more video is compressed (i.e. the more data are thrown away), the more artifacts appear. Compression of a video, therefore, is a compromise of throwing away as many bits as possible, while maintaining an acceptable image quality. As of this writing, the best codec is Sorensen. We'll look at aspects of this codec in more detail later.

There isn't the space in this book to delve into all the technical details, but there is one item to watch out for: DV has non-square pixels, while computers have square pixels. This means that a video clip in DV format will look distorted if shown on a computer screen. Fortunately, some software packages (such as After Effects and HEURIS MPEG Professional) can automatically do the required adjustments to pixel aspect ratios.

Production Process for Digital Video

The production process for digital video is shown in Figure 14 - 4. The digitization and compression steps have no equivalents in analog production.

Figure 14 - 4   Producing digital videos

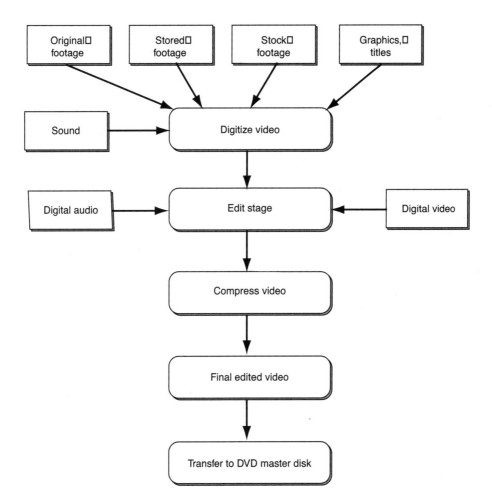

Capturing and Assembling Digital Footage

Previously, we discussed how to digitize video from an analog source. The other way to obtain digital video is to use a camera, which stores the sound and images in digital form while recording. There is an increasing number of affordable video cameras use digital video. The main formats of these cameras are miniDV, DVCPro, DVCam. Figure 14 - 5 shows one camera made by Canon: the Optura. This camera uses a miniDV minicassette. Most DV cameras can be connected to your computer via a standard protocol connection, IEEE 1394, called FireWire by Apple and i.Link by Sony. Note that the DV format is only for recording; it is not intended for playback at the user stage.

Many professional digital cameras use the miniDV cassette. One such camera is the Canon XL1 (Figure 14 - 5, courtesy of Canon U.S.A.). The advantage of the miniDV format is that cheaper "prosumer" cameras, such as the Canon Optura shown in Figure 14 - 5 (courtesy of Canon U.S.A.) also use this cassette. This means that you can get a professional team to shoot the footage on the more expensive XL1, and transfer for editing using a much cheaper camera.

Transfer of the digital video to a computer storage device is done either by taking an analog signal out from the camera and digitizing it (not the best way), or using the IEEE 1394 connection that all of these cameras have. The cable takes the video file directly to hard disk without the need to convert from analog to digital. When choosing a camera for your company, make sure it comes equipped with an IEEE 1394 cable to connect to your computer (which, obviously, must be chosen to have an IEEE 1394 input port), and that it can be computer-controlled by brand-name software. Premiere and Final Cut Pro, the most advanced digital editing software packages, both support control of cameras by IEEE 1394 cables.

Figure 14 - 5   **Professional XL-1 (left) and the "prosumer" Canon Optura camera**

Another way to get video into a project is to retrieve it from a stored library of digital clips. There exist software packages which store still images, video and audio files. Users can search for previously-stored clips by name or by keyword. These packages are generally called multimedia or asset managers. Two such products of note are Extensis Portfolio and Canto Cumulus. The source digital files can be stored on CD-ROMs or DVD-ROM/RAMs. You can set up a network-based system which makes these files available over a corporate Intranet. Figure 14 - 6 shows a typical interface for such a class of product. The figure shows the interface for Extensis Portfolio. Figure 14 - 7 shows the arrangement which allows digital files to be retrieved over a corporate intranet.

Figure 14 - 6   **Asset manager interface (Extensis Portfolio)**

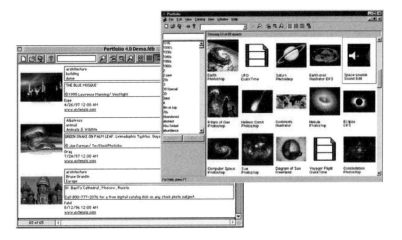

Figure 14 - 7   **Retrieval of digital files over an intranet**

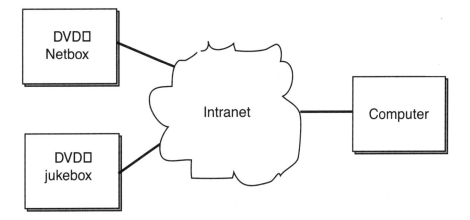

The device which holds many discs, each of which can be selected and files retrieved over a network is called a jukebox. There are currently many jukeboxes installed which use CD-ROM discs as the storage medium. Jukeboxes for DVD discs are just becoming available. ASACA offers a device called a TeraCart. Configurations of this device can hold up to 60TB of data stored on DVD-RAM discs. (That's Terabytes, ten to the power 12!) This system can hold 18 days worth of continuous video playback. Now that's storage capacity! The TeraCart uses multiple Panasonic LF-D101 DVD-RAM (rewritable) drives. You can get more information on this system at *www.asaka.com*.

A cheaper, more limited-space solution is available from LaCie (*www.lacie.com*) which makes a network device they call a Netbox that has a DVD/CD reader connected to a 36GB hard drive. DVD and CD data can be read and copied to the hard disk, and so the information is made available over the network. This is a fast and easy solution. The way this can be used is to archive video and audio files on CD or DVD, and copy them to the Netbox hard disk for access over the network. Thirty-six Gigabytes is not a lot of space for video files, but the ease and convenience of the device makes it worth consideration if your capacity needs are modest.

Cygnet Storage Solutions (*www.elms.com*) produce a bundle for addressing the network access issue. They sell a DVD jukebox, coupled with a special version of Canto's Cumulus. Their system holds 100 single-sided 2.6GB DVD discs. This system is called the "Digital Librarian".

Other vendors such as Pioneer (*www.pioneerusa.com*) and NSM (*www.nsmjukebox.de*) are also offering jukeboxes worthy of checking out.

Figure 14 - 8   **LaCie Netbox**

The Editing stage To do the actual editing of video files, you need a software package such as Adobe Premiere, Media100 or Final Cut Pro. The input video clips to be edited must be readable by the software package. You also need a package for producing special effects, such as the industrial-strength Adobe After Effects. Other software packages needed for prepping the files include Adobe Photoshop and a sound editor such as Sound Forge, SoundEdit or Bias' Peak.

The Compression Stage In advance of outputting the files to the final storage medium, you need to compress them and you need to make a significant choice: play back on DVD video or DVD-ROM/DVD-RAM. (This illustrates what we said earlier: The end mechanism drives the choices you need to make).

To play back from DVD video, the files must be stored in MPEG-2 compressed format only.

To pay back from DVD-ROM/RAM the video files can be stored in any format you desire, with any codec you desire (provided, of course your users can play back the format!). This flexibility, coupled with flexibility of size, makes this option very attractive for corporate use.

An excellent source of information about codecs and compression can be found at *www.terran.com*. Terran makes Media Cleaner Pro, an indispensable software package which is discussed later in this chapter.

Chapter 14

Getting the Best Compression

There are several things you can do at the filming stage to make the file easier to compress. Here is a checklist of things to do:

- Use a tripod. Any camera movement or shaking leads to frames which are different from each other, which means that the inter-frame differences are large, and the file becomes difficult to compress well.

- Keep the background simple. Any fussy wallpaper or lots of background detail causes problems with compression. If you cannot avoid detail in the background, set up so that the background is out of focus.

- Make sure people don't wear loud checks or stripes. These look to video as excessive detail in the frame and are difficult to compress.

- Use lots of light, either daylight or artificial lights. Video does not like low light. It leads to excessive noise in the video image. Noise is the most difficult thing to compress. Increasing the light level reduces the video noise, and makes the file easier to compress.

- Another aspect of the previous item is to avoid large shadowy areas. You do this by using diffuse lighting. Any film or video supply house can show you how to get diffusion lighting.

- Use video transition effects between scenes with restraint. A transition effect like a page turn looks to video like a lot of data passing in a short time (a data spike), and can cause stuttering or dropped frames.

- Sound quality also affects how much a clip can be compressed. Broad-spectrum noise (from computer fans, air conditioners, and so on) is difficult to compress. When shooting video, reduce the amplitude of such sources as much as possible. For example, turn off computers and air conditioners when shooting in an office location; stay away from exterior locations with lots of traffic noise. Reduce broadband wind noise by using windscreens on microphones when shooting location sound.

The story so far Armed with the technical background, we can now see how to choose options for use of video and DVD.

Delivery via DVD video

The second delivery option of Figure 14 - 1 is playback of a DVD video disc video using a DVD player (also called a "Set-top box") and a television set. (You can actually use a computer monitor to play back this option, but there is no good reason to do so: if you plan to use a computer in the delivery mode, you are far better to use DVD-RAM or DVD-ROM, as discussed later in this section.)

DVD functionality Why choose DVD video? There are several pieces of functionality which make it very attractive for use by a corporation.

- **Improved quality**: While DVD video offers excellent video and audio quality, the current television standard in North America, NTSC, only produces an image resolution of 300 lines. This makes it appear as though you're watching images projected onto a Venetian blind. Current cameras and monitors can actually display more than double this resolution, making the picture quality very much better. This is achieved by keeping the brightness information of the video signal ("luma") separate from the color information ("chroma"). Separating these signals reduces their interference with each other, a major source of problems with current television standards.

  DVD video also eliminates the severe bandwidth limits of NTSC, allowing viewers to receive the full resolution that cameras can capture, with colors that have the same dynamic range as the brightness information, and are therefore more vibrant.

  DVD video also has superb audio specs. Home systems have used Dolby ProLogic Surround Sound, an audio scheme that allows four channels of sound to be encoded on a normal stereo pair. The four channels are center, front left, front right, and rear (although the rear channel is typically divided between two speakers). While Dolby ProLogic provides a much better sound than stereo alone, DVD can do better. DVD video can play either Dolby's AC3 Surround Sound or MPEG-2 audio, with 5.1 channels (left, right, center, rear left, rear right, and the ".1" is a subwoofer).

  The combined improvement in quality of video and audio is significant. Presentations to corporate clients or potential customers using DVD are cutting-edge, and should create a favorable impression of your company.

- **Multiple camera angles**: DVD video provides the capability for creators of discs to use multiple camera angles for the same scene. The viewer can then select a particular camera angle. This opens up a whole new arena of creativity for the film-makers. For example, a DVD video compilation of, say, SuperBowls, might make available multiple camera angles, just like the broadcast show, but with the added advantage that you can choose which camera you want to use at any one time in the game.

- **Multiple screen shapes**: DVD video has the ability to play back video in different screen shapes. Current television screens are 4 units wide by 3 tall (a 4:3 aspect ratio). Theatre movies have aspect rations of 16:9. DVD video can play either, and, more importantly, a DVD disc can contain a movie in both aspect ratios, so you can choose which to view.

- **Security**: the DVD standard has built-in security capabilities. Originally conceived as a means of parental control over rented movies, this capability can be used as password protection for proprietary information in the corporate environment.

- **Navigation**: A DVD video disc has its own navigational system of *menus* and *chapters*. These chapters are just indexed places on the disc you can jump to directly. For a Hollywood movie, the chapters you can jump to are just pre-selected scenes or special trailers or added documentaries. Let's look at samples of navigation, and then see how the functionality can be used in the corporate environment.

A DVD video player behaves like a VCR on steroids. Navigating a DVD video disc is simple: you use a remote control (shown in Figure 14-9), to display and select choices on-screen. You can control playback via a remote control, just like a VCR. You can just hit the "play" button to start playback, or you can display a menu screen, which lists a series of places you can jump to or options for language, camera angle, parental control, and so on.

Digital Video

Figure 14 - 9   **DVD-Video remote control**

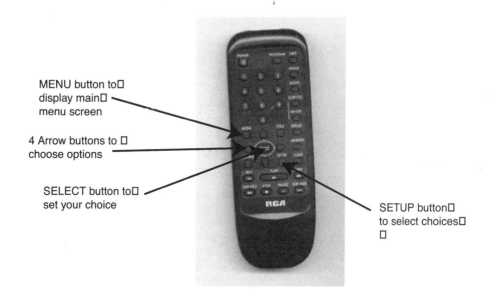

Figure 14 - 10 shows a typical screen displayed after you press the SETUP button on the remote.

Figure 14 - 10   **SETUP screen**

Chapter 14

Navigating this screen is easy: you just use the arrow buttons on the remote to highlight choice 1 to 6, and then press the SELECT button to select it.

The RATING choice in this screen refers to the level of parental control (for movies). In the corporate environment, this functionality would refer to a level of security: for example, level 1 could refer to movies only viewable by Research and Development teams, level 2 by sales staff, and level 3 to movies viewable by anyone. The security levels are password-protected, as shown in Figure 14 - 11.

Figure 14 - 11   **Password security in DVD-Video SETUP screen**

The AUDIO choice screen of Figure 14 - 12 allows you to choose the language of the soundtrack. This is independent of the language of subtitles. (You've never lived until you've watched *Das Boot*, a German movie, in Spanish with French subtitles. A bizarre experience.)

Figure 14 - 12   **Screen for choosing the language of the audio track**

The SUBTITLE choice screen of Figure 14 - 13 allows you to display or hide subtitles. For movies, this is simply the direct translation of the dialog. For DVD video use in corporations, this facility could be used for other-language prompts to support the main presentation movie. Or, it could be summary points displayed for a trade show kiosk, where there is lots of distracting noise.

Figure 14 - 13   **SUBTITLE option**

The TV SCREEN choice of shown in Figure 14 - 14 allows you to select the aspect ratio and method of display of the video image. The options are:

- **4:3**. This is just the usual aspect ratio of televisions

Chapter 14

- **4:3 pan and scan**. One way to display theatrical movies which use a 16:9 aspect ratio on a 4:3 screen is to display a 4:3 chunk and move the chunk around to accommodate which part of the 16:9 screen to show. This is called pan and scan, and is the reason for the message "This film has been modified to fit this screen" which occurs on many rental movies.

- **16:9**. This displays the full width of the movie, but leads to black bars above and below the image (the "letterbox" format).

Figure 14 - 14   **TV SCREEN selection**

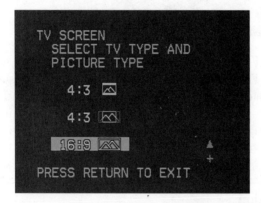

While quite common in Europe, widescreen televisions are only just beginning to appear in North America. The image quality is far superior to existing television sets, especially combined with the more-impressive 16:9 aspect ratio of the screen. If you want to impress clients, get a widescreen television.

The OSD choice means *on-screen display*. This is similar to regular VCRs which display "REW" when rewinding: the command selected is displayed on-screen.

MENU LANGUAGE means.... Well, let us leave that as an exercise for the student.

The menu screens shown above for SET-UP of most DVD video discs are depressingly like MS-DOS, and these form the bulk of early DVD movie discs. However, the technology allows much more creative uses of the remote control and on-screen graphics. For example, Figure 14 - 15 shows a mock-up of a main menu screen for *Chariots on Fire,* a non-existent movie that never saw the light of a projector bulb. For a real-world example of a good use of menu screens, view a copy of *The Matrix,* a title released by Warner Brothers. In the introductory menu screens for *The Matrix,* the movie scenes are actually playing behind the menu choices. This disc is a very impressive use of the DVD video format, and we recommend that anyone reviewing the format for use in a corporate environment examine this disc, and how the technology is used to best effect.

Figure 14 - 15 **Mock-up of a DVD Main menu**

Selecting SPECIAL FEATURES for most titles takes you to a screen with additional selections. An example of a Special Features screen appears in Figure 14 - 16. (By the way, that's my wife, dressed as a nun in high heels with a pogostick, leaving the grounds of the residence of the Governor-General of Canada. Don't ask.)

Chapter 14

Figure 14 - 16   **Example of a Special Features screen**

The SPECIAL FEATURES screen itself has more than one choice of added content: behind-the-scenes documentaries of the cast and of how the movie was made.

The DVD video disc for *Das Sub* (a mocked-up movie about a German U-boat in the Second World War—don't try to find this in your local video rental outlet), adds a nice touch: the cursor to indicate choices on screen is not a flashing square, but a reticule of a submarine periscope (Figure 14 - 17). This kind of approach shows that the creative freedom offered by the DVD video disc is large.

Figure 14 - 17   Das Sub main menu, showing creative use of cursor graphic

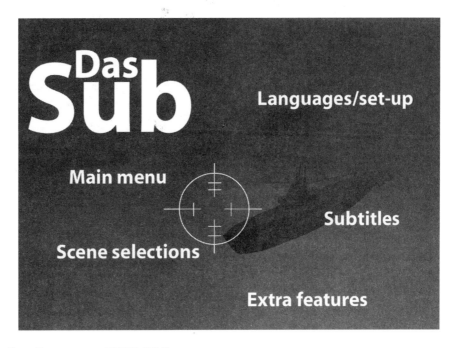

Playback Set-up for Corporate DVD Video

Playback of corporate DVD video discs could be in a boardroom or in a trade show or a point-of-sale display. The ambient lighting of each of these situations is very different. Choose a player that allows you to select the type of display device connected to it. The ideal set-up is to project the video onto a high-quality screen or large-format LCD. To do this requires a CRT or LCD projector. These devices have different display characteristics from a television, or monitor, and so the image must be adjusted for these display characteristics. One example of such a player is the Panasonic DVD-A320 which allows you to select from 4 different monitor types (Standard, CRT Projector, LCD Projector, and Projection TV) so that your DVD player can adjust its picture for optimum video performance when the correct monitor is selected.

Digital Picture Mode allows you to choose among 4 preset viewing modes (Normal, Fine, Soft, and Cinema) to best suit the varying image quality offered by differently formatted DVD discs.

The DVD-A320 also benefits from the inclusion of Digital Cinema Mode. This setting is ideal for movie watching in boardrooms where lights can

Chapter 14

be dimmed. It reduces glare and picture noise to give you a more realistic picture. Light and darkness are more accurately presented, while colors appear more natural; for example, flesh tones look like real flesh tones, instead of being overly saturated with color. A better picture, of course, presents your company in a better light (pun intended...) to customers and visitors.

Figure 14 - 18   **Panasonic DVD-A320 player and remote control**

There are also players appearing on the scene which can hold 5 discs at a time. SONY has gone the whole hog and announced plans for a 200-disc system, essentially a personal jukebox.

Advantages of Using the DVD Video Delivery Format As with all choices, there are advantages and disadvantages of choosing to deliver video on DVD video discs.

- Easy to use, like a VCR
- Familiar use of a remote control
- High quality video and audio, with widescreen option
- Simple, built-in navigation system
- Multi-language capability (via multiple audio tracks and sub-title tracks)
- Don't need to worry about computer horsepower or bugs and incompatibilities or upgrades or the latest version.
- Easier and more reliable than a computer to set up (an advantage at trade shows, for example)

Disadvantages of using this delivery format As you might expect, there are also some disadvantages to using DVD-Video for presentations.

- Complex, expensive process to produce a DVD video disc
- Video must be stored as MPEG-2 only
- High computing requirements to compress as MPEG-2

Should you produce DVD videos? As a corporate user, you need to decide whether to set up in-house capability to produce DVD videos (or DVD-ROMs and DVD-RAMs) or have the work done externally under contract. The fastest way to get into using DVD is to have it done externally under contract. This is currently expensive, because production houses are just buying new equipment and setting up shop to offer DVD services, and costs of such set-up are being passed on to you. For insights into the creative process of producing DVD titles, for both commercial distribution and corporate use, refer to Chapter 17, *DVD Creation*, which highlights some of the experiences of Zuma Digital, one of the pioneer title producers in the field.

Over the long term, costs of DVD production will come down (just as they did for CD-ROM). At some point within the next two or three years, DVD will probably surpass CD-ROM as the primary distribution medium for digital content.

You can actually do everything except duplication of DVD discs today, by choosing the right combination of hardware and software. The discussion below indicates the classes of hardware and software you need to produce DVD discs in-house, and mentions some specific examples. When researching to set up DVD capability in your company, you should be aware that new solutions are becoming available every month.

Producing a DVD video

Producing a DVD video disc consists of the steps shown in Figure 14 - 19.

Figure 14 - 19   **The DVD-Video authoring process**

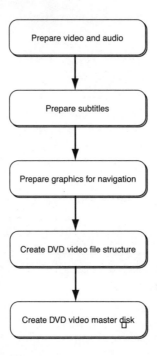

Preparing the Video and Audio Tracks

We've already discussed the step to get video and audio in digital form and to edit them.

To produce video files for inclusion into a DVD video disc, you need to compress ("encode") them in MPEG-2 format. Currently, the best solution is to buy Terran Interactive's Media Cleaner Pro, with HEURIS Power Professional MPEG export engine. Media Cleaner Pro is an indispensable tool for anyone using digital video and audio. It can batch process a group of files, which means that you can set up the compression of many files, and let it process away all night. Since compression of video files into MPEG-2 format takes a long time (up to 40 times the original length of the video clip!), you really need a hardware computer board which accelerates the MPEG encoding process. There is a special version of Media Cleaner Pro which comes with a hardware accelerator board. The HEURIS Power Professional export engine also works with non-linear editing systems such as Media100 or Avid.

To produce audio files for inclusion into a DVD video disc, you need to encode them into Dolby AC3 Surround Sound format, or MPEG-2 audio format.

Prepare Subtitles

Most corporate applications will probably not use this DVD video capability, but the process is easy: just create text-only files which will be used by the authoring program described below.

Prepare Graphics

The next step is to prepare the graphics used for menus and on-screen buttons. Some authoring programs can import Photoshop layers directly. This is a great time-saver, as all relevant buttons and their backgrounds for one menu can be produced together in one file.

Create the DVD Video File Structure

To produce an integrated file which will be burned onto a DVD video disc, you need to use an authoring program. *Authoring* simply means setting up the navigational links, subtitles, language selections, etc., and then creating an integrated file for writing to disc. There are many details involved in authoring. For example, the compressed video and audio must be combined into a single data stream, called multiplexing.

There are several authoring programs available, at varying levels of sophistication. The following list is a sampling which should allow you to get a feel for the types of solutions already existing.

Astarte (*www.astarte.de*) offers a series of authoring software packages. The simplest is DVDelight. This program, used in combination with a DVD-RAM burner, produces linear DVD videos almost as easily as preparing a slide presentation. Using a simple drag-and-drop interface, a DVD video project incorporating up to 99 video and audio tracks can be created. The tracks can be set up to play automatically whenever the DVD-RAM media is inserted. Videos can be set up to loop indefinitely, for use in trade shows, kiosks, and point-of-sale displays. You can create single-sided 2.6GB or double-sided 5.2GB DVD-RAM discs. This solution cannot produce menus or interactivity. For that functionality you need one of the other Astarte solutions, such as DVDirector.

DVDirector is a high-end authoring solution. Besides creating normal linear presentations, elaborate interactive projects with complex menus and navigational structures are possible. Using the integrated player, the

Chapter 14

project can be previewed at any time, without the need to "build" the whole project. Figure 14 - 20 and Figure 14 - 21 show the interface for DVDirector. As you can see, DVDirector presents a highly graphical user interface.

Once the project has been completed, DVDirector multiplexes the entire project and formats it so that it can be played from the hard disk, with a standard DVD player application. This file is then transferred to a DVD-RAM disc, using a burner application like Adaptec Toast DVD.

Figure 14 - 20   **Interface for DVDirector project**

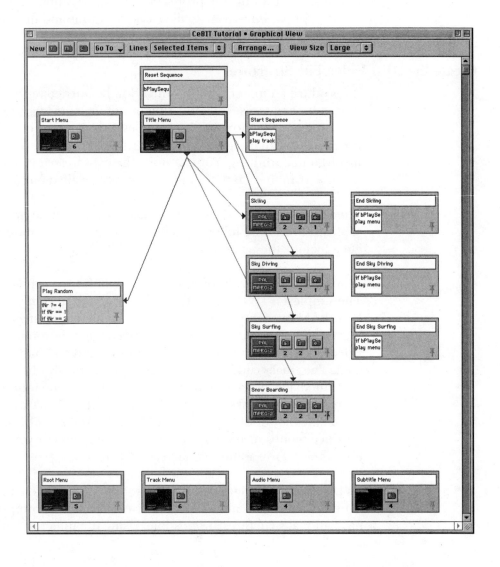

Figure 14 - 21 Interface for jump matrix for DVDirector project

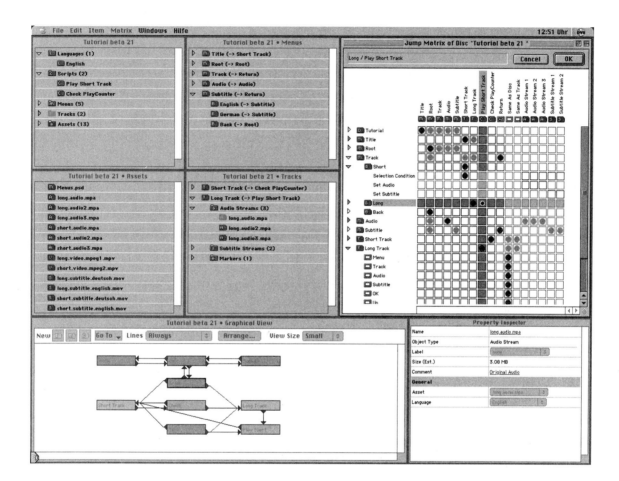

DVDirector Pro adds a hardware MPEG-2 encoding system, a Subtitle Editor, and Astarte's A.Pack Dolby AC-3 encoder.

One interesting and unique tool in the DVD arsenal from Astarte is DVDExport. This product is a Macromedia director Xtra which is used on conjunction with Director create simple menu-driven DVD-Video discs. Since director is used extensively in the corporate CD-ROM world, this product allows staff to migrate to DVD using a familiar tool. The Xtra writes a DVD video disc file structure to a hard disk, ready for burning to a DVD-R. This file can be transferred to a DVD-R, or can be passed to the high-end product DVDirector for further refinement or addition of func-

Chapter 14

tionality. (DVDExport cannot create adding advanced DVD features like region coding, subtitles and DVD scripting).

Astarte also offers utilities for encoding of original video and audio files. Astarte M. Pack is an easy-to-use application for encoding QuickTime movies into MPEG-1 or MPEG-2 formats. It produces very high quality MPEG streams, comparable to hardware systems costing thousands of dollars, albeit much more slowly, as it's a software-only solution.

Astarte A. Pack is a Dolby-licensed AC-3 audio encoder. It also has a pre-encoding process capability, allowing band pass filtering and de-emphasis filtering.

INTEC's (*www.inteca.com*) DVDAuthorQUICK software supports Panasonic's LF-D101 ReWritable DVD-RAM drive. Designed for Windows NT environments, DVDAuthorQUICK is a DVD-Video authoring toolkit that produces simple DVD video discs with menu and chapter selections.

Sonic Solutions (*www.sonicsolutions.com*) DVDit! is another basic easy-to-use solution. which supports Panasonic's LF-D101 DVD-RAM drive. With DVDit! and the Panasonic rewritable drive, Windows NT and 98 users can produce simple, linear DVD video discs. Navigational structure can be created with drag and drop simplicity. It has an integrated subtitle generator, removing the need for an external text editor.

One step up in power, Sonic DVD Fusion gives video producers and editors a comprehensive set of tools for encoding, authoring and proofing DVD-Video titles from projects created on Avid, Media 100 and QuickTime-based video editing systems. Up to eight audio streams and 32 subtitle streams can be assigned to each video clip.

Minerva Impression for Windows NT (*www.proh.com*) is a complete software-only DVD video authoring solution. The Smart PhotoShop support feature enables a user to import entire multi-layer image files with a single mouse click, which are then parsed and captured as backgrounds, buttons and button rollovers (highlights).

The step after producing a DVD video disc file structure is to verify the file and to test its operation before burning a DVD-RAM disc. This process is called emulation, and allows you to test the interactivity and how your disc will perform. When purchasing a DVD video authoring solution, make sure it has an emulation or premastering capability.

Producing a DVD video

Create a Master DVD-Video Disc

Once the file structure has been verified and tested, the last step is to create ("burn") a DVD-RAM disc. This disc can be a *one-off*, or can be passed to a replication plant for making multiple copies. Figure 14 - 22 shows the LaCie DVD-RAM burner.

Figure 14 - 22   **LaCie DVD-RAM burner**

Delivery via DVD-ROM and DVD-RAM

The third delivery option, as it appears in Figure 14 - 1, for playback of corporate video is playback of video using a computer and its monitor. As mentioned earlier, for this option, DVD-ROM is just like CD-ROM and DVD-RAM is just like CD-RW. The discs simply store files, which the computer reads and uses.

There are two flavors of this delivery mode for video:

- Having the DVD player connected to the local computer
- Having the DVD device connected to a corporate server

The delivery set-up can be such that the user views the video on the monitor screen, or (for large audiences) on a projection screen, using an extra device which projects the monitor image. This device is usually called a *data projector*.

Advantages/disadvantages of using this delivery format. The advantages of DVD-ROM and DVD-RAM in relation to using video are in essence the advantages of QuickTime itself. In principle, you could use MPEG-2 for playing back video from DVD ROM/RAM, but choosing QuickTime as your video for-

Chapter 14

mat opens many more doors for practical use in a corporate environment: QuickTime is a much more versatile tool.

To see why QuickTime is so effective, let's look at specific applications:

- **Lecturer-support presentation modules**. For presentations before a live audience, fullscreen display is not always necessary for video clips. For example, using 90% of full screen allows you to have a graphic border around the video being played back. This is a professional aesthetic touch which ties the video clip to adjacent all-graphic slides. The MPEG-2 codec does not allow you to choose such a size for your video, while QuickTime does. Figure 14 - 23 shows the use of a smaller size for video clips in a presentation on how to communicate with the news media. The "Channel 10" logo is followed by a compilation of news clips. (Thanks to McLoughlin Multimedia Publishing for permission to use this screenshot.) Notice the drop shadow of the graphic wooden motif on the video clip. QuickTime has the capability to use alpha channels (masking channels) for compositing on-the-fly in such programs as Macromedia's Director. This opens the door for such creative techniques as non-rectangular video clips. MPEG-2 has no such capability.

Figure 14 - 23   **Screenshot of a presentation module showing creative use of a video clip**

- **Self-paced training or corporate communication applications.** The ability to use video clips at less than full-screen size is also required for self-paced applications like training. Figure 14 - 23 could also be a screen for such a self-paced application.

- **Delivery over the Web.** Because the Web is so abysmally slow, video file sizes must be as small as possible. However, some users have fast cable modems and T-1 lines, while others still have 28K bps modems. One size of a video clip may be too much for one user's system to display properly, while the file may be too small (difficult to see) or too compressed (reduced quality) for another. To optimize the playback of video for all users, QuickTime has the built-in capability of using "alternate" video clips. This means storing on your DVD ROM or DVD-RAM server several versions of the same video clip. Which one plays depends on the speed of the network connection.

- **More programming control.** There are more tools that allow program control of videos stored in QuickTime format. For example, a software package like LiveStage can make a QuickTime movie with embedded control buttons, just like a CD-ROM project. Another, more subtle use of this programming control is to compensate for the fact that Mac monitors have a different gamma (brightness curve) than Windows monitors. The programming detects which platform a user has, and displays a video clip optimized for the gamma of the appropriate monitor. In addition, it is much easier and faster to make, say, a navigational change to a Director interactive module stored on DVD-RAM than it is to make a navigational change to a DVD video title, and re-burn the DVD video disc.

- **More choices for a codec.** QuickTime is a file format, and it can read and write many different codecs. As of this writing, the best codec to choose is Sorensen. This codec comes in a Professional version, which is used to encode (compress) video clips, while a user just needs the standard installed version of QuickTime to have the playback Sorensen codec available. The key aspect of Sorensen which makes it very good is that you can encode a video clip using Variable Bit Rate encoding. This means that the encoding algorithm adjusts the data rate to allow for varying difficulties of compressing different part of the video clip. (Actually, MPEG-2 has this capability also.)

Chapter 14

- **Can do VR**. QuickTime has built-in Virtual Reality capability. MPEG-2 has no such capability.

Summary Put succinctly: MPEG-2 is a one-trick pony, while QuickTime is a multi-trick pony. Corporations need multi-trick ponies, and so if you choose to deliver video using DVD-ROM or DVD-RAM, then choose QuickTime over MPEG-2 as your corporate file standard.

Summary of Recommendations

As you've no doubt deduced, there are many options available to bring DVD into the corporate environment. To help you crystallize your decisions, here is a short list of recommendations:

- If you can only choose one, choose DVD-ROM /RAM over DVD video. This choice gives you the most flexibility in producing and delivering your video material. (Computers are more flexible for deployment and use than set-top boxes. For example, PowerPoint presentations can be made using the same computer, whereas a set-top box does not allow this.)

- For trade shows or point-of-sale kiosks, choose DVD video over DVD-ROM or DVD-RAM. The DVD video player is much easier to transport and set up in remote locations than a computer. It is also more foolproof, something essential for a frenetic situation like a trade show.

- For DVD-ROM or DVD-RAM delivery, choose QuickTime over MPEG-2 as your file format standard. This choice also gives you more flexibility: QuickTime offers Web streaming, alternate movies for differing speed of user machines, alternate movies (with use of Director) to play gamma-corrected movies for specific platforms, VR and ease of production (because QuickTime native files are used by all the major editing and special effects software packages).

- For DVD-ROM and DVD-RAM delivery, set up a presentation computer in your board room or presentation room with a DVD-ROM or DVD-RAM player attached.

- Designate one person to act as DVD guru and techie. This person will be responsible for designing and implementing the specific hardware and software configurations your company requires.

Summary of Recommendations

These recommendations are to be viewed as a guide on what decisions you need to make and why one DVD format is better in one situation over the other. Perhaps the best advice is to research thoroughly in advance of any purchases or delivery mode decisions. With care and planning, you can make video delivery for your company impressive to customers and clients by being on the cutting edge of DVD technology.

This chapter was written by David Martin, an independent multimedia developer based in Ottawa, Canada. David can be contacted at: djfilms@home.com.

15

Interactive Multimedia on Disc

One of the bedrock uses for the CD-ROM and DVD-ROM is the creation of titles for the entertainment industry and for marketing groups in large corporations. This chapter showcases two such cases. One, a promotional CD-ROM title for Macromedia highlights, the techniques necessary to produce a Web-enabled CD. The other, a spirited journey through the birth, development, and release of a major game title for LucasArts, illustrates the process and issues involved in competitive niche market development.

Case Study: Macromedia *Add Life to the Web* CD-ROM

Goal: To create a Web-enabled CD-ROM

Tools: Macromedia Director, Flash, Shockwave

The staff at 415 Productions in San Francisco has a long history in producing CD-ROMs and an equally solid background in developing Web sites for clients. More and more today, clients are insisting on CD-ROM that provide direct links to their site. To accomplish this in as smooth and seamless a way as possible is the goal of most developers. I talked with Jeff Southard and Chris Xiques at 415 Productions about the techniques they employed in a recent project for Macromedia. We started out talking about the disc architecture that was used on their Add Life to the Web CD-ROM, a project that was developed for Macromedia to showcase the current crop of Web tools and technologies. The hybrid disc included an autostart function for both Macintosh and Windows users.

I know that on the Windows side to get the autostart it is a function of what you put in the AUTORUN.INF file, but I'm curious on the Mac side what actually kicks off the autostart.

Chris: It is a switch that you throw in the Adaptec Toast software for setting up CD burns. When you are treating the Mac file, you can enter a file that you want it to autostart.

In terms of the disc architecture, you do have a small Mac partition that is unique that you can't even see on the Window side. Is that true?

Chris: It works like this: You make a temporary partition on a hard drive in Toast. Then you copy all the files that you are going to use for both platforms into that partition. Then you drag the whole partition into Toast and that works for the Mac. You then have to go in separately and drag files that are in that partition into Toast again for the PC version. You get one CD where both partitions share the common files.

Jeff: Windows can read ISO 9660, and so can the Mac now. But, then there is a special partition that is a Mac-only partition.

Chris: Mac can read ISO 9660, but you do have to count on the user having certain extensions and managers enabled. It would be great to just be able to make ISO discs, but some projects you really do have to make a Mac partition. With every project, there are usually one or two files that you only want one partition to see.

Looking at the file contents of the disc, it looks as though it might be Level 2 ISO 9660. Some files appear to be longer than the 8.3 format.

Chris: This is done with another switch in Toast. You set the premastering so Toast will only burn files that conform to 8.3, or you can make the naming conventions looser. There is a Joliet standard that lets you do DOS plus Windows 95, or you can just specify Macintosh naming conventions. Toast will just burn everything according to your selections.

So, you used Joliet throughout this CD-ROM?

Chris: I think we did use Joliet on that CD.

It sounds like it is fairly safe at this stage of computer evolution to use those filenaming conventions for both Mac and PC users?

Case Study: Macromedia Add Life to the Web CD-ROM

Chris: Whenever you make a CD and you start talking about the specifications for the platform you are going to support, you're also immediately identifying the computers you're not going to support. These days we really don't support Windows 3.1 very often on projects. We can have a much looser naming convention.

You're probably safe on the Mac side with Joliet for everything that has been released in the last couple of years, I would guess.

Chris: If you name files so that they work in Windows, they generally work on the Mac, as well.

Do you want to talk a bit about the creative design decisions that were made when you first picked up this project from Macromedia?

Jeff: In this situation, Macromedia had a pre-existing visual style and pre-existing assets that we knew we could use. I'm speaking of their product home pages, which were made in Flash (by product home, I mean Director, Dreamweaver, and so on). Clearly, the two goals of this CD-ROM were: get people to the Web site but also give them the large product demo files. They call this approach "Try-and-Die" software.

It didn't seem smart to create some whole new aesthetic. We went with their look, and then we were able to reuse their existing Flash movies that appear on their main Web site. In some cases, we needed to tweak them a little to make them higher-resolution bitmaps.

So it sounds like you had most of the existing assets, and most of the decisions involved how you would present the material?

Jeff: Yeah, I'd say half of the assets were already designed. The other half we created based on the look they had made. Of course, the home page, the main menu, did not exist. We created that.

The circular selection menu was an interesting approach.

Jeff: We were pleased with it. The pages that appear after the products, including the product information, all lead to the Net. Those were the pages we made.

Chapter 15

Figure 15 - 1   **Circular menu used in the project**

Jeff: Of course, on many other projects, we do much more design work. But in this case, it didn't really make sense. It let us give them a lot of bang for their buck.

Chris: For me, one of the cool things that came out of this project technically was the blending of the Web and the CD-ROM. When people are running Internet Explorer on their machines, we have the ActiveX stuff coming straight out of their machines—you could actually see their Web content on the Director stage. Since it is nearly exactly the same as the content we were showing locally on the CD-ROM, you have this really nice transparent switchover to the Web content. I thought it looked really nice.

How did you handle the ActiveX stuff that is linked into the content?

There are Director Xtras you can use. Basically, you have an ActiveX Xtra that you invoke in Director; you just have to make sure that the Xtra is included on the CD-ROM. You first find the path to their browser and then check to make sure it is Internet Explorer. If that is true, then you can invoke this ActiveX Xtra and all the content will appear on the stage.

Case Study: Macromedia Add Life to the Web CD-ROM

Was there a way to deal with the ActixeX components on the Mac side?

Jeff: There really was no option on the Mac side. The contents just appear in a rectangle that is a Web browser. We worked around it. Whenever you go to Web-based content, you can launch whatever browser someone had installed. If someone didn't have a browser, we provided browsers that could be installed. It was really kind of a bonus that the display was seamless under Windows if you had Internet Explorer installed. At that time, the total numbers of IE installations were not as high as they are now.

What do you think it is now: 60% IE to 40% Navigator?

Jeff: I don't know the exact numbers, but I know that if our office is any indicator, Netscape's numbers have fallen a good bit.

Chris: 60% sounds right to me; I think that's the last thing I read a couple of weeks ago.

Were there any serious problems during development where you had to change direction?

Jeff: Mostly little technical problems. Communication with the actual Xtras offered some challenges. I think that maybe halfway through the project, we switched the Xtras used for detecting browsers. So, that was a glitch.

Chris: I don't think there was anything that was really a change of direction. It was more or less a normal development cycle. There is a lot of interapplication communication on the CD-ROM, because you're running Flash inside Director, you're launching HTML content, there may even be some Acrobat Reader files on that disc. Plus, you have all the ambient sound that is separately stored—AIFF files that are being played in sync with the Flash movie. The main thing was just getting all the components to talk to each other.

Jeff: Luckily, Director is quite good at it.

Chris: Yeah. And, of course, on this disc Macromedia had a lot of software that you could actually install right off the disc. And invariably you would go through some iterations with the installers, getting them just right.

Do you use external test facilities or do you do most of your testing inhouse?

Jeff: We used an external testing facility.

Chris: Macromedia was very concerned about this disc running on as many platforms as it possibly could. They hired external testing to handle the test process.

Did you run into anything during testing that required your attention?

Chris: Yeah, we did. We initially developed for an 800x600 screen, the size for the Director stage. Of course, when the testers came back, they said, "Well, we're running on platforms where you can only do 640x480." We had to develop an approach where we would do a system call and check to see what the resolution of the monitor was. Then we could launch the appropriate size application.

Most of the stuff we were running is Flash content and it is completely scalable. You set the size of the Flash movie that you want to play and it just sizes all the content immediately. The screen sizing actually turned out not to be that big a headache, once we figured out how to do it.

I guess it becomes more of a headache if you've got fixed bitmaps that you're displaying?

Chris: Very definitely.

It literally changes the aspect ratio of the images, doesn't it, if you force it into a new framework?

Chris: Yeah.

Jeff: You have to layout everything again. Or, you need to rethink everything. Of course, with the Flash movies the rescaling of all the vector art is handled automatically. Even with bitmaps being used within it, Flash does a pretty good job of scaling up or down.

Chris: That's one of the reasons that Flash is so popular with the developers.

That seems like a very big advantage...

Chris: Often, companies will have a logo that is fairly small in its original form and they want it blown up. If you get those kind of assets into Flash, you can make the logo any size you want at the touch of a button.

Jeff: The better tools let you stay in the vector format as long as possible. Primarily, I'm thinking of Flash, but Fireworks is also very good. At the time, we weren't really using Fireworks yet for CD-ROM development, but it is very good for creating everything from icons and interface graphics with rollover states to content artwork.

Does Fireworks handle both bitmapped images and vector images?

Jeff: Yes.

And you can manipulate any type as needed?

Jeff: They are basically all just objects within layers.

Chris: I think you can actually choose what format you end up saving stuff out.

Jeff: The great thing is: it slices up all your graphics for you. Although it was designed for making Web pages, most of the features directly transfers to making CD-ROMs. You can make a mockup of your whole screen with all the interface elements, and all the content, and then you just draw these little slices, little rectangular areas on top of that layer. You can also define whether you want to have multiple states, a basic state and a rollover state, or a mouse down or a mouse rollover and down state. It will define the different looks of each of those interface elements across the first two or four frames.

You can automatically export them all, using various kinds of compression formats or no compression. You can also export graphics with anti-aliasing channels. You can export PNG files with maps. Fireworks is just great for Director 7.

So, what is the relationship between Director and the Fireworks content?

Jeff: When you design stuff, you want to design the whole thing, but in the end you need little bits—pieces of graphics, for the interface or other areas. Fireworks is a tool that is great for handling that part of the graphic production process.

So it's the same kind of image slicing for rollovers that you do for HTML processing?

Jeff: Yeah, but it all directly applies to CD-ROMs.

Chris: Right. Most often, if you have some kind of navigation bar in your project, you're going to want to have three states for each button. Invariably when you show the prototype to a client, they will want to change certain things about each button. The nice thing about Fireworks is that you can go back in and make little graphics changes to your button and then just spit them all out again. The program saved the information about where you wanted the slices. That one feature is a huge timesaver for a graphics person working on a CD.

Jeff: There is sort of a handoff between the designer and the programmer.

So all this different content is getting integrated by Director.

Chris: Well, certainly you can make all your images in Director if you want to, but the paint window in Director is not very sophisticated. Almost all of the graphical content is being made outside of Director and then being cut up into usable elements and put in Director. All of the code in a CD-ROM project like this is done in Lingo, but of course you are using more and more Xtras (either those that you are getting directly from Macromedia or from other third-party developers). Many of those little chunks of code are coming from outside.

And some of these Xtras manipulate the HTML components?

Jeff: A lot of times the Xtras are doing system calls or operations that are a step behind the usual capabilities of Director. Director has always been a melting pot of media types, and it has gotten more and more elaborate. There are certain ways that make more sense to do things.

You will almost always do the whole interface for your CD-ROM in Director using Lingo. In some cases, you don't even need to use the programming language—you use behaviors, standard bits of code that you can drag and drop onto graphics elements and suddenly you have created a button, without knowing a line of code. Sometimes you may want to be pulling in all sorts of media types. That is where you might throw in a page of HTML or a Flash movie or launch PDF files.

Can you output things from Fireworks in a format that can be integrated into Lingo?

Jeff: The main thing Fireworks exports is little graphics. In one production process, you take all those graphics and make Web pages. You go

Case Study: Macromedia Add Life to the Web CD-ROM

from Fireworks directly into Dreamweaver. We're talking about doing Fireworks into Director. Director needs all these little graphics. Fireworks makes them in a coherent way.

Chris: Fireworks is not really spitting out any code. It is not enabling the rollovers. It is just making different states of the graphics that you can use to create rollovers.

Jeff: The button states end up in Director as cast members.

Once you've got those pieces you could go to Lingo or HTML rollovers?

Chris: Right.

Did Macromedia give you much feedback during development?

Jeff: There was a regular process of review. We weren't dependent on them for any expertise or technology. We did all the design work. Since we were designing things that were very similar to what they had already approved, it was a very simple process to get final approval.

Chris: Of course, they wanted to hear the sounds. It wasn't anything out of the ordinary, in terms of the client relationship. We had to make alpha and beta deliverables. Once we got to the testing lab, we had to send out different versions.

Jeff: Every night. Until the bugs were shaken out and it was ready to go.

Who did the sound design?

Jeff: We choose what we wanted and we decided to make it sort of fun. Since Macromedia is a hip company mostly pitching to people like us, they pretty much defer to us to decide those sounds and other elements that we like.

Do you happen to know if the sounds were from libraries?

Jeff: We hired a composer for that part of the project. We work with a number of different composers—we give them a spec and they make us maybe twice as many sound loops as we need and then we narrow them down.

What was the final resolution for the sound files?

Chapter 15

Chris: I think they are 16-bit, 22KHz. That's a safe selection. You like to keep it 16-bit if you can. In this case, we used stereo, because the files weren't that big—they were loops. They didn't have 64 or 128 bars of original music. You had maybe 24 bars that looped. So, you could keep the resolution pretty decent and not end up with huge monster files that Director would have to load.

Jeff: Since the sounds are being stored on CD-ROM, you don't really have any tight limits there.

Chris: It's mostly about the sound performance in Director and the amount of memory that it takes up. You run into problems if you use too many resources.

Were there any other things you did in particular, dividing up the content between the Web and the CD-ROM?

Jeff: At one point we considered putting the entire Macromedia site on the CD. Then, we decided not to. We could have. We dreamt up a lot of things we could do. The essence of the project was delivering the demo files. At its very core, this project was just a glorified menu for launching demo files, and launching the installers. We provided the basic information about the products and the minimum requirements for the demos. From there, the user just launches the demos.

So, the decision was to keep it really basic—and then have the Web site provide the real depth?

Jeff: I'd say that is a good approach overall. You want to have some take-home thing that people can launch and use—in this case, the Macromedia software demos. To go much further is kind of pointless. Earlier versions of the showcase CD had to encompass the entire experience, because the Web really wasn't around then and it wasn't commonly used. In previous iterations, we had whole worlds of information about all of the products and all the things that had been made with all the products. It was quite elaborate. For this version, it was trimmed down.

Are the demos on the CD-ROM primarily 30-day demos?

Jeff: I think they are save disabled, which just means you can't save any of the work. They run forever, but you can't really do anything useful with them, because they won't save files.

Case Study: Macromedia Add Life to the Web CD-ROM

Did you actually do the premastering for the production CD in-house or do you farm that work out?

Jeff: We do it in-house.

Chris: What do you mean by premastering exactly?

Creating the one-off that went to the replicator prior to production.

Chris: We created the one-off.

When you produce a one-off, there are a bunch of utilities around that will do sector-by-sector checking of content. Did you go to that extreme or did you rely on the replicator to run a check before they create a glass master?

Chris: Usually we rely on the replicator. If we're having some problems, I definitely have gotten to the point of doing sector-by-sector checks. But usually we do that kind of stuff to see where a file was written on a CD. A lot of times when you're using QuickTime media for example, it is crucial that the media is written as close to the center of the disc as possible—to get every bit of performance out of the media that you can.

That's interesting. I thought it was much less an issue with the faster CD-ROM drives, but it sounds like you still need to worry about file placement quite a bit.

Chris: You can't count on the end user having a faster CD-ROM drive. Even in a situation where you are developing for a 4x drive and you have some rich QuickTime media playing along with some other animation in Director, you still have to worry. You do every thing you can to get good performance.

Does Toast even have a feature that lets you control disc geography?

Chris: It does, but only works for ISO-9660. It's really frustrating when you are making a cross-platform CD.

How did you get around the placement issue then?

Chris: On this particular CD, it wasn't so much an issue. We just tended to put the files that would only be accessed once, like the software demos, towards the outside of the disc. The Flash content and the Director files we were going to be repeatedly accessing, we put in the middle of the disc. To answer your question: On this particular project, we really didn't do any sector-by-sector checks and we just let the replicator verify that the CD was in good shape.

It sounds like you normally do that for QuickTime assets.

Chris: Especially if you have a CD that is really pushing the limit. You have 630MB of content; you really want to make sure where all the files are getting placed on the disc.

Is Toast your application of choice when premastering?

Chris: It sure has been the last couple of years. It is really friendly. It has the features you need it to have and it doesn't have a lot of extra stuff that is confusing. It is easy to teach someone how to use Toast, if they are going to be doing archiving. We have been using it a lot.

Just out curiosity, do you have any idea how much Mac development took place compared to PC development in your project?

Chris: Most of the design still gets done on the Macintoshes. The Mac is the first choice and then each developer will check their work on a PC. Obviously, if it is a cross-platform disc, then the engineers are going to be working on both platforms as well. I think a lot of engineers these days are either working on applications that are PC only or they are working on back-end NT stuff that doesn't have anything to do with CD-ROM. Engineers tend to use the PC more because they end up doing more development on it.

It depends on which part of the development is being done. A lot of times the sound software is still better for the Mac, so many of the sound artists and even a lot of the After Effects people are still using Macintoshes to get their media churned out.

Jeff: At least in the final phases of any CD-ROM and Web production, we steer towards the PC of course, because that's where you really have to get real and that's where the numbers of users are.

Case Study: The Creation of DroidWorks

DroidWorks was conceived in the imaginations of staff members at LucasArts and designed from the ground up as the first release in a new division of the Lucas empire, Lucas Learning Ltd. The history of the design process behind this well-received children's title, based on the

Case Study: The Creation of DroidWorks

Star Wars series, is instructive to anyone contemplating a major CD-ROM project. Collette Michaud, the project leader, offers insights and perspective on the development process.

How many people were actually involved in the DroidWorks project?

If you count marketing and internationalization, if you count everybody, the numbers definitely add up. It takes a lot of people to do everything. But. the main part of the development, which took about 18 to 20 months, was done by the core team. That team started off with just me and the programmer, Jon. It then it ballooned out to about 14 people at the height of the project.

It's typically a bell curve during development. For a long time it was just Jon and I, plus three other artists. For almost ten months, it was just five of us. Then we hired some level designers and the team gained three more people. We got up to about 14 total at the peak period, which lasted about 5 months. Slowly and surely, as the product drew to a close, the artists dropped off and soon we're back to just me and Jon and the programmers at the end, trying to finish it up.

The original release of DroidWorks was in 1998, is that right?

Yes, it was October of 1998.

Can you remember back to when the idea for DroidWorks first materialized and how you got involved with it?

It was in 1997, when I helped start the company with Susan Shilling. Basically, George [Lucas] said at the time, "Come up with ideas for games for children that are entertaining as well as educational." He didn't have any specific ideas as to what he wanted to do. He just wanted to make sure that it was as educational as it was entertaining. Also, very innovative, technically.

While I was working at LucasArts, I had always thought that a good idea for a children's educational game based on Star Wars, would be to use the Droids and to give kids the opportunity to build their own Droids. Given that the technology continued to get better and better, in joining Lucas Learning we knew that we could do something like that pretty cool using 3D graphics.

Chapter 15

The idea almost sounds like the digital equivalent to one of the classic toys, the Erector Set.

Yes, the computer not only gives you the ability to make your own Droids, like an Erector Set, it also allows you to take them out and see them animate—which is something you can't do with a real-life erector set. We were able to do something on the computer that you couldn't do with the toys you've played with in the past, which was one of the goals of the project. We didn't want to just create something on the computer that you could go out and buy and do with your hands. We wanted the computer to be an extension of play. Not a replacement for what you can do.

So, that was the basic idea. I submitted two other ideas to George. But, this one was—by far—the favorite of everybody. George included. He had tinkered with the idea of doing something like a spaceship builder or a Droid builder. So, he approved it right away.

It sounds like George is directly involved in the approval process, rather than having a design committee.

He is. We usually present a high concept to him—maybe just a page of description. If he approves it, then it goes to the next step, the step that we call "Concept." That is where we brainstorm and flesh the idea out to ten or fifteen pages. We get together with the core team, which is a programmer and a lead artist. Now Lucas Learning has grown to the point that we have a director of development and a director of content. They get involved in that brainstorming process, as well.

So, where did it go from this rough idea that you wanted to do some animated computer droids?

From there, after George approved it, I moved into the Concept stage. At the time, we were trying to start Lucas Learning at the same time as were trying to get a new game going. We didn't have any artists on staff. We barely had any programmers. There were literally only about 10 of us on staff at the time. I immediately started looking to hire artists at the same time we were fleshing out the game design. Around February or March, I worked with an outside concept artist, Peter Chan, to come up with some Droid parts, beyond what we saw in the Star Wars movie. He drew a lot of different concepts around that idea. Simultaneously, I'm trying to find 3D artists and 2D artists, not only for this team, but also for another team that was starting up another project.

Case Study: The Creation of DroidWorks

Around May, Jon Blossom, the lead programmer, had a rough prototype of the Droid workshop (the place where you put the Droids together) working in the computer. Very rough. We had hired two artists by that time, so we had some 3D models to work with. From there, it just kept building. The final design was submitted around April or May, which was probably about 25 to 30 pages long. It included ideas for all the Droid parts and mission ideas—what you could do with the Droids after you built them.

Is it a formal development process or more of an informal one? It sounds like it is fairly formal, because you do go through design specs.

It is formal in that way, but it is informal in that the document is very fluid. The design doc that I presented to George and everybody else—it changes. Around October and November, when we started really designing the levels and the missions, we ran into all sorts of technological issues that had to be solved. This changed how we designed the missions. A lot of the missions that were proposed in the design doc never happened.

In the heat of development, do you actually continue to update the design docs? Or, do they get thrown out the window?

I don't. I pretty much throw them out the window.

Sounds like typical software development.

The skeleton is there. When I go back and look at the design doc, it is pretty much the same game. It is just the details of the missions are different. That is why I am not a big proponent of making detailed design documents. I think it is a waste of time. They inevitably change because of the technical issues you encounter.

There are a lot of people who will do these incredibly detailed design documents that script out every possible interaction. I just look at them and shake my head, thinking, "Why did you spend time doing this?"

I would rather spend the time getting into production and prototyping and finding out in the real-world what is working and what is not.

That makes sense. I think that works for a small dynamic team. I guess a large part of your job in coordinating a project team is to make sure that everyone stays on the same path.

Chapter 15

That's right. Stays on the right path and stays motivated.

What's the trick to keeping people motivated?

I guess admitting that you are wrong a lot of times, that you can make mistakes as the leader of the team. So often project leaders get this attitude of infallibility and then everybody gets disillusioned with them and the team sort of breaks off and starts doing their own thing, because they don't trust the project leader to make the right decision.

It sounds like the personalities involved have got to be a pretty key factor.

That's true. For me, it is communication with the team, keeping them up to date on the changes that happen—when they happen and why they happen. Because, so often, artists and programmers get the design doc and read it (or, they say they read the design doc, at least), but they never do. Instead, they concentrate on what they're specifically working on. When something changes, especially something that they have worked on, they want to know why it changed and they want to be part of the decision for making the change.

That, for me, is the biggest trick—just keeping everybody involved. Keeping the decisions on a flat plane as much as possible, but always ensuring that the final decision is mine.

In DroidWorks, when we started getting into the missions, things started changing quite a bit. It turned out that the technology that we decided to use (which was an engine that was used at LucasArts, the one that they used for Jedi Knight) didn't work as well as we had hoped. We thought that was going to be a very strong engine with a lot of capabilities. But, when we actually got into the programming and the designing of the levels, we ran into a lot of obstacles with that engine. It just was not as robust as we had thought. So, as a result, some things changed. The game changed from having 30 levels to 20 levels and it went down to 5 levels at one point. Then, we split the last five levels up into 13 levels. But, the number of missions was changing a lot for quite a while.

Are there multiple game engines in use at LucasArts?

Yes. It depends. It used to be that you could make two or three games out of a game engine. Nowadays, the hardware technology is changing and getting better so fast, if you want to stay current and on the cutting edge, you pretty much have to start over and do another game engine in order to take advantage of all the new hardware features.

So, if a new graphics accelerator comes out or Microsoft comes out with a new DirectX driver, you need a new game engine.

Now, they have new things like these 3D engines, like the ones they use to make Jedi Knight and Quake, where you run through hallways. Now they have a way to lay textures on the background so you get all these beautiful effects. It looks far, far more detailed than it used to.

With graphics accelerators, you can take advantage of those features. So, now they have to rewrite the engines to use the features. Before they didn't have that capability.

So, it seems like you've got to make a tough decision somewhere during the process, looking at the most current hardware platform and looking at what you're doing with the software, and deciding where to set the minimal system requirements.

Yes. For us, being a learning company, our minimal platform is never as high as the cutting edge. We don't try to be cutting edge. We do want to try to have great-looking games. So, it is a balance. We're not doing the lowest end stuff, but we're not striving to be cutting edge either. We try to take advantage of the technology that LucasArts develops.

Most people give their hand-me-down computers to their son or daughter to play educational games. Our minimal platform is usually much lower than the typical entertainment games.

Are you still actively supporting the Mac platform?

Absolutely. For kid's games, a lot of people out there have Macs for their kids. Or, hand their Macs down. We're big proponents of Macs.

Supporting both platforms must be a little complicated, too?

In DroidWorks, we had simultaneous development. We had a Mac programmer devoted to converting DroidWorks over. It was released on disc in Mac and PC versions. It was a pretty involved exercise to get the Jedi Knight engine onto the Mac.

The single disc had both Mac and PC versions?

Yes. It was bi-platform. You could put the disc into either a Mac or PC and it plays the version.

Did you use an outside testing service?

Because we are right next door to LucasArts, we were able to use their testing facility.

So, they really have a serious inhouse test setup?

About half of the bottom floor of LucasArts is testing. It used to be all of the bottom floor, but now it is shared with customer support. Customer support and testing are huge departments. They grow and shrink, depending on what is going through development to release. When they have something that is going to be released, they will hire on a bunch of testers. For us, we got up to about fifteen testers before it went golden master.

Creatively, as you got deeper into the program development, did you reshape and change look and feel of the program much?

Yes, one of the things that changed the most was the interface. Especially for the Droid workshop. There was so much that was happening in there, in terms of designing the Droids. You could color the Droids. You could learn about what each Droid part was capable of doing, what the Droid as a whole was capable of doing. So, there was a lot of information in this interface that we had to get across without making it too crowded. That took a tremendous amount of work and it was pretty much in development all the way up to the end—constantly changing. We got it working early on but to make it look good took a long time. So, we were always futzing with it.

So, once you build these guys you send them out on their missions, you actually get to see what they're doing? Can you control them?

Yes, you do. One of the things we do all the way throughout the development of the games at Lucas Learning is to do focus group testing with kids. From the very beginning even before the game was up on the computer in prototype when it was still just a concept on paper, we brought in our first group of twelve kids. One of the things we weren't sure we should do is program the Droids (script them to go left, take four steps, go right, take three steps) and then watch the Droids do what you programmed them to do. Or, should we do it more like the Jedi Knight

Case Study: The Creation of DroidWorks

model, where you get to directly control the droid and run around the environment.

We put those two ideas in front of the kids to find out what they would prefer to do, and unanimously they said, "No, no. We want to be a Droid. We don't want to sit back and watch these Droids."

So, you're a Droid, but you're limited by whatever parts and capabilities you picked up during the design in the workshop?

Yes. Exactly.

What sort of things can these guys do? Do they have weapons?

No, no, no. One of our mantras at Lucas Learning is no violence. Instead, we promote using your mind as your weapon.

Not a bad idea. So these are exploratory Droids then?

Exactly. Each mission is designed around a simple machine. On one mission, you have to use the droids to figure out how to use a machine with gears. On another mission, you have to use the Droids to figure out how to use the lever in a fulcrum.

On another mission, you have to catapult your Droid across a ravine. And, depending on the weight of your Droid, how fast your Droid is, how big or how tall your Droid is, you can complete the mission. So, the missions were tricky in terms of what you had to figure out. There were many possibilities for building Droids—we had eighty-seven parts total for building Droids.

When you do the math, there are 25-million different Droids. Early on we thought we would figure out what Droids could be built for each mission. After we found out how many different combinations there were, we knew there was no way to figure out exactly how many Droids could be used for each mission. However, we did implement certain filtering devices in each mission. For instance, you have to have a Droid arm that can lift a certain amount of weight or you have to have a Droid who can jump, as opposed to one that rolls. In some of the missions, you have to have a Droid that rolls, rather than walks or jumps.

Chapter 15

If you take the Droid on the mission, and he tries to jump the ravine, and his legs aren't long enough, you just take him back to the workshop and give him longer legs?

Yes. Exactly.

You can keep tinkering with him until he is equipped for whatever task you need to do?

We left it up to the kids to find that out for themselves. A lot of times they would start the game and they would just throw all these parts together, depending on what they liked or disliked, and then they would just take the Droid out on any mission. They would find out pretty quickly, "Hey, I can't get up this hill because my Droid isn't fast enough, or I can't jump because my Droid has wheels. They would take it back. Suddenly, the active use of the Droid parts starting becoming much more important to them. Which was great, on their own they figured out that they needed to pay attention to that rather than us walking them through saying, OK, you need to pay attention to these parts and what they mean." They would figure it out on their own.

We felt as though we accomplished one of the biggest goals, which was to let them have fun, but have them learning at the same time—without really knowing that they're learning.

That's the whole essence of learning. If you can do that successfully, you've really accomplished something. What ages were you aiming for?

We were aiming for 9 to 12 year olds. We tested those age groups.

That sounds a bit young to me, but then I'm probably underestimating how quickly kids pick up computers these days.

They do. They are actually very sophisticated at about 10 years old. It's a big jump between 8 and 9 and then by 10 they are cognitively very, very smart. They didn't have any trouble at all operating the Droids, putting them together. At age 8, it was a little bit challenging for them. The missions were definitely challenging. We've seen kids as young as 5 have no trouble putting the Droids together in the workshop.

Have you seen where older kids are getting into it as well?

Case Study: The Creation of DroidWorks

We've even gotten feedback that adults like it, which was tricky, because the marketing department wasn't sure where to market it at first. We had targeted it at 10 years old. When we finished, the marketing department thought that maybe it was attractive to older kids, and they felt like it should be marketed to older kids. The concern was that we really didn't design it for older kids. We compromised and marketed it to 10 and up.

When you've been involved with a project so intensely for some 18 months, do you sometimes go back and find yourself amazed at what you have accomplished?

So often you get done and you don't want to look at it. I got done and then I went on vacation for a month and a half. When I came back, I was doing a lot of demos and interviews between when we finished it and when it actually shipped, so I had to look at it a lot. But, by then, I was able to look at it more objectively.

Are you pleased looking back at it?

I'm very happy with it; it's the project I'm most proud of.

It sounds like you did some pretty amazing stuff with the Mortimer game as well.

Thanks. Mortimer was good, but DroidWorks really pushed the limit. I'm happy it has gotten a lot of great recognition and appreciation from the educational community. They appreciate it for what we tried to do with it, which was create a game that kids love to play, but that teaches them something at the same time. A lot of educators have come up to me or written me. They have all said the same thing. It has a great balance between education and learning.

That has to be satisfying knowing that you have set a ground-breaking path and accomplished something in a CD-ROM game that hasn't been done before?

The other thing is that when we first started out doing games for George Lucas and knowing that we had to use Star Wars, it was a challenge because one of the biggest fears we had was that people would look at these games and say, "Oh, they're just capitalizing on the Star Wars license. They're not doing anything with integrity." One of our biggest goals was to make sure that we kept the integrity of our mission in the forefront at all times. To be sure that the concept supported the license and vice versa and that we weren't just using the license to get the kids or to get more sales.

Chapter 15

We felt that a lot of other reviewers appreciated that. Not one of the reviews ever came back and said that our games were just using the Star Wars license to get more sales for George.

Being the first product out for Lucas Learning, it was absolutely critical that we did not start off on a bad foot and alienate the reviewers and our customers into thinking that we were just into fluff.

Did you ever get any feedback from George on it?

I've never really gotten direct feedback from him on it.

Is he kind of an intimidating figure, or are people relaxed around him? He seems fairly intense when you see him in interviews and such.

He is intimidating. He doesn't really interact with us at the development level. He pretty much talks with the General Manager and people very close to him at that higher level. At the end of the project, he does send out thank-you letters to everyone on the team, which is nice.

But he does play our games and he has a son who loves to play computer games. I believe George relies on him for feedback.

Summary

Interactive multimedia continues to evolve at the same accelerated pace that optical disc technology itself has advanced. The mega best-selling titles that define the medium have yet to appear on DVD-ROM, but literally every developer that I talked with mentioned that their clients are investigating and weighing the options involved in producing titles on DVD-ROM. The advantage of multimedia development on DVD-ROM is that the developer is not shackled with the many format considerations that apply to CD-ROM. All types of content can be accommodated on DVD-ROM without need for adopting a special format. CD-ROM developers who have struggled with CD-ROM XA, VideoCD, PhotoCD, Enhanced CD, and all the other variations of the compact disc family can relax and enjoy focusing on content development instead of format issues.

On the other hand, CD-ROM drives are ubiquitous throughout the industry, having achieved nearly universal penetration to every level of the computer marketplace. If your goal is to reach the widest possible audience, you can accomplish that end most successfully on the CD-ROM. With improved authoring tools and significantly enhanced playback performance on modern computers, there is a lot of life left in the CD-ROM as a multimedia platform. Whether your aim is to push the envelope or address the vast multitudes, DVD-ROM and CD-ROM offer a great way to present multimedia content.

16

Disc Replication, Printing, and Packaging

Most of the focus of this book has been on the desktop recording of CDs or DVDs. Optical recording from the desktop fulfills a very large range of business and personal requirements, and with the right equipment you can produce discs in quantities that make it unnecessary to rely on a replication service.

At some point, however, you may find yourself with a project where you want to produce 500 or 1000 or more discs for commercial purposes. Instead of an inkjet-printed label, you'd like the extra professional finish of a silk-screened label. You also want to bundle the finished product in a handsomely produced, environmentally sound package to enhance the overall image of your company and your content. These are all services that you may want to receive from a disc replicator. The popularity of optical discs for distributing everything from digital music to legal databases has resulted in a very competitive market for replication services. Services range from a corner shop where you can get a dozen discs produced in an hour or two to large-scale manufacturing plants that produce discs in the hundreds of thousands. Most likely, your requirements are somewhere between these two extremes. This chapter investigates the issues surrounding disc replication, printing, and packaging.

Chapter 16

Replication Overview

The magnitude of risk rises sharply whenever you're dealing with large-scale replication concerns. Instead of possibly wasting a few dollars by producing CD one-offs that contain errors, suddenly you're looking at hundreds or thousands of discs, each manufactured identically from a glass master that contains the files exactly as you presented them to the replicator. If you didn't run a virus check on your final master file set, you risk infecting your entire audience with a nasty virus. If you overlooked some filenaming conventions that resulted in the shortening of filenames required by a running application, your program may not work. If you failed to understand the physical limitations encountered when printing to the surface of an optical disc, your printed label may turn out completely different than expected. Failing to address packaging considerations early in the process may result in your disc being ready, but your packaging materials lagging behind schedule because of their additional production requirements. Many different individual steps have to be completed in the intricate process of producing a disc through a replicator.

You'll gain the best results if you have some insights into the processes that are necessary to create a finished disc out of a collection of files and some artwork. These issues are summarized in the following sections and then covered in more detail throughout the rest of this chapter.

Preparing Files for the Replicator

The various CD formats and the relevant DVD structures have been discussed in earlier chapters. At this stage, you should have a master disc ready for hand-off to the replicator. Nonetheless, there are several common problems involving files that can stop the replication process until they are corrected.

Keep these considerations in mind when delivering files to the replicator:

- Find out ahead of time what kinds of removable media the replicator accepts. Most replication services have certain media types they handle regularly—there may be a conversion fee or other charge if you use something other than standard media. For CD-ROMs at most facilities, a CD-ROM one-off is the media type of choice. Other media types, including Syquest cartridges, magneto-optical cartridges, Jaz or Zip disks, and similar media are often acceptable; check first to be sure. For DVD-ROM or DVD-Video, DLT is the favored medium at most locations, but a

Replication Overview

DVD-ROM, burned from a DVD-R disc, is generally acceptable at many locations.

- If you deliver files on a medium other than optical disc, make up a complete file and directory list for the entire project so that the replicator will have a means of cross-checking the assembled files for completeness. Make sure that all necessary support files are included with the project.

- If you create your own CD or DVD one-off, make sure that filenaming conventions appropriate to your selected platforms are maintained throughout the premastering process. If the premastering application renames any files, you may have problems with links (if including HTML content) or referenced drivers or other files, if you are including a software application on disc. If creating a DVD-ROM for playback on the widest range of equipment, choose the UDF Bridge format to ensure backwards compatibility with the ISO 9660 file system.

- For software applications being distributed on CD-ROM, perform extensive testing of all the software components before committing final files to master. Ideally, this testing should include full tests of the installer software for each of the targeted platforms. If you don't have the resources to perform this testing in-house, hire an outside service to perform it. The expense is far less than if you end up having to scrap an entire disc manufacturing run because of software errors or installation problems.

- Run a virus check of all of your files on the optical disc using a up-to-date virus application for each of the operating systems that you are targeting. You can be fairly certain that if you transmit a virus to a customer, you'll lose that customer's business for life. If you discover a virus after replication, you've just purchased a huge batch of Christmas tree ornaments that will never see the inside of an optical disc drive.

Preparing the Artwork

Production of a CD-ROM, DVD-Video, or DVD-ROM involves a substantial amount of printed material, including the inlay card insert and liner notes for a jewel case, or the cardboard packaging for a custom project. You also have the option of text or images silkscreened onto the disc itself, or text or images attached using a stick-on label. In all these cases, you need artwork. You can prepare it yourself, or contract a graphic artist to do it.

Chapter 16

Some replication services offer printing as part of their services, either performed inhouse or subcontracted to a local printer. If you have the materials printed yourself, check with your replicator to ensure that the method you have chosen won't require additional work (such as elaborate manual filling of packages), which will invariably generate extra expenses.

The supplied artwork must, of course, be designed to fit the precise space available on the disc. Artwork fashioned to be silkscreened onto your discs must be set up to avoid non-printing areas, such as the center of the disc. Specifications that define the exact available printing area, as well as graphics templates that can be used in graphics applications, can often be obtained from the replication facility. These templates offer a sure-fire method for getting the artwork prepared to necessary precision. Figure 16 - 1 illustrates the region on a disc that is printable. Figure 16 - 2 designates the appropriate size artwork required for a jewel-case insert.

To get a sense of the typical artwork template, Disc Makers offers free downloadable templates at *www.discmakers.com/cdrom/*.

Figure 16 - 1 **Printable area on a CD-ROM**

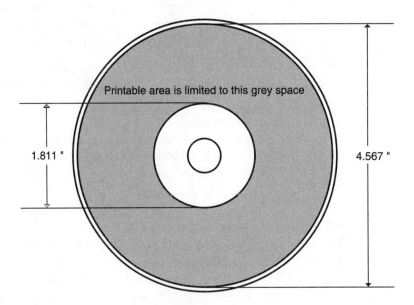

Figure 16 - 2   **Layout of a jewel-case insert**

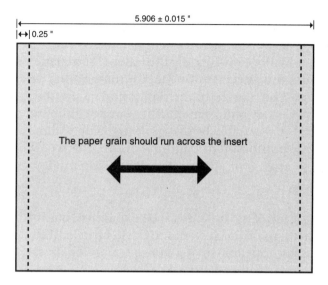

Jewel cases provide sufficient room for a booklet, the front and back cover of which displays the artwork printed on the booklet or other insert. The booklet size is about 4 inches square, consisting of a single sheet or multi-paged layout with a maximum of about 30 pages. Booklets are typically bound with a staple passing through the spine.

Whenever you're including artwork with a project, it pays to ensure that the recipient is properly equipped to read and process the files you submit. If you are using a Macintosh page layout program, and the replication facility only uses Windows NT computers, you may have to convert the artwork before turning it over. Versions can also be an issue. If you've produced artwork in Adobe Illustrator version 8.0, the replicator may not be able to read the file if they are running version 6.0. On the other hand, a properly generated Adobe Acrobat file can be structured to contain all the necessary information for printing and this may be the best cross-platform solution. Sometimes facilities like to perform final tweaks on the graphics files before producing printing plates—if so, you will need to use a format other than Acrobat for submission.

Four-color printing is available for optical discs, but is a less common approach. For more details about the printing requirements for various kinds of projects, refer to *Expert's View: Preparing Artwork for CDs* on page

399. Four-color printing generally requires that you use Pantone values to specify ink colors. To avoid problems, discuss any printing questions with your replicator before finalizing the artwork.

Printers use photographic film in the printing process—one sheet for each color used, to a maximum of five sheets (one each for the CMYK colors, and an optional, costly fifth sheet for a matte or metallic finish). Sometimes you may have the film for the printing process made by one facility and then passed to the replication facility. Be aware that there are certain necessary requirements that must be followed to ensure that the materials will be usable by the receiving facility: film can be positive or negative, emulsion up or emulsion down. The receiving facility will not be able to print your job correctly if you don't follow the specifications for the film.

Four-color process printing can be expensive, but it allows silkscreening high-quality color images onto the disc surface. Well-designed images enhance the appearance of a disc and can benefit sales of product.

Delivering the Files to the Replicator

Communication is the paramount concern when handing off files to the replicator. Make sure you address these concerns:

- If you deliver on a medium other than a fully premastered optical disc, you need to communicate to the replication service how to deal with the files that you provide. You should map out the file and directory structure, specify the intended format of the DVD or CD project, itemize the individual assets, and communicate all of this clearly to your account representative. Since most services have preprinted forms designed to accept all of this information, spend some time filling in the necessary data. If you need help with some of the requested information, don't guess at the proper response—ask someone.

- Whenever possible, hand over artwork for packaging and labels ahead of the disc master that is delivered. Extra time is generally required for processing these items, particularly if the replicator outsources the work to specialists. Packaging, in particular, can sometimes require an extra week or two for preparation.

- Be clear about all the scheduling issues and any proofing responsibilities or approvals that are required from you during the manufacturing. Replicators will do everything possible to make sure the job comes out right, but you will necessarily have to oversee some of the process and respond to their sign-offs and approval cycles as the work progresses. Take advantage of these opportunities to make sure that the manufacturing is proceeding according to your intentions—once the discs are pressed, you can't erase them and start again.

Manufacturing the Discs

Once you have turned all the necessary materials over to the replication facility, you are generally finished with your role in the process. At this point, you just need to sit back and wait for the manufacturing process to be completed, which can take anywhere from a few days to three or four weeks. If you have included artwork for printing, many replication facilities will offer a proof copy for your approval to ensure the design will print as intended. While this will approximate the final appearance of the completed disc, replicators will be quick to tell you that there are differences that will be apparent on the manufactured version, because of different materials and processes used. As long as you understand the nature of these differences and how they will affect the final results, you can avoid unpleasant surprises. Have the replication service explain the printing issues to you in detail before proceeding with the job.

Packaging the Discs

Discs can be packaged in a multitude of packaging materials and various types of containers. Your choice of packaging reflects the image of your company and provides the first impression of your title. Many companies are choosing minimal packaging materials and insisting on recyclable materials to reduce the environmental waste and responsibly deal with disposal issues. For more details on the packaging options, refer to *A Gallery of Package Types* on page 412.

Expert's View: Preparing Artwork for CDs

The steps involved in preparing the packaging materials for a disc and designing the artwork for the disc label should generally be accomplished before the disc is released for replication. These steps can require as much as an extra week or two beyond the disc manufacturing process, so it helps to begin somewhat sooner.

The characteristics of a silver or gold reflective disc are not the same as paper, so the printing process must necessarily be quite a bit different. The following Expert's View covers the issues involved in printing to CD

Chapter 16

as explained by Lance Svoboda, the Electronic Pre-Press Manager at Disc Makers, who counsels clients on the techniques that ensure the best results.

I imagine from your perspective, there are certain artwork preparation problems that you see all the time.

There are definitely certain common mistakes. Probably the most common mistake is supplying images that we can't use. Or, supplying images that we have to work with extensively (we generally always have to do something with the images). Some customers will supply their artwork in four-color, when they're looking to do spot color (which is what we use on the CD label). Or, they will supply artwork for true duotones, where they have one screen overprinting another screen. To convert those, we pick a lighter color and apply a base color; we do a flood coat underneath the halftone and actually only print a halftone over a solid area.

Why wouldn't a duotone work for a label?

There is a possibility of registration issues. Certainly, we can make film that is good and won't have any registration issues. When it goes to the plant, this is done in silk-screening. The registration can't move as much as a point in any direction. So, there is a possibility that a dot pattern will appear, whether it is a moiré pattern or a funny rosette—any of those things that happen when you lay a screen over a screen and it isn't in proper register.

The other issue is that there is no way that we can do a proof on a duotone. We don't have these inks. We can't predict how they're going to stick, how they're going to lay down. We can't show the customer anything that says, "This is how your label is going to look." It could come back looking very different from the laser proof that we supply.

So, you're really basing your judgement on how it is going to look on past experience?

That's right. There are a thousand different PMS inks and depending on how they are mixed at the plant, they are either more or less translucent. Two plants could mix the same color and one could come up with a very opaque ink and the other come up with a very translucent ink. The same disc printed at two different plants with the same duotone could look very different.

Expert's View: Preparing Artwork for CDs

Are there recommendations that you give people in terms of structuring the design?

Stay away from large areas of tone. Unless you're doing a halftone where the image changes as it goes across. If you wanted to do a 40% tint across the disc, just by the nature of silk-screening, you can get waves in it—where the ink is thinner or thicker in certain places.

You won't see that in a half-tone of an image, because the image will hide it. But, if you were to lay down a 40% tint and put it down over a disc, you would probably see some shift somewhere, just because it is impossible to pull it down evenly.

Large areas of solids aren't a problem—just screened areas. If you're doing a large area of solid, the printer will mix the ink thicker. So, the CD printer is going to mix the ink based on whether or not there is a half-tone in the image or it is a solid image and whether or not it is silver or applied over another color. So, if working with a solid area, the printer can mix a really thick ink. You can lay that down without a problem; you'll never see any waves with that.

But, if you're working with a half-tone, you need to mix a much thinner ink. The thinner the ink is, the more likely you're going to see any inconsistency in the thickness of the ink. Because the ink is translucent.

Do you have recommendations in terms of where colors are butting up against each other? Or, for bleeds?

We don't bleed the CD label. One hundred sixteen millimeters is all that it goes. It is a circle and that is that. However, we do trap very aggressively. Depending on what the colors are—if it is type on black—we will give it as much as a one-point trap, which is a very big trap as far as offset printing goes.

With silk-screening, the registration isn't that great, so big traps are pretty useful.

What work do you have to do on the artwork submitted for labels?

We check all the files to make sure they aren't using any duotones. And, to ensure that the files aren't using RGB colors.

With CD labels, we are more likely to take apart and then reconstruct whatever the customer did. We make sure that each piece of the artwork is good. This is the part of the job that takes the most time. We'll send it off to the plant and it will be a 10-day turnaround before the discs get back to us. With the print work [booklets and inserts], it will be printed and cut up in the next 3 or 4 days.

You really don't want to lose any time when you're dealing with the CD labels. We check them extra carefully.

Once we get the label checked out, we supply the customer a proof. They approve it and then we generate film.

Do you submit a positive or a negative?

A film positive. For most of our labels, it is an 85-line screens.

If somebody wants a higher resolution, is it available at extra cost?

Definitely.

I understand you can also get 4-color if you want.

You can get 4-color offset printing. It is actually 150 lines, so it is very high resolution. We use the same resolution that we use with our print work for our 4-color process work.

There are some interesting issues with it. It is a waterless system using a common blanket. So, when overprinting type and large solid areas sometimes it doesn't seem to be as opaque as you would predict. You overprint a black and it comes back kind of grayish. Whereas, with a silk-screen, you can overprint a black any time that you want, because it is a different process.

Can you get around that by using a knockout for the area that you would normally be overprinting?

Exactly. If you are going to be using heavy tones on top of another color, you should definitely knockout and not overprint on the process label.

If someone submits artwork where they haven't done that, do you make those modifications before they go to film.

Definitely. If a job came in and it was already trapped, we would still go back through and trap the job again. Even if we didn't make any changes, we would check every single setting. Just because the majority of the time, the labels are set up for a regular process trap. This is the default of most applications. With the silk-screening process, we're a little bit more aggressive.

Spot printing versus process printing—there are things that you want to do and other things that you don't want to do in terms of trapping. A lot of times you have common tints in process printing that just don't exist in spot printing.

It sounds as though you probably have to modify 98% of the artwork that you get in just to make it ready for the specialized CD printing?

Definitely. Most people don't do this every day. Most people deal with printing on paper or cardboard. Each material has its own characteristics. When printing on discs and when doing any type of spot color printing, you have special issues to think about.

Does it make any difference which the graphics program someone uses to prepare the artwork? Does Adobe Illustrator work better than Photoshop artwork, for example?

That is really determined by the customer's artwork and their comfort level with a particular application.

So you don't favor any particular application?

If I did, it would only be a personal preference. It wouldn't apply to the guy who is sitting next to me or across the room.

I understand Disc Makers provides standard templates in different formats on your Web site to help simplify the artwork creation.

When working with CD labels, this artwork is really an important thing. Because these labels have rings, stacking rings contained within them, and everybody's template is different—every plant uses a slightly different template. They each have their stacking ring in a different place. Chances are, you're going to have art going across that stacking ring. If you're not using our template, you can't predict where it is going to be. If you haven't allowed for it, we're going to have to make room for it when we modify the artwork. That can affect the appearance of your art.

Whether you are preparing a job for us or you're going to have it printed somewhere else, you should always find out what the specifications are first. Disc Makers offers these templates free of charge and they are a gigantic time savings to the customer. If you use them, the job will be closer to being in spec, so we don't have to change things in a unpredictable ways.

Can you give any guidelines in terms of workable designs for CD?

An important thing to remember is that the design is being laid over silver and not over white (unless you are laying down a flood coat). A lot of times people lay down a halftone on their LaserWriter that prints on white paper and they think it is going to look exactly like that. But, when it goes onto the silver, some of the lighter tones disappear because the gray value of the silver is similar to the value of those screens.

Another point is that, although your package allows for 2- or 3-color printing, you can use the silver of the disc as a design element. So, it almost becomes a third or a fourth color.

Are there other media types that you work with other than silver?

There is one type of disc where the customer can still opt for a gold surface. But, that is only for the short-run CD-Rs. As far as I know, all the replicated discs are now on silver.

Most of our runs are pretty short. Most customers want the advantages of a specific replication package. Those are all done with the silver.

When you say pretty short, you're thinking in terms of 1000?

I'm thinking 300, 1000, maybe up to 5000.

Can people use metallic inks and fluorescent inks on their discs?

If it has a Pantone number, a PMS number, you can use it. But, once again, with the metallics and with the fluorescents, when we give you a proof, it will be a color laser proof, and it won't show those colors exactly as they will print on disc.

If you're doing a black disc with some gold lettering, it is pretty easy to tell what is going on. But, if your layout is more complicated, it might be difficult to see on the proof what it is actually going to look like when it is printed.

Expert's View: Preparing Artwork for CDs

One of the things that we always want to do is to show a reliable proof to the customer, so they're not surprised when they open the box and see their discs. This is why we stay away from duotones and designing things with fluorescent colors or metallic colors.

Can you use a simple inkjet disc printer to show what a design would look like?

We don't. We give them a color laser proof. I know that from time to time, Disc Makers has looked at different proofs that would be over silver, and we still haven't found one that we like. This proofing problem is just something that we live with.

Looking at the color laser print, can you generalize as to what you'll see on the actual CD? Is there some color shift?

There is some color shift. If you're printing over silver instead of white—especially if you've got a halftone over silver instead of a halftone over the white, you just have to try to visualize the appearance over gray.

A lot of times an artist will put the design over a gray color (on the computer) so he can get an idea of what it is going to look like on the disc. But, we can't show that as a preview, because it isn't really that close either. It can makes things look worse, when it really won't look that bad. Sometimes it works OK on your monitor, but it doesn't work as well when you print it.

Sounds like a difficult problem. I can see where there are probably customers who don't understand why they're not getting back exactly what they submitted.

I have seen jobs bounced because of it. We put notes on the proof and we try to warn the customer ahead of time what is going to happen. Especially in our department where we work with the design issues, where a customer will do something that isn't going to look very good and oftentimes they are in a big hurry. The job goes through anyway. Maybe we give them a call, but they don't have time to talk about changing it or getting proofs. When they finally get the CDs, sometimes they're disappointed because the design elements don't stand out like they did on their laser proof.

Any other design advice?

Here are some good points. When doing halftones, a lot of contrast is a good thing. These are big dots. These are 85-line dots. When you've got images that are faint, with the silk-screen process (where it could have waves in it from the ink anyway), subtle images won't show up nearly as well as a large image with a lot of contrast. Also, the plants generally say that their screen tonal range is 15 to 85%. When you think of offset where the tonal range is 3 to 97%, that is a big constriction.

I don't know how many jobs I've seen with pictures of clouds, where they've got a range from 15 to 40%. It looks OK in the proof. When it goes to print, the contrast isn't enough, so it is hard to see the shape of the clouds. Sometimes you don't even see the clouds unless you are holding the disc at the right angle.

High-contrast and solid colors work well. Tonal shifts get lost in the lighter ranges.

Gradients are good to stay away from too. With gradients, you'll get banding. The same kind of thing, the tonal range is so short it can give you what appears to be a ring for a band. The way the printing picks up the ink and picks up the dot, it puts almost a line.

Sometimes the customer will supply something that seems like it is going to be OK. But, then when it gets to the plant and it prints a little light, all of a sudden it isn't OK. You thought that 40% to 25% would be enough contrast to show something. It ends up not being enough.

What percentage of discs do you seem to run into artwork problems with?

Well, with many of them there are no problems. Because the customers are just doing type on a disc. Then, once they get into doing a halftone on a disc or a gradient, it gets more complicated. I'm looking at two gradients here that just look beautiful, that worked out very well. You know, you really can't tell when you are doing it what it is going to look like. Some look very nice, but I've seen some that print terribly.

If you have a halftone on a disc, this is one of those things, we run through and do a series of checks. We check a whole bunch of different things, every element that is in the artwork for a whole range of potential problems.

A lot of disc art goes through with no changes. There are other designs where we change everything. That is one of the reasons that we always send out a proof—we don't make design changes and then go to print. We change the artwork and show you what we have done, or at least talk to you. A lot of times with the CD-ROM customers, they are in such a hurry they don't get a proof, but we do discuss it with them. I would guess that there are some changes needed on 75% of the jobs.

Most of the stuff you can fix just by tweaking the artwork a little bit?

Yes, I would have to say that there have maybe been two discs in the last 2 years where we asked a customer to do a redesign. The overwhelming majority of what we do, we come up with a solution that is close enough that we are willing to present it to the customer.

Summary of Disc Artwork Tips

The following tips, some of them gleaned from the previous interview, can improve your results when setting up the artwork for a disc design. These guidelines were adapted from the Spring 1999 issue of *CD-ROM Access*, published quarterly by Disc Makers. A free subscription is available by signing up at: *www.discmakers.com/cdrom* or by calling 1-800-237-6666.

- Keep your design simple. The standard line screen is 85 lines per inch (lpi) or about the same screen ruling as you find in newspaper print. Also, the tonal range allowed is 15% to 85%, which means that tints less than 15% may not show up and tones darker than 85% may close up. These factors limit the amount of detail that you can present in an image.

- Work from a manufacturer's template. Each manufacturer's CD has a slightly different printable area depending on the mold they use, so only their templates will be perfectly accurate for their CD work.

- Bleeds, design areas that run past the outer edge of the printed media, are not permitted.

- CD label proofs usually show more detail than the actual on-disc silk-screen print and do not always give an accurate impression of what these kinds of problem areas will look like.

- As many as four or five spot PMS colors can be printed, but this is not the same as full-color process printing. If a white area instead of a silver background is desired, a "donut" of white ink can be laid down first. Any of these on-disc print options involve additional cost.

- Full-color process printing is also available for additional cost. There are two varieties of this: silkscreened (85-line screen, but with improved alignment accuracy to permit process blending of colors) and offset printing (150-line screen that approaches the quality of full-color process printing on paper. Offset is the most costly method of on-disc printing and it can add several days to turnaround time, but the quality difference between full-color silkscreen and full-color offset is significant.

- Trap aggressively, depending on what colors you're using. Trapping addresses possible registration problems that can occur between adjacent colors in a printed design by slightly expanding one color region into the other. If the design includes type on black, give it as much as a one-point trap. Many graphics applications programs include support for trapping. If you are supplying electronic files to the replication service, a pre-press specialist will generally check and trap all of your files for the best results when printed.

- PMS inks have varying levels of opacity, so the color combination from printing the inks on top of each other may be different from what you expect. Other screened blends of two or more colors can also present problems for these same reasons.

- True duotones cannot be done predictably on disc, since registration in silkscreening is difficult and a moiré pattern can easily occur. Also, for the same reason, avoid using fine type knocked out of a screened background or screened type in small sizes.

- When proofing, keep in mind that printing your label proofs on white paper will be misleading, because your design will be printed over the silver surface of the disc (unless you are laying down a flood coat of white). This may cause some of your lighter tones to disappear due to the gray value of the silver being similar to the value of those tones. You can get around this by using the silver as a design element—so it becomes an extra color.

- Short of printing the ink on a CD-ROM, there is no totally accurate means for viewing what the actual printed surface will look like. Skilled professionals can anticipate many of the effects that may occur, but even they are surprised from time to time. The fewer unpredictable elements that you have in your design, the less chance you will be surprised by the outcome.

Packaging Options

Optical discs are durable, but if you're going to ship them or store them you need some type of package or storage container. Depending on your intended use, the package might be anything from a single fiber sleeve to an accordion fold multi-disc package with an accompanying full-color printed booklet. Disc titles that will share shelf space with books or other software titles generally require more elaborate packaging and professionally designed artwork. Titles that are sold or distributed over the Web may require nothing more than a protective shipping container. If you're producing an audio CD, the jewel case format is almost inescapable, since the vast majority of music stores favor this approach in their display bins. If you're producing a CD-ROM to accompany a book (as we have with this book), the disc is typically delivered in an envelope that is glued inside the back cover of the book. Discs that will be mailed require a sturdy package to survive transit during postal delivery.

Packaging is one part of a three-part process. It is a step that should be considered in concert with distribution and advertising. Different distribution channels have different requirements and considerations for packaging. Advertising initiatives differ for the different distribution channels, and this in turn affects the choice of packaging. If you are independently marketing or distributing a title, refer to Chapter 18 for details on techniques to integrate your packaging approach with your marketing plan.

Steps in the Packaging Process

Points to consider in your individual approach to packaging are as follows:

- Clarify the target audience and identify precisely the intended use of the product. Will the package itself be used in marketing the product? If so, the artwork should be coordinated with any printed ads. If you need a mailer, will a plain one suffice, or will you need a custom approach designed to integrate with your total marketing plan? What other materials (such as user manuals, brochures, reference cards, and so on) will be packaged with the optical disc?

- Find a replication service or production house that specializes in package design and execution. If you can find a single facility that offers both design and replication, this can speed the process and offers fewer chances for miscommunication. Obtain samples of their previous packages to examine the quality of workmanship. It is a major bonus if the company has a client-services person who is articulate and understands the needs of the marketplace. This person can then assist you in making the proper choice of regarding a package design, and even advise you on marketing issues, since they've been through the process many times with other clients. Such assistance is invaluable, especially if it's your first venture into CD-ROM production.

- Choose a package design. An effective package design attracts attention (if it's for a retail sales outlet), communicates the key information about the disc title, and provides protection for the enclosed disc or discs. As you're selecting an appropriate package, you might want to qualify the selection based on your budget, the package's intended use, the protective characteristics of the package design (some containers, such as quad-disk jewel box holders can come apart if they receive rough handling.

- If the package is intended for retail sales, the form factor of the container must not be too large or too small—if it's too large, the store may be forced to place it in an out-of-the-way location. If it's too small, adjacent products will overshadow it.

Of course, the overall appearance of the package should be evaluated by a true artist and a marketing mind. For example, glossy, laminated packages not only catch a person's eye, in theory, they are supposed to increase the perceived value of a product. A counter trend has developed that deals with packaging in minimalist terms. If you can achieve a distinctive design using fewer materials and recyclable fibers, a wide segment of the market will be attracted to this approach, particularly if your audience includes socially responsible businesses or individuals.

Watch the trade press for new box designs. For example, LaserFile has recently released a jewel box variant in which the tray slides out and folds back for easy removal of the disc. The figures in this chapter show several examples of the more popular types of packages.

Note: At this stage, you need to decide about details such as placement of stickers on the outside of the shrink-wrap, whether hang tabs are needed, size of shipping cartons, and so on. The replication service or production house usually has a form for you to fill in that steps through each of these details.

Packaging Options

- Prepare (or have prepared) the artwork and collateral materials. As always for such work, make sure that the file format for the artwork can be read by the production house or printshop. It can help avoid problems if you obtain electronic templates of layouts from the production house or printshop before you begin designing your artwork. That way, you can be assured you are working within acceptable guidelines.

    You should use CMYK colors, not RGB, for the file format, as printers use CMYK, and color shifts can occur in the if your art has to undergo translation from one format to the other.

Note: At this stage, you need to arrange for the writing of the manual, warranty card, and any promo materials that you plan to include with the package, as well as the design of these materials. In this MTV era of attention spans measured in nanoseconds, the writing should be forceful and to the point.

- Arrange to have a prototype of the complete package made. This step allows you to verify that all the materials will fit inside the package, that the package is sturdy enough for the contents, and so on. Think of this as a "beta test" of the finished product, and try it out in the way that it's intended to be used. If it's for a retail sales shelf, pick it up, drop it, pick it up, bang it down. Open it up, take out all the items, and then try to put them all back properly. Is it easy to get the jewel case out? Is it easy to replace the manual in the box?
- Go ahead with production. The packaging and replication house should take care of this step for you. If you have properly addressed the design issues, packaging contents, and so on, there should be no problems at this stage. A couple of issues to think about:

If the packaging house and the replication house are not the same company, then you must take extra care to make sure there is frequent contact and discussion between them, to minimize the chance of problems arising due to miscommunication.

If you expect reorders, order more packaging than you need for the first run of discs. This will help you fulfill the reorders more rapidly, and it will save you reprinting costs. Many replication services and production houses will store extras for you, and act either as a storage facility or a fulfillment center.

Chapter 16

A Gallery of Package Types

This section comments on the figures which show samples of various packaging options which are available for CD-ROMs, DVD-ROMs, and DVD-Video discs.

The standard jewel case, shown in Figure 16 - 3, has a booklet insert for a cover, a plastic tray and a tray card with the title on the spine, and which displays graphics and text on the back. You can leave out the booklet or insert to show off the silk-screening, and also select colored or clear trays.

Figure 16 - 3   **Standard jewel case**

The package shown in Figure 16 - 4 has a cardboard fold-out that includes five surfaces for holding graphics and text, as well as a plastic tray for the CD itself. The large number of surfaces lets you add extensive promotional or informational material for the prospective buyer.

Figure 16 - 4   Cardboard fold-out with a plastic tray

The package shown in Figure 16 - 5 has a cardboard fold-out with five faces for holding graphics and text, a plastic tray for the CD, and a flip lock to hold the package closed. There is also room for a booklet insert. The five surfaces and the booklet allow even more information to be presented to the user than the option shown in Figure 16 - 4.

Figure 16 - 5   Cardboard fold-out with plastic tray and flip lock

The cardboard package shown in Figure 16 - 6 holds four discs and has eight surfaces for holding graphics and text. This package is quite delicate, and so it should be stored inside a more robust container, such as a mailer envelope or a box.

Chapter 16

Figure 16 - 6 **Four-disc cardboard holder**

The package shown in Figure 16 - 7 is a plastic tray which holds the CD, with a cardboard slip cover. The front and back surfaces of the sleeve can hold graphics and text. This is another type which must be packaged inside another container. It is very convenient for a user's shelf beside a computer, since the disk is easily removed.

Figure 16 - 7 **Tray with a sleeve cover**

This package (shown in Figure 16 - 8) is similar to that of Figure 16 - 4, but with a cardboard slipcover, rather than a plastic tray. There are four or six surfaces to hold graphics and text. This type of package is suitable for inclusion inside a box with a manual and other collateral material.

Figure 16 - 8   **All cardboard fold-out**

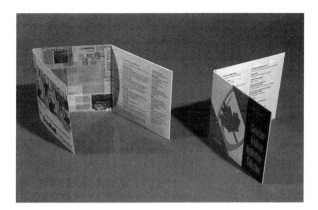

Figure 16 - 9 shows a simple slipcover package, commonly known as an LP cover. There are two surfaces for graphics and text. Clearly, it must be packaged within a more robust container for distribution.

Figure 16 - 9   **Cardboard slipcover ("LP cover")**

The package shown in Figure 16 - 10 is a simple sleeve made of vinyl. It is not intended to have graphics or text, but merely acts as a distribution holder inside another container.

Chapter 16

Figure 16 - 10   **Vinyl sleeve**

This "package" (shown in Figure 16 - 11) is simply a paper sleeve to hold the disc. It is used within other packaging options, such as a box.

Figure 16 - 11   **Paper sleeve**

Figure 16 - 12 shows a simple box with the materials and disc placed inside through a flap on the thin top edge. The disc can be held by anything from a jewel box to a paper sleeve. The figure shows the disc in a sealed envelope. The buyer breaks the seal to accept the licensing conditions printed on the envelope.

Figure 16 - 12   **Slim-line cardboard box**

This package shown in Figure 16 - 13 can hold a thick manual plus the CD-ROM and collateral materials, or (as shown here) 25 CD-ROMs, each in its own paper sleeve, and wrapped with cellophane in two bundles. The package shown is part of Corel Corporation's 25-CD-ROM photo stock library.

Figure 16 - 13   **Front-loading cardboard box**

Chapter 16

This package shown in Figure 16 - 14 is similar in opening to that of Figure 16 - 13, but with the addition of a die-cut holder for a jewel case. There is room below the die-cut for a manual or other materials.

Figure 16 - 14   **Front-loading box with die-cut holder**

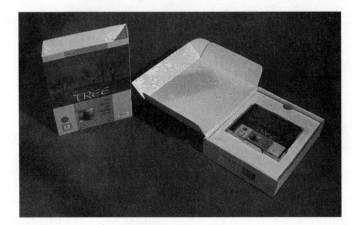

Figure 16 - 15 shows an envelope with a sticky back, which is pasted into the back of a book or manual. The envelope is printed with a licensing agreement, and the buyer signifies acceptance of the licensing agreement by cutting open the envelope.

Figure 16 - 15   **Envelope in back of a book or manual**

Summary

This package shown in Figure 16 - 16 is a simple mailer, with no graphics or text. It is sufficiently robust to withstand the rigors of snail mail.

Figure 16 - 16   **A simple mailer**

There are many variants of each of the packaging options shown in these figures. Your replication service or production house can work with you to find the one most suited to your application, budget or needs.

Summary

As you can see from this chapter, there is a fair amount of specialized knowledge that is required to successfully prepare artwork for CD and DVD printing. The requirements should be clearly described by your choice of replication services. Any issues involving file transfers and disc formats should be clearly identified in advance of the turnover of materials to the replication service—items can often be overlooked or mistakes can be made in haste when deadline pressures arise. With some planning and forethought, getting discs replicated, printed, and packaged can be a straightforward and simple process.

17

DVD Creation

If you want the best perspective on the process involved in authoring and premastering DVD titles, turn to those who are on the front lines of the DVD creation process. This entire chapter is devoted to an interview with Blaine Graboyes of Zuma Digital, one of the leading DVD design and creation facilities in the world.

Expert's View: Developing Titles for DVD

Interview with Blaine Graboyes

While the industry continues to tinker with DVD copy protection schemes and the DVD-Audio format undergoes countless revisions, Blaine Graboyes and the crew at Zuma Digital have been busily developing titles in a variety of DVD formats. Not content to wait until there is a consensus on the direction of the newest optical storage media, Zuma Digital has been steadily working with the pioneers in the industry to take advantage of the impressive storage capabilities of DVD discs and bring titles to market.

Recent projects have earned the development group notoriety, particularly a project completed for the Guggenheim Museum titled *Premises: Invested Spaces in Visual Arts, Architecture, & Design from France, 1958 - 1998*. An exploration of the artist's interpretation of space, this project stretches the limits of DVD presentation capabilities with sequences that play back on forty-five individual displays.

Chapter 17

During their short but prolific history (Zuma Digital has been around since 1997), some valuable DVD development lessons have been learned, as discussed in the following interview with Blaine, the COO and Creative Director of Zuma Digital (*www.zumadigital.com*), conducted in January, 1999.

I understand that Zuma Digital typically handles projects from start to finish, including development, authoring, and premastering?

Yes. I come from a multimedia background, so when it comes to DVD authoring, we're in a pretty good position to help our clients.

Are people using Macromedia Director to author DVD-ROM discs?

There are a number of Xtras for Director that allow you to incorporate VOB files [Video Object Files], which contain MPEG-2 video and AC-3 audio, into Director. You would need another application to handle your premastering and the creation of a disc image (since Director doesn't have the ability to create a DVD-legal disc image).

In reading some background material, I noticed that you also have a background in art. How did you get from there to your current work in DVD?

At Zuma Digital, we only hire staff members that have background in other areas—fine arts, literature, history. I think this kind of background is essential for working in any aspect of multimedia or DVD production.

I went to Bennington College in Vermont as a Photography Major and left as a Mathematics, Sculpture, and Philosophy of Science Major. The first term of my Junior year, the college won a grant. The New Media Center was a consortium of hardware, software, and technology companies who were interested in setting up advanced facilities at colleges and universities. Bennington was the 21st school and the only liberal arts school. The rest were schools like NYU, Columbia, Duke, Stanford.

A fellow by the name of Reuben Pentadura from Bennington College was put in charge of running the media center. He was my thesis tutor in chaos theory, thermodynamics, and relativity. I had actually never touched a computer before, but I was very interested from a photography class that we had; my photography professor had shown us how to scan images. He asked me if I wanted to take a job managing it.

I said yes. He had already hired a man named Bill McCabe to run the center from a senior management point of view. I was put in charge of running it from a staffing and upkeep point of view. I just literally dove in and started working every day, seven days a week there, learning about computers.

Before that time did Bennington have much going on in the way of interactive media?

We didn't have any computers, to speak of. There was a little computer lab that was mostly for writing papers. There was one scanner that my photography professor had installed. There was nothing, though, that could be described as interactive media at Bennington.

Were you using something like Director at the time?

Actually we were using AuthorWare at the time—one of my favorite tools.

Do you still work with Macromedia products?

Yes and no. Right now I'm working on developing a board-room presentation product and the front end for it is written in Director. But the product is not computer-based at all. This product simply uses the computer front-end as a way to author presentations that are then downloaded to a hardware device that we are building that allows you to control an industrial DVD player and play back various video and audio elements on the disc in a custom order for a presentation.

When you say you're "building" a hardware device, it sounds as though you're maybe getting into some logic design and that type of thing.

Absolutely. We're designing the hardware. Just last night I finished the specification for the cabling that will tie it all together. It's a proprietary cable that allows the device to function as widely as it does.

This sounds like fairly heavy-duty engineering type work—have you trained yourself in digital logic design along the way?

We work with outside designers, but—yes.

As projects come up where you're working with DVD, at some point you must have to pick a branch and decide to go either with DVD-Video or go with DVD-ROM. Do you have certain criteria to decide when one makes sense over the other or is it more driven by the client's needs?

It is very much driven by the clients, but there are a few clear-cut areas. If you need to have database access, you need to be on DVD-ROM. If you need to do anything that is platform-specific, or something that needs to be handled by a computer processor, it seems like at that point you're going to need to go with a computer.

Although we are working on some solutions right now—I have something that is up in the air where we're doing a kiosk for the Federal Law Enforcement Training Center. For that project, we are handling all aspects of the production, from the design and fabrication of the kiosk itself to the production of the media to the DVD production, as well as the support.

Their specification calls for tracking the user. They want to know what screens they access, how many times they access it. I have two proposals I'm working on right now. One is for doing it in AuthorWare. It's funny, it's the first project I've done in AuthorWare since I did that one for the college. AuthorWare is great for tracking users while you're going through an application and using AuthorWare with a DVD-ROM solution, playing back MPEG-2 VOB's, using the Tibianni plug-in. The other way to do it is to use a proprietary input device that we built here from a technology that Pioneer calls JAMMA technology, which is a different way to use the RS-232-C port on their industrial player, making a new category of devices that access it. So, the second proposal we put together (and the one that I am really going for if I can finish the last piece of the puzzle) is tracking the user through there.

Since they're activating contact-closure buttons, there is some way to track that. Now it is just a matter of coming up with the logic of determining how you know what screen they are at, if they press the down button 17 times, they end up on this screen. We're working on developing another device that would fit between the JAMMA control unit and the DVD player, and actually track how many times and what it accesses.

I get the sense that maybe you consider the DVD-Video a more stable platform to work on—maybe because of the hardware?

Without a doubt. In a week and a half I'm heading out to California. Microsoft and Intel sponsor these plug fests. On January 11th, Microsoft is going to be releasing their new Windows Media Player, which actually gives DVD-Video support at every level of the Windows 98 operating system. I'm a Mac guy, but I have to say that Microsoft is really leading the pack when it comes to integrating DVD-Video and DVD-ROM playback into their operating system.

With this new Windows media player, you will literally be able to embed any or all of a DVD-Video project anywhere in the Microsoft operating system. From a Web page. To a PowerPoint slide. To a Microsoft Word document.

[*Updated comment from Blaine*: It is interesting to note that a year later, the promise ofered by Microsoft has not come true and that this is what prompted Zuma Digital to create our ActiveDVD software for PowerPoint.

The downside, of course, is the potential instability of the Windows 98 OS, or even NT. It seems that for multimedia, the Mac OS is more solid.

Absolutely. I talk to the guys at Apple every day and I'm really sad that they're not moving along a lot faster. I think they have made a lot of mistakes early on in supporting DVD, in not supporting DVD soon enough. Also, taking a very hands-off approach up until very recently in developing products for DVD.

Don't you think the potential is there within QuickTime?

Supposedly the potential is there. We're a total Mac-based shop here. From every side of the fence, Apple has done a poor job in helping us. All the way from PC Exchange to their UDF support to their DVD support. With Apple, there is no way to play back a VOB file standalone in a DVD-ROM application on a Macintosh computer.

So they really are lagging behind?

We've had that capability on the Windows side for 12 to 18 months. That's way too far behind for Apple to be on anything that is multimedia related. We send every product that we make to Apple for them to test for DVD-Video compatibility. At first, we were having what I consider to be a fairly high failure rating.

Now I do have to say they have been incredibly responsive. From what I understand, they are really scrambling and really pushing hard to get some good products to market very quickly.

Chapter 17

Do you think that some of the problems may be staff related, since they've been through several upheavals recently and are still rebuilding?

I can certainly imagine that they don't have the resources, either qualified or unqualified resources in an area like this which certainly if you looked at the DVD-ROM market 6 or 8 months ago you could have said, "Well, I don't really know where it is going to go." Nowadays, I don't think that there is a question. My accountant just asked me this morning—she just bought a new home theatre system and right on the back of her receiver there is a spot that says DVD. She was going to buy a CD... She needed a new CD audio player. She asked, "Do I really need to buy a new CD player cause I want to buy a DVD player as well?" No, you don't need a CD player any more. Go buy a good DVD player. You'll have all the functionality of a CD player plus you'll be able to play these new discs.

The same thing is happening in the computer market as well. I just bought a new laptop last night. It was no more expensive to have the DVD installed.

Did you get a PowerBook?

No, unfortunately I bought an IBM ThinkPad 770X. It comes with the DVD installed in the bay. One of the reasons I use that: we have a number of PC DVD systems here. There is a type of playback called PES–Packetized Elementary Streams. If you decode using PES and playback using PES, there is a great deal of functionality that Windows 98 is supporting via DirectX (or ActiveX or ActiveShow or whatever they call it now).

There is an incredible ranch of functionality for the playback of standalone VOB files in the DVD-ROM type environment. In the same way that I was describing before that the Windows Media Player will allow you to incorporate DVD-Video authored content anywhere, the PES playback in DirectShow will allow you to incorporate VOB playback in the same way.

For someone who hates computers, you've really become immersed in the technological side of it.

I have. I really like to think that it is just another hammer in my tool belt. It is one that is a very, very integral part of what we're doing.

Whenever it comes to trying to develop presentation, playback, installation show system, you need to have a computer at least somewhat involved, but it has really been my goal to have it minimally involved.

Expert's View: Developing Titles for DVD

Within the DVD-Video spec, there is some built-in interactivity. Is that something that you are able to take advantage of?

Absolutely. We've developed some solutions here, for example, for an advertising agency. The solution I described for you earlier is a way for—in a board-room situation—for an advertising agency to be able to re-order spots and make a custom reel. We've developed that same exact technology on the DVD-Video side from a regular, standard, set-top player. Basically, you can go to a screen and you just start picking things. You pick the first spot, the second spot, the third one you want to play. When you're done with your presentation, you hit "end" and it goes and plays back your presentation for you.

There is a great deal of interactive potential in the DVD world. It is just that sometimes the limitations and the expenses related to those limitations make it easier to develop other solutions. The one limitation we found there is, after you have picked the order of spots for your presentation, since everything is hard-wired and there is no data input, the way there would be on a computer, there is no way for me to tell you what you have just picked. Without you watching it.

This is something that, for most of the agencies, they weren't very comfortable with. They find that their executives are so nervous when giving a speech that to not know what they're getting into and what order everything is going to play in—it's just too much for them.

It sounds like you've fairly quickly become one of the foremost companies using DVD technology. Were you focused on the DVD aspect in your growth plan?

It was our goal from the very beginning to focus on the more difficult projects, the more complex projects, the tight time-frame projects, the large creative projects. The staff and facility that we've built, while we are in a position to handle a mass amount of very straightforward work, some of which we do take, we do have some clients for whom we do four or five films a month for them. They're always all the same except for minor artwork changes.

But, we've always been very aggressive going after the more complex projects, the more difficult projects. We've had numbers and numbers of projects that other houses have turned down, either because of time, expense, difficulty, complexity. We'll spend a day or two here in the studio. We'll get all the stuff together. We'll brainstorm how we can go at it. We'll give it a shot. I think that at this stage of DVD, the greatest advantage that anyone producing any project for DVD can do is to be innovative.

If you're innovative now, you're going to stand out and eight months from now, your products are going to be up to par when compared to the new stuff that is being done. As opposed to some of these film distributors who are literally just shoveling out anything they can get.

I talk to film distributors every day. Their concerns are getting products on the shelves. If they can get two up the shelves versus one with half the features, they're going to do it. It's the folks that take a second and say, "Wait, if I really take my time here and put a little more expense into this title; if I take a chance, and put my money into one title instead of two..." (which I don't think is taking a chance, I think the other way is taking a chance), without a doubt, they're going to have a pay-off now and they're going to have a pay-off later, too.

Within the corporate environment, it seems as though companies are forced to go outside for development just because the tools aren't there to let them do what they are used to. Do you think tool availability will changes things?

I think so. We've already seen some $5000- to $10,000 authoring systems. For some corporate work, I think these tools do a great job, actually. I get a lot of calls, people from various vendors and people that are looking to get into the fray talking about tools. And, mostly, the thing to do is: when a company needs to make a decision as to what type of work they're going after (and if you want to go after, there's a lot of very straight-forward corporate work), I'm all for companies getting their own system. Actually, a huge agency that we've been going after for ages just got their own system.

I would think that would just help grow the entire industry? Which would help your business, as well.

It already has. Without a doubt. Now that they have a system in house, I don't need to go after them any more about authoring their little DVD projects—their $3000 to $10,000 DVD projects. Instead, I think that we're going to be better off going after larger scale solutions for them. How they can implement DVD in a variety of ways and ways that their tool may or may not let them do, but that their expertise and experience doesn't allow them to do. That is really most of it.

Expert's View: Developing Titles for DVD

What we have built here at Zuma, if you go out into my loft right now, we'll have a music project running, a corporate project running, a film project going, an archiving project going. What we gain from working across all of those disciplines... Yesterday, I was in two meetings with DVD-ROM clients. I was in meetings with DVD-Video for film. I was in an industrial one. I quoted two museums. Working across all of those areas, it is really just about the experience and expertise that you get to add. There are a million things... Actually, on the DeVry Project, part of the specification for the project was that they had to use the Panasonic portable Palm Theatre. They absolutely had to use them. They bought 25 of them before they even called me about the project. They really had to use them.

Isn't that kind of the industry standard anyhow?

Yeah, there aren't many other choices. The Sony doesn't have a screen. I have one sitting on my desk. I think it is a pretty OK little unit. I wish that they in particular would have been more responsive to some of the industrial markets needs. Since, they had really imagined it to be a rich person, leisure device (rather than an actual business tool). Part of the spec was to use the Panasonic Palm Theatre.

Another part of the spec was that, basically they have one long video and all the presentation does is pause every so often. It is playing through one 15-minute video and every so often it is broken up to pause. Maybe 150 times. When it pauses, they need to have a full-screen pause. Not a half-field pause. Which is what the Panasonic player will do by default.

So, this is kind of a passive kiosk type product?

It's for presentations. DeVry uses this for road-show presentations. They have 20 some sales folks who go out on the road to high schools, colleges trade schools, universities, to make presentations about DeVry Technical School.

So they've got a captive audience who sits there and watches these little segments.

Exactly.

During the pause, is someone speaking?

Yeah, there is a speaker up there. A lot of it is silent and he is speaking over it. But when it paused, it was essential to them that it was a nice crisp clean pause. I spent literally three-and-a-half weeks working about at quarter time figuring out how to do that pause on this player.

It ends up being the most obscure, random, unconnected setting in your DVD-Video encoding that Panasonic is looking for to let you do that full-frame pause. I think that is where Zuma has something special to offer the corporate market. In terms of authoring their projects so that their 15-minute video plays and pauses 150-some times, certainly we'll provide that service to them, but certainly in another year or so they're going to do that for themselves in house.

But, when they ran up against the problem that every time they pause no matter what they do, they can't get this full-frame pause, then I'd love for folks to give Zuma a call and we can work with them on some really critical issues.

Actually, as it turns out with this one, it is an encoding feature that is generally not supported with the lower-end encoding system. Across the board, it is unsupported by authoring tools under $100,000.

You actually build in that pause as part of the encoded stream on the disc?

This is one thing that is kind of interesting. In DVD, you don't have frames. You can't just put a chapter mark where you want. You have to put a chapter mark at a GOP header. GOPs are anywhere from 4 to 16 frames. One of the things that we do is work with them on creating a specification. This is another area that I think that Zuma has a lot to offer folks. Even if we don't make a project for someone, talking with us to specify your project can save a lot of headaches later. We work with them from the beginning so that we can guarantee that places that they'll want to make a pause will be on a GOP header. By structuring their code and authoring properly beforehand. And again, this is something that just comes from experience.

Now, when they are making their content two months before we're ever going to see it, they're making it knowing what GOP structure is, what encoding is going to do to it, what all of the settings I'm going to do, we've been through: don't use very slow, horizontal fly-in's of text. It doesn't look very good when you MPEG encode it. MPEG doesn't handle horizontal changes in text very well. It will get all jumpy and weird.

So, by being familiar with the media, you know what works and what doesn't, and you can advise the client along those lines?

Exactly. I think that without a doubt, that is what I want to be helping people with. Those issues. We will certainly author anyone's disc for them, and we'll do the best job in the industry, but at some point people are going to want to bring their stuff in house. I'm all for it. I'm all for it because it creates markets for folks to push it. If I don't have to spend my time all day producing the same thing that I did last week, we're going to be able to move twice as fast. And it is just going to help the industry.

I think that film distributors should buy their own systems, if they want. Or, make deals or buy DVD authoring companies. But, when it comes time for them to produce the absolute killer best-of-the-year, award-winning DVD title, I think that without a doubt every time they are going to want to go out of house and they're going to want to get experience and expertise and ideas from folks who have been working in other areas. Because that is really where it comes from.

So, you think it's better to have more of the work done in-house by a staff that is used to working together.

Absolutely. And create relationships where it is smart. There is definitely an advantage between the authoring facility and the replication facility, because there is a lot of things that happen in between that are very confusing. We just had a title for Goodtimes Entertainment go out and it went to JVC for replication. JVC was having trouble getting the CSS encryption keys to work. Now, at that point a huge conversation starts between the replicator, my studio, Sonic, and other parties that we might bring in. In this case, Doug Carson Associates and Interra. And a huge conversation starts with email and phone calls and faxes flying around about who did what and how it is set up and it's here or there, and this person did this or that or the other thing. Having a nice comfortable relationship in that area, is a real advantage to everybody.

We have tried from the very beginning to be open and helpful with the replicators, because I've had replicators call me with questions and problems, I've called them with questions and problems. I think that a real collaborative outlook on the whole thing is the only way.

There are undoubtedly many issues for any individual project.

And ones that you think there couldn't possibly be an issue here. I have a client, Shannacie Entertainment, they do what has to be the absolute simplest titles in the world. One main menu, one scene-access menu, usually under an hour of content, that's it. We had a title that had a really obscure weird problem with Toshiba players. You couldn't author a simpler Video disc. Anyone who has been sold by a tool manufacturer thinking that this is easy, I do easy things, they were sold. There is no such thing. It is not easy. There is no part of it that ever easy. Every time that we think we've got something nailed.

That seems to be the nature of interactive multimedia. Even when dealing with so-called stable platforms.

I think importantly that is something that as you keep pushing is never going to go away. I talk with my partner a million times (because I handle the technical side of it), and he says, "When are we going to stop having these problems with Sonic?" Because we will often have a problem with the toolset. Working on a project, this never stops. If you want to have new features, you're going to have new problems. It is not that Sonic is not doing a good job. Sonic is doing an excellent job. It is that there is no possible way that you can innovate without creating issues. We've evaluated every tool out there. I've had every one in here. People tell me they have the easiest system in the world and then I have an MPEG-1 encode that is not working and I have to go in and edit lines of the BIOS code on an NT machine. There is nothing simple out there.

At the moment, you favor the Sonic Solutions stuff?

The reason I do is that it is a totally simple approach. They are the only ones who have totally embraced the workgroup solution for DVD. No one else has even come close.

It's great working now that we have educated the client about how to put this together. We get documentation back that I have to say is really great on these projects. Now that we're on the third version of this; for a lot of clients, we develop custom paperwork for them. Just for their jobs. For Shannacie, we have one single form that Shannacie delivers to use. From that comes everything that anyone needs to author their title.

Coming from the multimedia world, that is something that we are huge on. Naming conventions for assets. Proper paperwork when everything comes in. We do not accept tape without timecodes, for example. If someone sends me that kind of tape, I'll send it right back if we don't have timecodes for it.

So you're trying to eliminate the serious problems up front.

Absolutely. And every time we run into something, we have a set of terms and conditions, every time we run into an issue we generally make some addition to our terms and conditions.

Do you accept DVD-RAM for source files?

We haven't had any requests yet, but we have been testing a few systems.

Do you think that may become another viable alternative for distributing content?

I think there are a few things that are going to be interesting. Like Pioneer's DVD Designer. And DVD-RAM tools. Who was it? Minerva, their Impression tool lets you import QuickTime assets and AVI assets as placeholders. I think these tools are really interesting. There will soon be something analogous to the Apple Media Tools and the Click Works and Icon Authors of the DVD World. A lot of what I used to use Apple Media Tools for was to prototype something. I then I would pass it along to my C++ programmers or Lingo programmers or even the Apple Media programming environment programmers. From there, they would be able to run the show.

These tools that allow simple yet robust prototyping of projects are really where it is at.

Prototyping really seems to help people visualize and get on the same playing field.

My only advice that I ever give clients is: call us first. If you wake up in the middle of the night and decide that you want to do a DVD project, call me. Right then. Because your project is going to be that much better. If you don't call me, call some other DVD company, because your project is just going to be so much better.

I've had a million-dollar corporate client who made just a simple mistake in producing his artwork. So, we had to make some concessions in the end product. In this case, I don't think it affects the product at all. But why not make it perfect?

I understand that Adaptec has a version of Toast that support DVD mastering.

Adaptec has a version of Toast that does quite a few things. We've been testing it.

What do you think of it?

Chapter 17

It's a version of Toast that lets you control the DVD burner, which is how we burn most of our images. Additionally, it lets you make disc images, which is incredibly helpful for the DVD-ROM developer. I wish they would add support for DLT (digital linear tape) to be honest, and I've discussed it with them and they're looking into it. If you supported DLT as well, then multimedia developers that wanted to get into DVD-ROM—all they would have to do is contact a service bureau like us to create VOB files for them from their video assets, just in them same way that they might have sent out tapes before to have QuickTimes made. Then they can do everything in house. All of my DVD-ROM clients, we go back and forth. We make the VOB files, we send them back. They program them. They send us 18 CD-ROMs to master a DVD for them when really they could just be doing that themselves.

Then Astarte has made another authoring tool which is the first low-end Mac-based tool. Which is really interesting. They make a tool that falls in with the Spruces and Minerva Impressions of the world, while very helpful for a certain end of the market, they certainly aren't applicable to the highest end of the market. I think that while it is helpful to have the concept of extracting the DVD specification from DVD authoring, there is no way that you can do anything innovative without knowing what a PGC, and a part of title, and a cell is. Because you've got to make stuff that plays in players.

It sounds like there is a whole new level of education even for people who have been doing a lot of work in CD-ROM development over the years. With a whole new vocabulary.

And also this is something that multimedia developers will be the most comfortable with and this is the thing that kills video and audio professionals: In multimedia we were really used to system requirements and we were used to the fact that sometimes you would take it and play it back on one computer and it would play back in a certain way and the next time on another computer it would play totally differently. You would be hard pressed to figure out what the hell was going on.

I don't know how many times I've been wracking my brain trying to figure out what was different between two different systems, both of them running the same system software. We have so many projects, basically, if you take any DVD-Video disc and any two DVD players, even from the same manufacturer, you will be able to find some aspect of them that functions differently. Sometimes that will lead to a fatal error. I have projects right now that are being evaluated that on some systems just don't play. On other systems they play perfectly.

Do you use an outside test facility?

We use a number of outside testing means, as well as a great deal of internal testing. I'm sure that you've heard of Interra. I think Interra is jumping in to play a really important role in this market. They are building a tool called Surveyor which is a third-party tool that does two really important things. It inspects your project for specification verification, to make sure that your project is DVD legal, according to the spec. Additionally, it's going to run it through a database of known player issues and lets you know that you might have a playback issue on certain players. Now, because of obvious reasons, they can't say that you're going to have a problem on this Toshiba or this Sony player. Cause they're going to get sued by Toshiba or Sony pretty quickly. Since that is information that people don't like to have out there.

It does sound like the early days of CD-ROM. That was one of the headaches back in the early development process. You couldn't get things to play back consistently.

You end up with the lowest common denominator approach. We did the Olympics for Panasonic. Because we needed to support many of the first generation players, we had to scale back on features in some areas. There are products that are going to be shipped (I've shipped a few) that will recommend a certain player and say that they don't play back on other players. At some point, people stopped making for 2x CD-ROM drives and starting making for 4x, 6x, and 8x.

Seems like typical technical evolution. It must make your job frustrating at times.

It can be frustrating, but it is something that someone has to expect and I think also at the same time, it seems to me that it is inevitably part of anything like this. Maybe I'm just willing to accept it because of my background. I think that when they make new parts for cars, they probably run into the same kinds of issues, I'm sure.

Chapter 17

It's still kind of fun working on the edge.

There is a certain responsibility that I think is often overlooked by manufacturers in providing quality-assured products that function.

It's the classic battle between the marketing department and the engineering department. In many companies, it's like a war. The marketers want to get the stuff out there, get the sales going. The engineers want to get the bugs out and get everything working perfectly. Somewhere in between you're going to get the product released in a relative state of completion.

I think that is exactly the issue.

18

Independent Marketing and Distribution

One of the most appealing aspects of optical disc recording is the ability to create your own interactive multimedia titles or audio CDs from the computer desktop. The analogy to the revolutionary breakthrough of the printing press is often used to characterize the incredible communication capabilities of this medium, but, in many ways, optical disc recording is even more revolutionary. During the hundreds of years following the invention of Gutenberg's movable-type press, only those involved in printing as a business could typically afford the expense of a printing press—it was not the kind of equipment that someone could set up in a spare room for occasional use.

In comparison, CD recorders are among the least expensive computer peripherals. For $150 you can buy a recording device that lets you create a disc for mastering hundreds or thousands of CD-ROMs or audio CDs. Replication costs have dropped to levels that allow anyone to drop off a CD one-off and get a small-volume run completed for anywhere from $1.00 to $3.00 a disc. CD duplicator costs have come down to the point where for less than the cost of a high-performance computer you can buy equipment that will let you generate 10 or 20 discs an hour. For a small business, this capacity could easily support the release of an independently produced title marketed through a variety of channels.

Authoring tools for all forms of digital content—audio, portable document files, cross-platform help systems, interactive rich media productions—have also become more accessible, allowing anyone conversant in typical computer operations to become skilled at creating polished productions in a very short period of time.

Similarly, DVD-R technology makes it possible for an independent film company or video studio to create original masters and release them for replication in DVD-Video format. While the authoring tools and DVD recorders are still quite high in relation to CD-R levels, prices continue to drop as acceptance for this new medium grows. Software tools, such as Sonic Solutions *DVDit!*, bring impressive capabilities to the mainstream market, making it possible for anyone to turn digital video content into a published DVD work.

Opportunities abound for independent developers to produce successful works and realize a profit with unit sales in the hundreds, rather than the tens of thousands, as is the case with many commercial titles. The challenge, of course, is finding an audience in a market where many of the retail outlets are geared to large mainstream publishers of software, games, audio CDs, and educational titles. This chapter offers some suggestions for creatively marketing your original title that has been published to optical disc. It also discusses developing unique sales channels that you may not have even considered.

Targeting Niche Markets

As an independent title producer or developer, you probably don't have the funds to compete head-to-head with the well-established corporations that dominate the industry. On the other hand, you don't have their overhead to burden your operation—you can realize a significant profit on a small percentage of the sales that would be considered a success by one of the major players. While it is not impossible to release a general-purpose title and achieve success with it, you may find that by targeting your work to smaller audiences with niche interests lets you capitalize on market opportunities that the large-scale producers don't even consider worth trying.

Some examples of narrowing your market are:

- The CD-ROM market has been saturated with general-purpose encyclopedias that require tens of thousands of hours of research and production time. Why not create a specialized reference that mirrors your interests? Develop an encyclopedia of dog breeds to distribute to the Humane Society of the U.S. shelters across the country. Produce a compendium of British motorcycles from 1960 to 1980. Create a recipe book for people on sodium-limited diets. Develop an investment guide for those who only want to invest in socially responsible businesses. By choosing a narrow segment of the overall market for a CD-ROM or DVD-ROM title, you can reach an audience that is untapped by the mainstream title publishers.

- Resource databases are a popular and effective use of optical disc publishing. Find an appropriate vein to mine in this area and you can probably create an incoming-producing title. Many people are actively looking for resources to serve their power needs off the grid. Why not develop a CD-ROM containing worldwide resources for solar power and wind power tools and technologies? If you have an interest in genealogy, you could develop a disc that lists the best genealogical research resources in the country. If you want to aid in the recycling movement, you could produce a disc that provides all the recycling centers for different types of materials in a specific geographic location. CD-ROMs and DVD-ROMs make great vehicles for conveying this kind of information in a very accessible way. If you take advantage of abundant hyperlinks and indexing of terms, you can make such a resource more usable than an equivalent printed directory or resource listing.

- If you're working on an audio CD, follow your music interests to an extreme. A compilation of Native American flute music might not go platinum the first week of release, but to those who love the music, you can find an audience. Do you like imaginative experimental percussion sounds using found objects? Turn your passion into an album. Do you enjoy poetry read over a classical guitar backdrop? Start recording. Odds are that if you like a musical form well enough there are thousands of other people out there who might enjoy the same music.

- Training and educational titles are perennial sellers. Pick a subject that you know something about and can share with a community. If you enjoy historic architecture and know something about acquiring property and restoring it, produce a title on that subject. If you've mastering the latest sound-editing and processing software and want to help other do the same, create a training CD on the topic. If you've mastered an oriental martial arts program that no one else has even heard of, you might have the material for a successful optical disc title.

The range of material available to those working with digital content is virtually unlimited. You have the potential for building a sustainable business around supplying information that is needed in specialized areas or delivering entertainment to those with specialized tastes.

Applying Permission Marketing

The term *permission marketing* has entered the consciousness of the modern marketer in a profound way. Partly fueled by the emerging capabilities of the Internet, partly a response to a market so oversaturated with commercial messages that people filter out everything, permission marketing is built on a simple premise. Respect your potential customer, only give them information that they request, and build a trusting relationship over time. The Internet, of course, provides mechanized ways of doing this through email, but the basic ideas apply to any kind of direct marketing techniques. For example, the information that someone requests from you could be delivered in the form of a Web-enabled CD-ROM containing a catalog of your products and links to an FAQ about your manufacturing processes.

The Flaws in Interruption Marketing

The opposite of permission marketing, as defined in Seth Godin's book *Permission Marketing: Turning Strangers into Friends and Friends into Customers*, is interruption marketing. Interruption marketing relies on intruding on someone's conscious attention, either through a commercial television message, a full-page ad in a magazine, or a direct-mail piece that you receive in the mailbox. Since we are all bombarded with intrusive messages of this type every day of our lives, people have become extremely adept at ignoring them. The more advertisers fight for our attention, the more they get tuned out by an audience that is saturated to the limit with these kinds of messages.

The fatal deficit of interruption marketing is that the more companies that are clamoring for your attention, the more expensive it is to thrust through the noise and communicate to anyone. The approach is inefficient and notoriously wasteful. Direct mail campaigns that get a 2% response rate are considered successful—despite the fact that 98% of the people receiving the message completely tuned it out. As more and more companies compete for increasingly smaller shares of each person's attention, it becomes more expensive gaining new customers and equally expensive to keep them. The spiral of costs with very little return is an arena that an independent developer can little afford, particularly when other more effective techniques exist for building a customer base.

Permission Marketing for Independent Developers

The concept behind permission marketing is particularly meaningful to independent developers who are interested in promoting and marketing their title. It relies on the tendency of people to seek out areas of personal interest and to actively solicit information on subjects or products that are relevant to their life. For example, someone who loves Gypsy swing music will spend a good deal of time seeking new music in that style or bookmarking Web sites that cater to that specialized taste. If you produce music CDs in this genre, you can find ways of politely fulfilling requests from people interested in hearing about your latest releases. You might offer an electronic newsletter with articles about Gypsy swing bands and performances, email announcement of new album releases, providing a free compilation album with tracks by several different bands, and so on. The significant difference between this form of marketing and other forms (in the interruption marketing approach) is that each person *opts in* to receiving regular information from you. If you abuse this trust, by bombarding a person with a constant stream of messages or selling their email address to others without their permission, you've violated the underlying substance of permission marketing.

Relationships with customers are deepened over time by offering them the kinds of updates and information that they want. You are catering to the customer's tastes by supplying interesting and focused content. If you trivialize the responsibility or betray the trust, the customer reserves the right to turn off the spigot of information and sever their relationship with you. If you do right by your customers, you are rewarded with more sales and a long-term relationship.

The approach offers many benefits for an independent developer or small business with an optical disc title to promote. You can build a customer base slowly and steadily with a very high probability of repeat sales and long-lasting customer loyalty. You don't need to take out an ad during the Super Bowl to get the attention of your audience. You can use carefully focused channels, taking advantage of the Internet (as described in the following section) to reach your unique customers and nurture a long-term relationship.

Using the Internet as a Leveler

The Internet offers a variety of well-established channels for reaching an audience, but if you want to compete at the levels of highest visibility, it can be expensive. For example, Yahoo!, one of the most popular search sites, instituted a fee structure to gain more rapid attention from their staff prior to including your site in their search engine. While this offers an avenue to avoid having to submit your site to Yahoo! multiple times to gain a listing, it does favor the larger, more well-heeled sites at the expense of those sites with few dollars to spend on promotion.

To effectively leverage the Internet as a promotional tool, sometimes the best way is to skirt the expensive paths and fit approaches that fit your budget. This might include:

- Trading Web articles or other content in exchange for links from another site to yours. Larger Web sites are often desperate for fresh content to keep their site interesting to visitors, so you have a powerful bargaining chip at your disposal. Choose sites that match your specialty interest for the best results.

- Producing a monthly electronically published newsletter made available through the Internet, using a simple sign-up list. You can build the subscription list through postings in newsgroups or special-interest forums, sign-up forms on your Web site, or through opt-in mailing lists for people who have expressed an interest in receiving information about your topic of interest.

- Performing an opt-in mailing through a service such as YesMail (*www.yesmail.com*) to reach a targeted segment of the market. This can be an effective way to announce a new product release or stimulate interest in a title that has been available independently for some time.

- Using services, such as GoTo.com's search term bidding, to selectively choose those terms that you want to use to identify your business on the Internet. Clever use of terms and skillful bidding can bring you to the top of the list in many select areas, as described in *Bidding on Search Terms* on page 446.

- Offering training in a specialized area and posting your training course on one of the Web sites that provides both the tools and the content storage. For example, if you have a product that facilitates 3D character animation, you could create a beginner's course on character animation and post it on *www.learn.com*. You not only gain attention from prospective customers, but you can achieve a level of credibility for your product or title.

- Offering some extremely valuable resource on your Web site that will become a focal point for the kinds of people you want to become aware of your product. For example, if you have an optical disc title that explains how to convert a conventional home electrical system to a solar-powered system, your Web site could provide an interactive tool that lets someone determine how many solar panels they would need for their particular geographic location. If you give something of value away on the Internet, you can often persuade someone to purchase something from you—if it reflects their interests and inclinations.

Imagination, creativity, and innovative communication can often make up for the sheer power of dollars when it comes to reaching an audience on the Internet. The more specialized your audience, the better the chances that you open a communication channel to them through the mechanisms available on the Internet.

Constructing a Web Storefront

The entry requirements for constructing and maintaining a Web site enabled for ecommerce can be steep—too steep and too time consuming for many independent developers and small businesses. The solution is using services that provide the site, the commerce mechanisms, and the visibility to get you quickly online selling your titles or other products. Setup time and maintenance are usually a small fraction of what they would be if you maintained your own server with custom software, or even if you implemented a Web store through an ISP.

Most Web store services provide a browser-based form for entering product information, uploading accessory files (such as graphic images), and handling the storefront configuration. You are typically charged a set monthly fee based on the number of items that you have for sale at any given time. Many of these services include close links with online credit card processing services, such as Authorize.net, that will enable you to automatically validate credit cards based on a customer's entries and perform any necessary transaction entries automatically or through browser-based forms.

One popular service of this sort—Merchandizer (*www.merchandizer.com*)—offers a highly configurable storefront model. You can use their templates to display your goods or you can upload your own HTML pages at any level of complexity. Many added benefits to the Merchandizer approach make it easy for customers to interact with your site. Shipping rates through UPS are calculated automatically and appended to orders (if UPS shipping is requested). Customers can assign themselves a password and come back to check order status at any time. Discount clubs can be set up to offer a select group of your customers special offers. A variety of activity reports can be generated to get a sense of customer page-viewing patterns from month to month. If you're looking for a relatively trouble-free way to enter the world of electronic commerce, Merchandizer is one very good provider of this type of service.

Other options for turnkey storefronts are available from different sources. Yahoo! offers an electronic storefront that is easy to setup and use. By the time this book reaches the bookstores, many new options will probably be available. The Web storefront can be the focal point of your marketing effort and your primary sales channel. You can direct potential customers to it through email campaigns, links from other high-profile sites, search terms that are designated through bidding, newsgroup postings, and other techniques. If you have a good title and can reach your audience, the sales will generate themselves in time.

Online Auctions and Other Sales Outlets

Online auctions have proven to be an extremely popular way to sell things on the Internet. For some mysterious reason, people respond to the excitement of auctions in a way that they rarely do for other forms of Internet commerce. If you follow the bidding on items at popular auction sites such as eBay or Amazon.com, you may be amazed at the prices that odd items sell for—such as old transistor radios from the 1950's, out-of-print books, and vintage guitar effects boxes. The secret is that these auctions are able to filter and categorize the auction items in such a way that small numbers of people searching for very specific things can sort through literally millions of items to find things that interest them. The market segment of people looking for a 24mm lens for a 1960's Yashica 35mm camera might be extremely slim, but if you have a dozen people who have been looking for such a lens suddenly bidding against each other on eBay, the price of this item can rise quickly.

For those who want to take the simplest approach, Amazon.com offers their zShops, which allow merchants to list items for sale for a very low weekly fee (typically in the range of $10 to $.25 per item). If you don't want to go through the tedious process of applying for a merchant credit-card processing account, you can sign-up to use Amazon.com's One-Click payment acceptance mechanism. Your sales item will include the One-Click purchase button, Amazon.com will track sales and deposit the receipts (minus a small processing fee) directly into a bank account that you designate. You get the advantages of rapid online credit-card processing without incurring the periodic monthly charges and other overhead associated with maintaining your own merchant account. For established Amazon.com customers, the One-Click purchase option is an extremely popular way to buy items; you can gain the benefit of this feature through one of the most powerful online presences.

Through zShops, you can also designate a variety of search terms that will link to your product when Amazon.com visitors search for books or other items. The skillful use of search terms can be a sure-fire way to reach specialized audiences. For example, if you're selling a CD-ROM title about how to perform maintenance on British motorcycles from the 1960's, you could list terms such as Norton, BSA, Triumph, and Vincent to cause your product link to pop up whenever these terms are entered by site visitors. As with all these approaches, the more tools that you have to finely filter the areas of interest of Web visitors, the more likelihood that you can find exactly the audience that will respond to your niche-market title.

A variation of a standard auction, known as the Dutch auction, is open to eBay sellers after they have established a track record based on a specific number of positive feedbacks from their winning bidders. Through a Dutch auction, you can sell multiple items through one auction entry. This offers a way to sell copies of an optical disc title, maybe offering it in terms of 20 items at a time through a Dutch auction. Amazon.com offers a similar option through their auction offerings.

Even if you don't generate immense sales through online auctions, you can include links back to your primary Web site to attract those potential bidders who might want to learn more about your products. The attention you gain in this manner can often be more valuable than the actual item you are selling through the auction, since sites such as eBay attract hundreds of thousands of visitors each week. You are almost guaranteed to see a sharp rise in your traffic following the posting of an item of interest on eBay that includes a discreet link to your Web site.

Targeted Press Releases

Press releases have long been an effective tool for large corporations. New product releases, merger announcements, patent filings, trade show announcements, and similar kinds of information often find their way into the mainstream press—including newspapers, trade magazines, radio, television—through targeted press releases. These are distributed to all of the listed press facilities and frequently incorporated into news pieces in a variety of formats.

You can compile a list of key press facilities and perform your own distribution of press releases. You can also rely on a service to do this for you. One company, Digital Works (*www.digitalworks.com*), offers inexpensive press release distribution to your choice of geographic locations, making it possible to achieve notice in a select market or location with your product news.

Writing successful press releases is an art form in itself. Ideally, you want a newsworthy item that will be picked up and reproduced in some form by the media, but, obviously, a blatant product description will not be likely to garner much coverage. The trick here is to develop an angle, look at your product information in the way that the media folks will and presenting it in such a way that it offers technological information, human-interest perspectives, or some other interesting angle.

For example, if you've developed a training CD-ROM that teaches people how to effectively compost their kitchen scraps into rich soil, you might try to work the announcement into a news story format or a case study. "The Recycling Center in Middlebury has been using the *Compost your Kitchen Scraps* CD-ROM to instruct their clients in the most-effective methods for using the composters that they sell." The right angle can help you gain some attention.

If you're not totally comfortable trying to write your own press releases, Digital Works, as well as other similar companies, provide specialists who can guide you through the process or write the release for you based on information that you provide. The press can be a powerful information channel when it is working to your advantage—learn how to get your titles showcased in the mainstream and trade press and you'll see sales take an uphill curve.

Bidding on Search Terms

Fighting to get yourself noticed in the major search engines can be an enormously difficult task. The sheer volume of companies that are engaged in some form of Internet business has become overwhelming. An easier way might be to selectively place yourself in front of the audience that you want to reach by bidding on specific search terms.

GoTo.com originated this interesting approach to the search term circus. They provide two complementary components in their approach:

- Free search engine interfaces that can be posted on your Web site, so that your customers can take advantage of the GoTo.com listings. You get paid a small amount whenever someone utilizes the GoTo.com engine, so this can be a revenue producer, as well. This also ensures that many sites will be utilizing the engine; this wide distribution will be to your benefit once you start bidding on search terms.

- A bidding system where you can choose the amount of money that you will pay for each click-through on a search term that links to your site. You only pay for click-throughs, not for views of the search term, so this can be an effective way to steer customers to your site.

The trick is in selecting search terms that will attract the unique interests of those customers you want to reach. You also need to steer away from the more common terms—such as sex, money, and MP3—since these become the terms involved in bidding wars, driving the prices upwards.

Let's say you've developed a CD-ROM that provides an illustrated history of the use of pewter and its value as a commodity in colonial New England. You run a few test searches using the GoTo engine and discover that there are no current bids on the term *pewter*. You place a minimum bid of $.01 on the term, and also select several other terms, including *colonial trading, New England history, oil lamps,* and so on. As part of the search term, you write a description that will appear when the search term is brought up; perhaps: Learn the history of pewter use in Colonial New England in a unique CD-ROM title. You then add your Web address. Each time someone clicks through to your site, after running a search on pewter, you pay GoTo.com $.01. If you get 1000 visitors in a month, you pay $10.00. Keep in mind that because you are the high bidder on this term, each time someone searches for pewter, your listing comes up at the top. For a small company competing with giants, being at the top of any search list can be a major competitive advantage.

Current bids on search terms typically range between $.02 and $1.00. At $1.00 per click-through, this becomes a bit more expensive means of generating traffic, but, otherwise, it is probably the least expensive way you can significantly boost your Web site traffic in a very targeted way. Through the selective use of different search terms, you can precisely identify interests in your prospective audience, which should generate some fairly good response rates as new visitors enter your site.

Chapter 18

Summary of Techniques

This chapter covered a lot of ground in a few short sections. The basic points worth remembering are summarized in the following list.

- Niche markets offer great opportunities for small-scale title producers. Find a niche that interests you and you can very likely find an audience that shares your interest.

- Permission marketing is a means of politely supplying information that a customer requests and then working to build a long-term relationship with that customer. It is the opposite of interruption marketing, which tries to sell things to you by jarringly intruding on your attention and ramming home an indelible message.

- The Internet offers a variety of ways to reach people with very specialized interested and to communicate with them inexpensively. Opt-in mailing lists, free newsletters, useful Web site tools that attract a specialized audience, and search-term bidding can all help you reach your audience.

- If you don't have the time or expertise to construct an entire ecommerce Web site, there are pre- built services that can have you online accepting credit cards in a couple of days. Other methods, such as online auctions and services such as Amazon.com's zShops, can also work in some situations.

- Press releases can get your product or business in the news. Write the yourself and distribute them to key news outlets, or hire a service, such as Digital Works, to do this for you.

- To avoid the search engine scramble, use a service, such as GoTo.com, that lets you bid on selected search terms. Targeted use of particular terms can attract the attention of exactly the kind of interests that your prospective customers will most likely have.

Look for ways to make the Internet work for you that give maximum benefits from the dollars invested. You don't have to spend a lot of money to get noticed if you use your imagination and creativity in bright and interesting ways.

19

Responsible Media Use

We live in a society that has grown accustomed to abundance, so much so that we have become slothful and wasteful in our habits. Despite landfills that are overflowing capacity at an incredible rate, we continue generating solid waste at a record-breaking pace. We denude forests to produce newspapers, toilet paper, and disposable diapers. We package our foods in plastic so that items can be shipped without breakage or spoilage, and then toss the wrappers in the garbage. Packaging of many types of products is done with little regard to the recycling of the packaging materials, and, often, our methods make recycling difficult or impossible. We generate cheap, easily breakable household goods that last only a short time before they end up in the dump.

Optical discs offer the promise of both material conservation and potential product lifecycle benefits that encourage the cradle-to-grave approach that is being practiced by many European companies. The idea is simple. Companies and organizations bear a responsibility for the goods that they produce. These goods should be designed in such a way that they do not generate toxins during the manufacturing process and the product design should include a method for recycling the spent materials or disposing of them in an environmentally sound manner. From conception to the end of a product's life cycle, the potential harm that a product can inflict on the environment is factored into the manufactured process and reduced or eliminated whenever possible.

This chapter considers ways that optical recording can help businesses and individuals reduce solid waste, minimize paper consumption, and adopt more responsible everyday practices.

Chapter 19

The Revolution That Never Came

Electronic publishing and the personal computer were supposed to initiate a revolution: the advent of the paperless office. More than 20 years have passed since predictions and forecasts touting the benefits of the paperless office have come and gone. Despite the proliferation of the Internet as a communication medium, near universal use of email within businesses, and electronic storage media such as tape and optical discs, paper use is as high or higher as in previous years. In fact, studies at companies such as Xerox found that when email was introduced to all employees, paper consumption internally went up as people enthusiastically printed the scores of messages that they received.

Change only comes about when sufficient numbers of people insist on challenging the "business-as-usual" model. Sometimes it is a matter of ideas taking hold so firmly that they become the prevailing paradigm—the deeply entrenched belief system that influences a person's day-to-day behavior. Excessive paper use continues because it is still not widely perceived as being a problem. Alternative fibers that could provide a direct substitute for tree-based paper—such as kenaf and industrial hemp—are ignored by the large-scale pulp industry, much as the oil industry has used their muscle to discourage development of alternative fuels and solar power.

Change also often comes about from the ground up. If you change your individual habits, you can frequently lead those around you by example. By finding ways of reducing paper consumption at home and in the office, you set an example that will be noticed by your co-workers, your children, and members of your community. Don't print your email messages; save them to disc every few months. Encourage your company to receive phone bill itemizations on optical disc, rather than voluminous printouts. Look towards distributing short-lived materials, user guides and references for products, on disc rather than on paper. Make electronically published content easy to read onscreen and easy to access, with well-indexed sections and skillful design. Most users will prefer this approach to the paper-based equivalent.

A single optical disc can hold the equivalent of hundreds of thousands of pages of printed material. You can generate enormous savings by using optical discs to replace paper as an information medium. As more and more individuals lean towards electronic publishing as a viable alternate to excess paper consumption, the more society as a whole will adopt more sustainable practices.

A Model for Sustainability: The Natural Step

From Sweden comes a model for sustainable business practices known as The Natural Step (TNS), founded in 1989 by Swedish oncologist Dr. Karl-Henrik Robèrt. Frustrated by the inability of government and industry to squarely address growing environmental problems, Robèrt, a scientist at a cancer research institute in Sweden, brought together 50 of his fellow scientists and worked out the fundamental principles that should form the basis of a sustainable society. These principles were further expressed as a series of one-day training programs that were attended by many business and community leaders throughout Sweden. The approach has spread to many other countries, as well, and training sessions are now offered in the United States. Branches of this movement also exist in the United Kingdom, New Zealand, Australia, France, Brazil, Taiwan, the Netherlands, and Canada.

The principles call for organizations to examine and redesign their activities and processes, using a system theory approach. As spelled out in their original form, the four system conditions are:

- In order for a society to be sustainable, nature's functions and diversity are not systematically subject to increasing concentrations of substances extracted from the earth's crust. In other words, those resources that we extract from the Earth's crust—including minerals, fossil fuels, and metals—must be redeposited back into the earth at an equal rate, rather than increasing in the biosphere.

- In order for a society to be sustainable, nature's functions and diversity are not systematically subject to increasing concentrations of substances produced by society. This condition addresses the need to reduce the use of persistent man-made materials so that they do not accumulate in the biosphere. It also states the requirement to avoid producing substances faster than they can be broken down in nature.

- In order for a society to be sustainable, nature's functions and diversity are not impoverished by physical displacement, over-harvesting, or other forms of ecosystem manipulation. This condition addresses our use of renewable resources and states the need to remain within levels of consumption and land-use that do not threaten other life forms or reduce the regenerative capabilities of ecosystems on the planet.

Chapter 19

- In a sustainable society, resources are used fairly and efficiently in order to meet basic human needs globally. This condition expresses the need for just resource distribution and the use of resource-efficient techniques to support the other conditions.

As an example of how these practices are put to use by companies, as CEO of the Fortune 500 company Interface, CEO Ray Anderson has changed the entire process by which his company operates. Interface trades in floor coverings and carpet tiles. Rather than simply selling carpet tiles to business, Interface leases them, providing a service through which they install new tiles in a business, replace them when they wear out, and recycle the old worn tiles. The tiles move through a natural life cycle and Interface maintains responsibility for the materials being used throughout this life cycle. The company continues to overhaul its practices and to work towards the adoption of sustainable practices, including the use of solar power in company plants. Anderson has won numerous awards for his vigorous support of environmentally benign business activities and has inspired many other corporate leaders to follow suit.

While the use of optical discs can dramatically reduce paper consumption, these polycarbonate-based objects also involve a number of manufacturing and chemical issues, packaging considerations, and disposal concerns. Discs can be recycled and packaging can be selected with sustainable principles in mind. Refer to Recycling Discs and Packaging on page 453 for additional details.

Further information about The Natural Step can be obtained from their Web site: *www.naturalstep.org*

The Board of Directors of the U.S. branch of The Natural Step includes numerous individuals with significant backgrounds in sustainable practices, including author and businessman Paul Hawken, Interface CEO Ray Anderson, and the organization's founder, Dr. Karl-Henrik Robèrt.

More information can be obtained by contacting:

The Natural Step U.S.

P.O. Box 29372
San Francisco, CA 94129
Phone: 415 561 3344
Fax: 415 561 3345
Web: *www.naturalstep.org*
Email: tns@naturalstep.org

Recycling Discs and Packaging

Because of their compact size, minimal mass, and relatively benign material content, audio discs, CD-ROMs, DVD-ROMs, and DVD-Video discs do not represent a severe threat to the landfill problem. However, with a bit of planning, discs and their packaging can be presented to customers in a form that makes it much easier to handle their natural life cycle. By minimizing packaging, reducing the amount of printing materials that accompany a title or a product, and providing a destination for disposing of worn, broken, or obsolete discs, the use of optical discs becomes a much more sustainable practice.

Follow these guidelines when producing optical disc titles:

- Current trends disc publishing favor minimalism. Avoid gratuitous package-fill items, such as flyers and brochures. You can include the same material on disc as Acrobat files or HTML content.

- Use packaging materials that are easy to recycle, and, ideally, tree-free. Substitutes for paper-based packaging can be found—materials such as kenaf, industrial hemp, banana, jute, recycled cloth waste, and recycled animal fodder. While more expensive than conventional packaging materials, these kinds of materials can add a distinctive appearance to your package and communicate to your customers that you follow sustainable business practices.

- Avoid plastic jewel cases when possible. These fragile, difficult-to-recycle components are not very environmentally friendly. Avoiding jewel cases can be difficult when you are introducing a product in a marketplace, such as the audio CD retail outlets, where the standard strongly favors the jewel case because of shelving requirements and other issues. Nonetheless, many successful artists introduce their works in non-standard formats and use this as a selling point, rather than a negative factor. Exercise whatever persuasion you can with distributors and retail outlets to overcome the resistance to non-jewel-case packaging.

- Offer your customers a way to return a broken or obsolete disc. Just as Hewlett-Packard provides a shipping container and free shipping for their laser printer toner cartridges, disc producers can provide a prepaid shipping envelope to allow discs to be easily recycled.

- If you're producing a software application or educational title that normally includes printed manuals, consider substituting electronic documentation on disc. Computer hardware and software manuals have a notoriously short lifespan and aren't good for anything within a year or two (after the next software upgrade is released). A small installation guide printed on tree-free paper and a similar quick reference card might be all the printed materials that your customer needs.

- Encourage others in the industry to adopt similar practices. Small independent companies may be able to influence the behavior of larger corporations by their examples and success in the marketplace.

Recycling Locations

The following facilities perform recycling of compact discs:

Digital Audio Disk Corporation (DADC) Recycling Program

300 North Fruitridge Avenue
Terre Haute, IN 47803
Phone: 812 462 8100

Plastic Recycling, Inc.

2015 South Pennsylvania Avenue
Indianapolis, IN 46225
Phone: 317 780 6100

Summary

With some forethought, you can release titles on disc that follow sustainable cradle-to-grave product principles. Optical discs, by their nature, help conserve other resources, particularly paper, and they can be designed to be a totally recyclable and environmentally sound communication tool.

20

Entrepreneurial Possibilities

As new technologies compete for customers and marketshare, opportunities spring up and sometimes vanish overnight. To those willing to test the waters, the potential for entrepreneurial ventures is extremely tempting and also genuinely hazardous. Producing interactive multimedia titles on CD-ROM once seemed like the sure road to fortune, but dozens of companies that invested in releasing expensively produced titles found they couldn't find enough customers to justify their big-budget expenditures. Many of these companies no longer exist or they have sold their catalogs to other. Voyager, one of the best respected producers of multimedia titles, has turned over their entire catalog to The Learning Company, marking the end of an era of bright optimism over CD-ROM development.

DIVX proved an expensive investment for Circuit City and other partners, who discovered that customers weren't interested in this approach to watching movies.

On a more positive note, optical recording continues to offer many bright possibilities to prospective developers. Whether you have a project suited to CD-ROM, DVD-ROM, or DVD, the optical disc can hold many different types of content and intelligent entrepreneurs will continue to find new ways to reach an audience with entertaining or enlightening content.

Small teams of developers, by exploiting microtrends and moving more quickly than the large corporations, manage to release titles on disc that have the power to stir the imagination and dazzle the senses. Newly

Chapter 20

emerging technologies, such as Digital Video cameras and editing tools, bring incredible levels of communication to anyone who can afford the reasonably inexpensive equipment. Digital audio processing hardware and software are revolutionizing the music world. Electronic publishing tools let you turn a computer and a CD duplicator into a communication tool that lets you produce and distribute thousands of pages on a disc that costs under a dollar and can be shipped for the cost of first class postage.

Many areas still look promising for those who are investigating new ways to share information or entertain an audience. Some of the areas that are still good bets for enterprising communicators include:

- Education and training are two areas that are wide open to the benefits of digital content on optical disc. The interactive capabilities of DVD-Video offer many options for authoring content that might cover any conceivable subject, from *How to Administer CPR to a Drowning Victim* to *How to Obtain Your Pilot Instrument Certification.*

- The increasing availability of digital video tools combined with the lowering costs of DVD-R equipment make it possible for homegrown movie companies to produce and distribute their own files, bringing new life to independent film production. *The Blair Witch Project*, created digitally for $40,000, went on to earn millions in the theaters and opened the eyes of many Hollywood executives. George Lucas has announced that he is abandoning film and turning to digital video and computer-generated imagery for future versions of Star Wars.

- The MP3 music revolution is in full force and independent musicians can now find an audience on the Internet and sell independently produced discs. Creating and burning masters in a home studio is commonplace, and replication costs have come down to the point where any band capable of making decent music can afford to release their own CD. Creating and marketing independent titles can be a liberating experience in comparison with the autocratic control that is often applied to musicians through large record companies.

- Resource lists and specialized directories are a natural application for the optical disc. With the increased capacity of DVD-ROM, the capability is available to include large amounts of rich media to complement other resource information. For example, a Bed and

Breakfast guide to North America could include two-minute video clips of each establishment. A catalog of independent music labels could provide audio clips of tracks from each library. A compilation of 3D authoring tools could include trial versions of every tool discussed. Movie guides on DVD-ROM could include theatrical trailers for hundreds of films. Used wisely, this extra capacity could open up entirely new product possibilities.

- The low cost of CD duplicators makes it possible to inexpensively open up a small title publishing venture. The ability to produce small quantities of discs on demand eliminates the need for large inventories or warehousing of products. Equipped with a single CD duplicator, an enterprising publishing business could venture into niche markets and exploit subject areas that could not be profitable for larger publishing concerns.

This industry is new enough that other possibilities may emerge or be developed overnight. The following case study illustrates how the characteristics of DVD suggested an entirely new model for video rentals and how a company was built around this simple premise.

Case Study: The Entrepreneurial Possibilities of DVD

Back while vendors were still hashing out encryption standards and arguing over product issues, Marc Randolph tossed his hat into the entrepreneurial ring and founded Netflix (*www.netflix.com*), an online DVD rental company. The foresight proved fortuitous, as finally in the latter part of 1999, sales of DVD players began accelerating and the industry ignited after years of growth predictions and speculation. Lightweight DVDs ship easily and Netflix devised a system where customers can queue up their movie selections, and then receive up to four DVD titles at a time, which they can keep as long as they want for viewing. As soon as a movie is returned, the next one in the queue is shipped. The approach has worked well; Netflix now ships some quarter-million DVD titles a week and business is growing strongly. Focusing on connecting customers to movies that suit their tastes, the Web tools being developed at Netflix find parallels in customer's movie tastes by means of their volunteered ratings and then offer recommendations based on similar ratings. More and more movies are being converted to DVD format. At this time, early in the year 2000, about 5200 movies have been converted to DVD format, as compared to 85,000 movies now in VHS. The total number of feature-length movies that have been made is around 250,000.

Chapter 20

For potential entrepreneurs who are wondering how you can gauge emerging technologies in a chaotic industry, this case study illustrates that the process is often more intuitive than it is analytical.

One of the noteworthy aspects of the DVD technology is the entrepreneurial possibilities of the medium. When did you first realize you could create a business around this medium?

Clearly the thing that we have tapped into with DVD anyway is the storage capacity. It just holds so damn much. What makes it the right time for it is that simultaneously there has been this huge interest in this whole concept of video-on-demand. It has unbelievable hype and it has had it for years and years and years (and through hundreds of millions of dollars of trials), but they still can't seem to nail it because movies are so huge. There is no easy way to get them down over an Internet connection or over a phone line or over cable, in terms of one to one delivery.

One of the things that DVD has enabled for us is a near-video on demand alternative. What we do: rather than trying to solve the problem of how do we get a movie from a central location to a customer, we don't try to solve that with a phone line or try to solve it with satellite or with any of the other alternatives that everyone is playing with for video on demand. We choose to ship it to them on a piece of plastic. That, I think, is the biggest advantage that DVD gave to us.

Originally, the idea for the business was DVD agnostic. It was just a way to make the concept of video rental better. And, it wasn't really until we began to hear about DVD's imminent launch, we realized that this was an enabling device for something even bigger. Which is: getting people any movie they wanted any time they wanted. And having this thing be small enough and light enough that they could actually use a truck to get it that last mile rather than a phone line.

It sounds as though you were exploring a number of different opportunities and it was just an ideal time for DVD. Was that a tough decision considering that the DVD market has been in flux for years?

Well, certainly. Part of the nature of being entrepreneurial is that you need to leap off a cliff before you have a very clear view of what lies at the bottom of it. And certainly the leap that we took for DVD was extremely early. When we launched, DVD was still in test markets in just a limited number of cities and even when we had officially incorporated the com-

Case Study: The Entrepreneurial Possibilities of DVD

pany and raised money, probably with less than 200,000 DVD players sold nationwide. So that was certainly a big leap of faith.

And then the second year, 1998, was no picnic either. That was also clouded by the arrival of DIVX and there certainly was a question of whether this was going to be a viable format. And it was really not until 1999 that it became obvious that was going to be successful—it wasn't a question of "if," it was more a question of how much and how fast.

I've heard recent figures stating that DVD has eclipsed the growth rate of CDs at the same point in their introductory period.

By a wide margin. By almost any measure, units sold in the first number of years or household adoptions in the first number of years, but it is now the fastest launching consumer electronics device ever.

You must feel pretty comfortable with your current corporate positioning.

It is very, very solid. This is the heir-apparent to the VCR. The VCR is in upwards of 90% of American households. Although on the horizon there are other technologies that will allow us to watch movies in a different way. They will not stop DVD from becoming a ubiquitous household item. They will be perhaps a second wave which then comes over that later. There is a need for a device in every household to be able to watch movies on, the same way there is a VCR right now, and I'm sure that will be the DVD player.

Do you have any concerns over the durability issues of the media, as you're shipping them back and forth multiple times?

We now have a tremendous amount of experience shipping DVDs back and forth. We do in excess of 250,000 discs a week now. We used to pay a lot of attention to how durable they were. Now, we don't worry about it any more.

So, people don't abuse them, obviously. Do you have an anticipated lifespan per rental disc?

Initially, we though the lifespan would be 30 to 40 roundtrips, but in reality we have some discs that have made 50 to 60 roundtrips and are still going strong. What we have found, interestingly enough (and we've shared some of this with the studios), it is very dependent by title. There is something in the manufacturing process that renders some discs much

Chapter 20

more fragile, both in terms of playability and also shatterability. When we go to analysis of breakage, it is usually skewed towards certain titles. In a given week, for example, if we have 100 broken discs, it would not be surprising to have 60 of them be exactly the same title.

So, that suggests that slight differences in the manufacturing process are the cause of the problem?

There is no question about it. The plastic somehow reacts to a bad plot and it curdles the plastic or something.

From those numbers, it really is a very minimal concern.

Very minimal.

Having used your service, it clearly is a very easy way to view top-notch movies inexpensively.

Amazingly enough, it works. It is a guess. Will people tolerate a day or two delay in exchange for all the other benefits we provide—all the selections, all the editorial tools. And the answer at this point is pretty clearly "Yes."

Can you give me a quick capsule history of how the business got started and what your background is?

My background is not in the movie business. I am not a videophile. I mostly have been with startups of all stripes. My background is primarily in direct marketing. I was the founder of a magazine called MacUser magazine, which we sold to Ziff-Davis. Then I was the founder of a mail order company called MacWarehouse and MicroWarehouse. Then I was at Borland Software for about 6 years, doing direct marketing until I became the General Manager of one of their divisions for awhile. I did a startup called Visioneer that makes a little scanner called the PaperPort. Then I did a small high-tech startup in the Silicon Valley, which did Quality Assurance software, which we eventually sold to Rational Software. That gave me about 6 months off where I said, "What do I do next?" I decided the time was right to do something in ecommerce. That was what lead us to examining what large categories are ripe for being exploited by the power of the Internet. We decided the video rental was a potential one. And, DVD would enable that.

Case Study: The Entrepreneurial Possibilities of DVD

Have you considering doing rich media previews on your site? Would the studios present an obstacle because of copyright issues?

No, not at all. There are two things. One, we could use the pre-canned previews, which they provide. They are, of course, eager for us to do that since it is a form of advertising for them. But, in effect, you can editorially excerpt brief segments from a DVD and show it as an editorial clip. And, we've played with the idea of doing both of those things.

The success of your model seems tied very strongly to the ease-of-use of the Web site and ordering through the Internet.

We've done two migrations on the site in the last month or so. One is moving most of our customers over to our Marquee program. The main innovation that we have done is to change the whole dynamics of video rental. Currently, for most people who rent movies, there is this level of spontaneity or same-day commitment. We say, I want to watch a movie. You need to get in your car and drive the video rental store to get the movie you want to consume then. We could never compete with that. No matter how much I spend, I could never get it so that I can get a movie to you faster than you get it at the rental store. With Marquee, we let you receive up to four movies and you keep them in your house until you're ready to watch them. I flipped the whole thing around. Now, it's eight at night and it's raining and you want to watch a movie. You don't need to get in your car and drive to Blockbuster. You've got four movies sitting on top of your TV or in your closet that you have prepicked that are ready to go. That one user dynamic changed everything around. All of a sudden, we have eliminated spontaneity as being an issue. And, of course, it is the same thing in reverse. You no longer have to drive to Blockbuster and drop it off. You can just drop it in the mail and you're done.

The second thing we have done, which has totally changed the business around, is that we've launched a service called CineMatch. This is very recent. CineMatch is a personalized movie service. Essentially, you go to the site and rate 20 movies and that allows us to build a taste profile for you, giving us a very strong ability to predict what movies we think you will like and dislike. And, then we can use that in all kinds of ways. Going from the most explicit, which is having a page of movies that we recommend for you. When you combine these two features, we've created huge amounts of convenience by having movies on top of the TV all the time with the ability to do very, very strong taste filtering for you, so you don't waste time watching movies that you are not going to like.

Chapter 20

I imagine that must be pretty tricky making predictions for people who have eclectic tastes. Is it a fairly complex algorithm you use to categorize movies to match people's tastes?

It is not a movie categorization service, because that is very complicated. What we do: let's say that I ask you to come in and rate 20 movies. The movies that I'm going to ask you to rate are movies that have pretty strong vectors of taste associated with them. If I said, let's rate *Titanic* or *Saving Private Ryan*, odds are that a lot of us would rate them the same. If I said, let's rate something more controversial, such as *Pulp Fiction*, or let's rate *American Pie*, you're going to get very strong conflicts. Some will love it. Some will hate it. Some are neutral. Now, if I take 20 movies like that and have you rate those 20 movies, I've got a pretty multidimensional sense of who you are as a movie liker. If we rated 40 or 60 movies, it gets even stronger. Now, what I do is take all those movie tastes you have told me about and I look into a large database of other people who have done the same thing, and I find people with movie tastes exactly like yours. In a nation of 280,000,000 people, there are a lot of people with movie tastes exactly like yours.

Is this anything like the User Circles that were featured on Amazon.com? Didn't they take heat for exposing the user data?

They did take heat for exposing the data. Telling the world, for example, that people at Boeing are reading this. And, of course, saying that people at Boeing are reading this doesn't really expose it; the only problem is when you begin to have organizations that are only five or six people. I am opposed to that. In our case, we would never tell anybody what you did or did not like.

What I am doing is matching you up with someone who has tastes just like you. And then you are going to come to my site and wonder, Am I going to like *American Pie*? What I do is look at the 60 or 70 or 80 people that I've found in my database who have tastes exactly like yours, I look at all of them who have looked at *American Pie* and rated it, and then I try to see if I have confidence that all of them have rated it the same. So, I find 17 or 18 who, like you, all hated *American Pie* and then I can report back to you with a high degree of confidence, You're not going to like this movie. We can do that in both dimensions: they hate it, they love it, they are neutral. What is remarkable is that, as scary as it seems, we are reasonably predictable in matters of taste when we are aggregated into very large communities. It doesn't work with 20 people or 2000 or 20,000. But, with 200,000 or 2,000,000, then you are able to pull out very, very interest-

Case Study: The Entrepreneurial Possibilities of DVD

ing results, because you are finding people who are very similar to you. Across the span of the Internet, we can find people who are separated from you physically, but who are very, very close to you emotionally. You create a community.

Some of this data I would guess you would have to tie in with your purchase and distribution system for particular titles. Would this determine what kind of depth you would have in your catalog of titles?

Yes, we do integrate the two. It does give us a very good ability to predict how many copies of each individual title we need to buy. Also, our ability to help present people with movies that are right for them on two dimensions: one is movies they are going to love and the other is movies that we have in stock.

Do a lot of people rely on the recommendations of reviewers who are featured on your site?

The problem with reviewers is that they represent reasonably the lowest common denominator. Most people, when they read reviews, they will read their local reviewer. Let's say you read the *New York Times.* You have one or two reviewers for millions of people. To think that millions of people all share the same tastes is crazy. A good reviewer is able to expose enough about the movie to allow someone to determine, is this for me? But someone saying that I love this movie or hate this movie is a dangerous thing, unless you have a long history with that reviewer and trust that their tastes mirror your own. We, of course, offer that because a good reviewer can help you determine if you might like a movie. A more powerful thing that we are coming out with shortly is saying there are 175 or 180 professional, widely circulated movie critics in the United States. If I can point you to a reviewer whose tastes are dead on with yours, that is a valuable thing.

Do you support an approach where users rate movies, similar to what Amazon.com does with letting users comment on books and other products?

It's great. I love it. We do that a little bit. We are expanding that feature, as well. The common ways that people figure out what movie to watch are: word of mouth (such as user reviews) and critics. If you are going to help steer people to movies they love, which is our objective, you have to do both of those things.

Chapter 20

What do you consider the next challenge for the business?

What we have done for almost two years is nail getting people movies by shipping them pieces of plastic. By doing DVD distribution as a near video-on-demand alternative. We are in the process right now of taking the next step, which is moving up the value chain a bit. Right now we do two things well. We help you find a movie that you're going to love and we get it to you quickly and conveniently. We look at those two things and we are much more interested in the former than the latter. So, the thing that we really want to continue to become phenomenal at is helping you find movies that you love. We will then become more agnostic about how you choose to get the movie delivered to you. We will say, Hey, we're great about getting you the movies you love. If you want to use the Marquee program, fantastic. If you want to buy a DVD, what tools can we give you to help you buy the right DVD. If you want to stream it to yourself, let's point you to where you can stream it. If you want to download it, you can download it. If you want to see it in the theater, you can see it in the theater.

Do you think your service would even be possible without the Internet?

The valuable part of what we do is by matching your tastes with millions of people, we can help you avoid watching a dog. Or help steer you to a gem that you never would have found otherwise. That can't happen without the Internet. I can't imagine that happening in your local video rental. It isn't conceivable. There is a network effect. The more people who do it, the stronger the recommendations get, the more people do it. That is the power of the Internet.

Resources

This appendix includes a number of resources that should prove helpful to anyone interested in optical recording technology.

Trade and Standards Organizations

Optical Storage Technology Association

The Optical Storage Technology Association (OSTA) is a trade association with a mission to promote various uses of recordable optical technology. The organization has an international presence with members throughout North America, Asia, and Europe. While OSTA does not actively define the standards that apply to the optical recording industry, they actively work with industry participants to ensure compatibility among devices and media employed in writable optical technology and to educate all interested parties in the potential uses of optical technology.

Optical Storage Technology Association Contact Data

311 East Carillo Street
Santa Barbara, CA 93101 USA
Phone: 805 963 3853
Fax: 805 962 1541 Fax
Web: *www.osta.org*

Appendix A

SIGCAT

The Special Interest Group on CD/DVD Applications and Technology (SIGCAT) began as a small user group back in 1986 devoted to U.S. Geological Survey operations. It has since expanded into an international organization with over 11,000 members. The organization's mission is to further the understanding of CD and DVD technologies and assist organizations and individuals with utilizing the technology in practical ways.

SIGCAT Foundation

11343 Sunset Hills Road
Reston, VA 20190
Phone: 703 435 5200
Fax: 703 435 5553
Web: www.sigcat.org

DVD Forum

The DVD Forum, formerly the DVD Consortium, is an industry group with members from various hardware manufacturers, software developers, and other parts of the industry. Their mission is to create an open forum for the discussion of ideas, advancing of capabilities, and resolution of any issues that might impede the growth of the DVD industry.

Their site (*www.dvdforum.org*) offers information and insights into many of the specifications and current industry plans for expansion into new areas. There are also numerous links to DVD-related topics.

DVD Forum

Toshiba Building, 15th Floor
1-1, Shibaura 1-chome
Minato-ku, Tokyo 105-01
Japan
Phone: +81-3-5444-9580
Fax: +81-3-5444-9436
Web: *www.dvdforum.org*

Appendix A

Digital Video Professional's Association

Providing resources for professionals involved in the tools and technologies of digital video, the Digital Video Professional's Association (DVPA) offers many benefits to members, including reduced costs for reference materials and access to breaking news on digital video topics.

Digital Video Professional's Association

1603 Main Street
Dunedin, FL 34698
Phone: 813 738 0656
Fax: 813 738 0659
Web: *www.dvpa.com*

PCFriendly

The merging world of computer DVD-ROM and DVD-Video is addressed by PCFriendly, an organization that provides resources on how to tap into the full DVD-ROM capabilities of your computer, including desktop playback of video.

The key element that you need to successfully play back movies on your computer desktop is a DVD decoder, which can be either a hardware decoder that was included with your DVD-ROM or a software decoder that uses the processor to perform the conversion of the video data, uncompressing the MPEG-2 content for streaming display on your computer. Clearly, hardware-based solutions to DVD decoding are faster and relieve your computer's processor of having to do the extra work required to stream the video. You need a very high performance processor to successfully decode through software; otherwise, your video playback will experience dropped frames and less than fluid performance.

For both consumer level guidance and information for companies that want to make their titles widely compatible. PCFriendly offers a valuable information resource.

PCFriendly

Phone: 408 436 6700
Web: www.pcfriendly.com
Email:questions@pcfriendly.com

Appendix A

Stock Photos, Fonts, Media Assets

EyeWire Studios

Offering a variety of CD-ROMs containing maps, digital artwork, movies, photographs, and fonts, EyeWire provides the resources that a busy developer needs to find that missing element for a project. Single items can be purchased online and downloaded. Medium and high-resolution content is typically delivered on CD-ROM. All EyeWire content is royalty free. Styles range from the conservative business model to outrageous funk.

EyeWire

1525 Greenview Drive
Grand Prarie, TX 75050
Phone: 800 661 9410
Fax: 800 814 7783
Web: www.eyewire.com

The Stock Market

Providing a sophisticated online selection service, using a lightbox metaphor for choosing images, The Stock Market also distributes catalogs on disc containing themed image packages. Individual downloads and CD-ROM collections are both available.

The Stock Market

360 Park Avenue South
New York, NY 10010
Phone: 800 999 0800
Web: www.stockmarketphoto.com

Authoring Tools

Authoring tools for producing digital content cover a very wide range of capabilities and encompass many different computer platforms. The following tools were described or mentioned in the text of this book.

Blue Sky Software Corporation

7777 Fay Avenue, Suite 201
La Jolla, CA 92037
Phone: 858 551 2485
Fax: 858 551 2486
Web: www.blue-sky.com

Appendix A

InstallShield Software Corporation

900 National Parkway, Suite 125
Schaumburg, IL 60173
Phone: 847 619 2266
Fax: 847 240 9138
Web: *www.installshield.com*

Sonic Desktop Software

P. O. Box 3205
Chatsworth, CA 91311
Phone: 818 718 9999
Fax: 818 718 9990
Web: *www.smartsound.com*

Sonic Foundry

754 Williamson Street
Madison, WI 53703
Phone: 608 256 3133
Fax: 608 256 7300
Web: *www.sonicfoundry.com*

Terran Interactive

15951 Los Gatos Boulevard, Suite #1
Los Gatos, CA 95032
Phone: 800 577 3443
Fax: 408 356 7373
Web: *www.terran.com*

Zuma Digital

59 West 19 Street, 5th Floor
New York, NY 10011
Phone: 212 741 9100
Fax: 212 741 1605
Web: *www.zumadvd.com*

CD and DVD Recorders

JVC Professional Computer Products Division

5665 Corporate Avenue
Cypress, CA 90630
Phone: 714 816 6500
Fax: 714 816 6519
Web: *www.jvc.net*

Appendix A

Microtech Systems

2 Davis Drive
Belmont, CA 94002
Phone: 650 596 1900
Fax: 650 596 1915
Web: *www.microtech.com*

Pansonic Document Imaging Company

2 Panasonic Way, 7D-9
Secaucus, NJ 07904
Phone: 201 348 7000
Fax: 201 392 4504
Web: *www.panasonic.com*

Pinnacle Micro

140 Technology Drive
Irvine, CA 92618
Phone: 800 553 7070
Fax: 949 789 3150
Web: *www.pinnaclemicro.com*

Pioneer New Media Technologies, Inc.

2265 East 220th Street
Long Beach, CA 90810
Phone: 310 952 2111
Fax: 310 952 2990
Web: *www.pioneerusa.com*

Plasmon IDE, Inc.

9625 West 76th Street
Eden Prarie, MN 55344
Phone: 612 942 3006
Fax: 612 946 4141
Web: *www.plasmon.com*

Plextor Corporation

4255 Burton Drive
Santa Clara, CA 95054
Phone: 800 886 3935
Fax: 408 986 1010
Web: *www.plextor.com*

Ricoh Corporation, DMS-C

One Ricoh Square, 1100 Valencia Avenue
Tustin, CA 92780
Phone: 714 566 3244
Fax: 714 566 3266
Web: *www.ricohdms.com*

Sony Electronics, Inc.

3300 Zanker Road
San Jose, CA 95134
Phone: 800 686 7669
Fax: 408 955 4771
Web: *www.sony.com*

TEAC America, Inc.

7733 Telegraph Road
Montebello, CA 90640
Phone: 323 726 0303
Fax: 323 727 7672
Web: *www.teac.com*

Toshiba America Electronic Components Disk Products Division

35 Hammond
Irvine, CA 92618
Phone: 949 457 0777
Fax: 949 588 7845
Web: *www.diskproducts.toshiba.com*

Yamaha Systems Technology, Inc.

100 Century Center Court, #800
San Jose, CA 95112
Phone: 408 467 2300
Fax: 408 437 9741
Web: www.yamahayst.com

CD and DVD Recorder Applications

Adaptec, Inc.

From their headquarters in Milpitas, CA, Adaptec specializes in I/O products, including SCSI host adapters, RAID technologies, and CD/DVD recorder software.

Appendix A

Adaptec, Inc.

801 South Milpitas Boulevard
Milpitas, CA 95035
Phone: 408 957 2044
Fax: 408 957 6666
Web: www.adaptec.com

CeQuadrat

From their headquarters in Germany, CeQuadrat has become recognized for their professional line of recorder applications and their packet-writing driver, Packet CD.

CeQuadrat GmbH

Dennewartstr. 27
D-52068 Aachen, Deutschland
Phone: ++49 241 949 020
Fax: ++49 241 949 0211
Web: www.cequadrat.de

CeQuadrat (USA), Inc.

691 South Milpitas Boulevard
Milpitas, CA 95035
Phone: 650 843 3780
Fax: 408 957 4544
Email: marketing@cequadrat.com

GEAR Software, Inc.

One of the pioneers in the CD-recordable field, GEAR Software (now a subsidiary of Command Software) was also one of the first companies to introduce software for recording to DVD. GEAR offers both consumer-level and professional-caliber recording applications.

GEAR Software, Inc.

1061 East Indiantown Road, Suite 500
Jupiter, FL 33477
Phone: 800 423 9147
Web: www.gearsoftware.com

Appendix A

Computer Output to Laser Disc

Computer Output to Laser Disc (COLD) applications consist of software, and sometimes hardware/software combinations, designed to use an optical disc directly as the target for information storage. Medical imaging systems, document scanning and storage applications, and video archiving often utilize COLD techniques to provide inexpensive storage and automated workflow. The following companies offer various products designed for COLD applications.

Bell & Howell

6800 North McCormick Road
Chicago, IL 60645
Phone: 800 646 3672

ALOS Micrographics

118 Bracken Road
Montgomery, MY 12549
Phone: 914 457 4400

Binnary Research

7100 East Valley Green Road
Fort Washington, PA 19034
Phone: 215 233 3200

Bluebird Systems

5900 La Place Court
Carlsbad, CA 92008
Phone: 800 669 2220

Elms System

2 Holland
Irvine, CA 92618
Phone: 714 461 3200

I. Levy and Associates

1633 DesPeres Road, Suite 300
St. Louis, MO 63131
Phone: 314 822 0810

Appendix A

Image-X International
1950 Stemmons Freeway S 5001
Dallas, TX 75207
Phone: 214 712 8500

Image
6486 South Quebec Street
Englewood, CO 80111
Phone: 303 773 1424

Network Imaging
500 Huntmar Park Drive
Herndon, VA 22070
Phone: 800 254 0994

Orion Systems
491 Maple Street
Danvers, MA 01923
Phone: 508 777 4747

Siemens Nixdorf DMS
6375 Shawson Drive
Mississauga, ON L5T 1S7
Phone: 800 565 5650

TDF Technologies
P.O. Box 458
Blue Ridge Summit, PA 17214
Phone: 717 794 5859

VUCOM
1120 Keystone
Lansing, MI 48911
Phone: 517 393 8610

Thin Server Technologies

Thin servers offer access to various network devices, including CDs and DVDs, in a seamless and thorough manner. The following companies offer products in this area.

Appendix A

Axis Communications

100 Apollo Drive
Chelmsford, MA 01824
Phone: 800 444 2947
Fax: 978 614 2100
Web: www.axis.com

Boffin, Ltd.

2500 West County Road 42, Suite 5
Burnsville, MN 55337
Phone: 612 894 0595
Fax: 612 894 6175
Web: www.boffin.com

Excel Computer

3330 Earhart Drive, #212
Carrollton, TX 75006
Phone: 800 995 1014
Fax: 972 980 0375
Web: www.excelcdrom.com

Microtest, Inc.

4747 North 22nd Street
Phoenix, AZ 85016
Phone: 800 526 9675
Fax: 602 952 6401
Web: www.microtest.com

StorLogic

498 Palm Springs Drive, Suite 100
Altamonte Springs, FL 32701
Phone: 877 786 7564
Fax: 407 261 8983
Web: www.storlogic.com

TenXpert Technologies

13091 Pond Springs Road
Austin, TX 78729
Phone: 800 9222 9050
Fax: 512 918 9182
Web: www.tenxpert.com

Appendix A

Library Systems, Jukeboxes, and Towers

Serious optical disc storage systems and recorders designed for network operation occupy the highest level of sophistication in the optical disc family tree. The following companies make optical disc products for network use.

CMS Peripherals

3095 Redhill Avenue
Costa Mesa, CA 92626
Phone: 800 327 5773
Fax: 714 437 1476
Web: www.cmsenh.com

Kodak Business Imaging Systems

343 State Street 2/20
Rochester, NY 14650
Phone: 800 243 8811
Fax: 716 724 2342
Web: www.kodak.com/go/businessimaging

JVC Professional Computer Products Division

5665 Corporate Avenue
Cypress, CA 90630
Phone: 800 488 4353
Fax: 716 816 6519
Web: www.jvc.com

MediaPath Technologies, Inc.

125 F. Gaither Street
Mount Laurel, NJ
08054
Phone: 800 357 0697
Fax: 609 222 0552
Web: www.mediapath.com

NSM Jukebox

1158 Tower Lane
Bensenville, IL 60106
Phone: 630 860 5100
Fax: 630 860 5144
Web: www.nsmjukebox.com

Plasmon IDE

8625 West 76th Street
Eden Prairie, MN 55344
Phone: 612 946 4100
Web: *www.plasmon.com*

Procom Technology, Inc.

2181 Dupont Drive
Irvine, CA 92715
Phone: 714 852 1000
Fax: 714 852 1221
Web: *www.procom.com*

SciNet, Inc.

268 Santa Ana Court
Sunnyvale, CA 94086
Phone: 408 328 0160
Web: *www.scinetcorp.com*

Sony Corporation of America

3300 Zanker Road
San Jose, CA 95134
Phone: 408 436 6300
Fax: 408 432 0253
Web: *www.sony.com*

DVD-Video Tools

Avid Technology

1 Park West
Tewksbury, MA 01876
Phone: 800 949 2843
Web: *www.avid.com*

Astarte USA

364 Wildwood Avenue
Birchwood, MN 55110
Phone: 651 653 6247
Fax: 651 653 6495
Web: *www.astarte.de/dvd*

Appendix A

Blossom Technologies Corporation

5555 West Flagler Street
Miami, FL 33134
Phone: 305 266 2800
Fax: 305 261 2544
Web: www.blossomvideo.com

C-Cube Microsystems, Inc.

1778 McCarthy Boulevard
Milpitas, CA 95035
Phone: 408 490 8017
Fax: 408 490 8590
Web: www.c-cube.com

Minerva System, Inc.

1585 Charleston Road
Mountain View, CA 94043
Phone: 800 806 9594
Fax: 650 940 1450
Web: www.minervasys.com

Sonic Solutions

101 Rowland Way
Novato, CA 94945
Phone: 415 893 8000
Fax: 415 893 8008
Web: www.sonic.com

Media

Creative Data Products

47654 Kato Road
Fremont, CA 94538
Phone: 510 668 4800
Fax: 510 668 4803
Web: www.cdpinc.com

DisksDirect.com

200 San Mateo Avenue
Los Gatos, CA 95030
Phone: 800 557 1000
Fax: 408 399 7671
Web: www.disksdirect.com

Eastman Kodak Company

460 Buffalo Road
Rochester, NY 14650
Phone: 800 235 6325
Web: *www.kodak.com/go/cdr*

Maxell Corporation

22-08 Route 208
Fair Lawn, NJ 07410
Phone: 201 794 5922
Fax: 201 796 8790
Web: *www.maxell.com*

Mitsui Advanced Media, Inc.

2500 Westchester Avenue, Suite 110
Purchase, NY 10577
Phone: 914 253 0777
Fax: 914 253 8623
Web: *www.mitsuicdr.com*

TDK Electronics Corporation

12 Harbor Park Drive
Port Washington, NY 11050
Phone: 800 835 8273
Fax: 516 625 0651
Web: *www.tdk.com*

Verbatim Corporation

1200 W. T. Harris Boulevard
Charlotte, NC 28262
Phone: 800 421 4188
Fax: 704 547 6609
Web: *www.verbatimcorp.com*

Packaging Materials

Bag Unlimited, Inc.

7 Canal Street
Rochester, NY 14608
Phone: 800 767 2247
Fax: 716 328 8526
Web: *www.bagsunlimited.com*

Appendix A

Calumet Carton
16920 State Street
South Holland, IL 60137
Phone: 708 333 6521
Fax: 888 333 8540
Web: www.calumetcarton.com

Compact Disc Packaging Corporation
320 Broadhollow Road
Farmingdale, NY 11735
Phone: 516 752 0750
Fax: 516 752 1971
Web: www.compact-disk.com

Information Packaging Corporation
1670 North Wayneport Road
Macedon, NY 14502
Phone: 315 986 5793
Fax: 315 986 4585

Multi-Media Publishing and Packaging, Inc.
9430 Topanga Canyon Blvd.
Chatsworth, CA 91311
Phone: 800 982 8138
Fax: 818 341 2807
Web: www.mmppinc.com

Univenture, Inc.
4707 Roberts Road
Columbus, OH 43228
Phone: 800 992 8262
Fax: 614 529 2110
Web: www.univenture.com

Replication and Production Services

Accurate Bit Copy
6 Otis Park Drive
Bourne, MA 02532
Phone: 800 696 0500
Fax: 508 759 5550
Web: www.accuratebitcopy.com

Appendix A

Acutrack, Inc.
3109 Castro Valley Boulevard
Castro Valley, CA 94596
Phone: 510 581 4536
Fax: 510 581 7386
Web: *www.acutrack.com*

Catalogic
990 Richard Avenue, Suite 103
Santa Clara, CA 95050
Phone: 408 486 0800
Fax: 408 486 0809
Web: *www.catalogic.com*

Disc Makers
7905 North Route 130
Pennsauken, NJ 08110
Phone: 800 237 6666
Fax: 856 661 3450
Web: *www.discmakers.com/cdrom*

Diskcopy, Inc.
P.O. Box 8197, 39 Shelley Road
Haverhill, MA 01835
Phone: 888 347 5267
Fax: 978 521 5300
Web: *www.diskcopy.com*

Green Solutions
13798 NW 4th Street #309
Fort Lauderdale, FL 33322
Phone: 954 846 8555
Fax: 954 846 9156
Web: *www.greensolutions.com*

Glossary

3DO An early high-performance 32-bit RISC platform, primarily for games and entertainment, that accepts only 3DO-compatible CD-ROMs or Kodak Photo CDs. Despite backing by Electronic Arts, Time Warner, and Matsushita, the 3DO format never gained wide acceptance. The hardware unit was sold off to Samsung and the software division lives on as The 3DO Company, producing games for Playstations, Nintendo machines, and the Internet.

8/16 modulation A technique used to store channel information on a DVD disc.

a-characters Fundamental ISO 9660 character set (compare to d-characters). Includes the uppercase alphabet, A through Z, numerals 0 through 9, and the following symbols: ! % " & ' ( ) = * + , - . / : ; < ? > _

Absolute-time The time elapsed since the beginning of a recorded Red Book digital audio program; also known as A-time. A-time is calculated by reference to an internal clock that monitors elapsed time starting at the beginning of the innermost track.

AC-3 The original name for the encoding scheme now officially termed Dolby Digital. The AC-3 term still frequently appears in reference to descriptions of the process.

Access Time The time required to position the laser read head over a specified sector and begin retrieving the data. For CD-ROM drives and CD-R devices, the access time generally ranges between 120 to 600 milliseconds, depending on the performance of the drive. Access time is also used in reference to other storage devices, such as hard disk drives and diskette drives.

Adaptive Differential Pulse Code Modulation A technique for compressing audio data by storing the difference between signals rather than the actual signal. Usually abbreviated as ADPCM.

Advanced SCSI Programmer's Interface (Usually shortened to ASPI. A set of ANSI-defined commands for application-level communication with SCSI host adapters. Most operating systems have ASPI drivers that are used to communicate with CD recorders included on a SCSI chain.

AGC Shortened form for automatic gain control. An electronic technique for boosting an incoming signal level to meet minimal acceptable recording strength.

Glossary

Analog to Digital Converter A hardware component that converts an analog waveform to a succession of digital values by sampling the waveform at periodic intervals. Often abbreviated to A/D converter or ADC.

ANSI-Labeled A tape specified to ANSI X3.27-19778 standards that includes details of the file structure, volume name, and a file header referencing the contents of the tape. Replication facilities and disc manufacturers generally prefer ANSI-labeled tape submissions.

ANSI-Unlabeled A tape formatted without ANSI label information.

Artifact A data abnormality that can appear in an audio or video file as the result of certain kinds of signal processing, including compression, data transfer errors, signal noise, or electrical interference.

ASCII Abbreviated form of American Standard Code for Information Exchange. Refers to a basic set of control characters and alphanumeric characters assigned to 7-bit values.

Aspect Ratio The ratio of the horizontal size to the vertical size of a picture. In television, the aspect ratio is 4:3. In widescreen DVD, the aspect ratio is 16:9.

ATAPI Abbreviated form for Advanced Technology Attachment Packet Interface. ATAPI provides a layer of commands used to manage devices connected through an IDE bus, including CD-ROM and DVD-ROM drives. ATAPI was introduced as part of the Enhanced IDE standard.

Authoring The integration of the individual elements—sound, video, text, graphics—within an interactive presentation. A wide variety of software tools are used for authoring, including search and retrieval programs, indexing software, multimedia development tools, SGML editors, and so on. Authoring is sometimes confused with the term premastering.

AVI An audio/video format introduced by Microsoft. Typically requires a high-performance computer platform with a fast storage device. No longer in common use.

Autoplay Discs encoded with the autoplay option will begin immediate playback when inserted in a DVD player that supports this feature.

Bandwidth A measurement of the data-carrying capacity of a bus or other data transmission medium.

BCA Short for burst cutting area. An area located near the center of a DVD disc that is reserved for ID codes and manufacturing data. The BCA is imprinted as bar-code data.

Bit Short for binary digit. A bit is the basic element representing digital data. Bits are combined into groups to form bytes (8 bits), words (16 bits), and double-words (32 bits).

Bi-refringence A term applied to the refraction of a beam of light in two different directions. This phenomenon occurs as an undesirable aspect of the compact disc manufacturing process resulting from residual stresses in the polycarbonate substrate introduced during injection molding. Excessive bi-refringence results in laser read errors.

Bit Error Rate Often shortened to BER; indicates the number of bit errors that occur in proportion to the number of correctly processed bits. Bit Error Rates for most CD-ROM applications are in the range of 1 error per 10^{12} bits.

Bits Per Pixel Sometimes referred to as color depth, the number of bits per pixel defines the maximum color variations available for each pixel that appears onscreen. 8-bit color allows 256 individual colors. 16-bit color allows 65,536 colors, and so on.

Block Error Rate Often shortened to BLER; indicates the number of blocks in which an error was detected in proportion to the overall number of blocks processed.

Book A The basic specification for the physical format that applies to DVD discs; this specification forms the basis for the DVD-ROM.

Book B The specification that defines the format of DVD-Video discs.

Book C The specification that defines the format of DVD-Audio discs.

Book D The specification that defines the format of DVD-R (write-once).

Book E The specification that defines the format of rewritable versions of DVD.

Buffer A temporary storage area used to compensate for differences in the data transfer rates of two devices. The buffer holds a quantity of data that ensures a continuous flow to the faster device while the slower device works to keep the buffer full. CD recorders usually feature a 64Kb to 2MB internal buffer for this purpose.

Bus Mastering A technique where a peripheral device, such as a bus master SCSI host adapter, takes control of a DMA transfer, moving data to or from the host computer's memory without direct processor intervention. By freeing up the host computer's processor during data transfers, multitasking operations can be performed more effectively. A SCSI host adapter featuring bus mastering significantly improves performance in a system that includes a CD recorder.

Cache A storage area created in high-speed RAM that contains a collection of the most recently accessed data, as well as anticipated data (in the case of a look-ahead cache). The computer processor can access information more quickly from the cache than having to retrieve that information from a storage device (such as a hard disk drive or CD-ROM drive). Effective use of caching techniques can make a slow storage device appear much faster.

Caddy A plastic enclosure with a sliding access window that holds a compact disc for insertion into a drive. Caddies are required with some, but not all, drives. CD recorders often use a caddy system to ensure that the blank recordable media remains free of dust and dirt that could hamper the recording process.

CAV Shortened form for constant angular velocity. In this type of data storage system,

Glossary

the rotation speed of the disc is kept constant as the read/write head is positioned over different points on the disc.

CD-DA Shortened form for Compact Disc–Digital Audio. The original compact disc specification as detailed in the Red Book standard. CD-DA discs contain audio data recorded in PCM format.

CD-I Shortened form for Compact Disc–Interactive. A standard pioneered by Philips to improve the play of interactive multimedia material. The specifications of CD-I appear in the Green Book standard.

CD-I Bridge Shortened form for Compact Disc–Interactive Bridge. The standard that defines how CD-I information can be written to a CD-ROM XA compact disc. Commonly known as White Book, this standard also encompasses PhotoCD.

CD-R Shortened form for Compact Disc–Recordable. A variation of CD-ROM standards that supports the use of recordable media and includes multiple recording sessions to a single disc. CD-R technology has improved dramatically over the last year; it now represents one of the least expensive methods for archiving and transferring data.

CD-PROM Shortened form for Compact Disc-Programmable ROM. A technology developed by Kodak that offers both a recordable region and mass-replicated content on the same disc.

CD-ROM Shortened form for Compact Disc–Read-Only Memory. A compact disc that meets the specifications defined in Yellow Book, supporting storage of computer data (as well as audio data in Mixed Mode applications).

CD-ROM XA Shortened form for Compact Disc–Extended Architecture. A standard that originated from Yellow Book enhancements to improve the playback of different data types and to support multisession operation. Photo CDs employ this physical format.

CD-WO Shortened form for Compact Disc–Write Once. An alternate term used to describe recordable CD media.

Cell As applies to DVD-Video, a single unit of video content that can vary in length from less than a second to several hours. This structure allows video content to be grouped in various ways for interactive playback.

Channel In audio terms, a division of the audio content that is typically directed to one speaker. For example, stereo signals include two channels of audio content.

Challenge Key Part of the encryption process used in DVD-ROM content presentation, the challenge key authenticates an exchange between the drive and host computer.

Circular Buffer Read-Ahead A method by which data are retrieved and loaded into the CD-ROM drive's buffer before it is actually requested. Done properly, this technique speeds throughput.

Glossary

Clamping Area The region close to the central hole in a disc that is gripped by the drive mechanism to rotate the disc.

Close Disc Closing a multisession recordable disc makes it impossible to perform further write operations to it. When the disc is closed, the last session's Lead-In does not contain the next address to which recording can take place. This address is required to accomplish multisession writes.

Close Session Closing a session on a recordable disc results in directory information being written to the disc's table of contents and a Lead-In area being created in preparation for the next write operation. New sessions can be written (and closed) until a Close Disc operation is performed or remaining disc space is consumed.

CLV Shortened form for constant linear velocity. Refers to the varying of the disc rotation speed to ensure that the laser read head encounters data at a constant linear rate (1.2 to 1.4 meters per second). Disc rotation speeds increase for inner tracks and decrease for outer tracks to maintain this constancy.

Composite Video A standard video signal in which the red, green, and blue components are combined with a timing (Synchronization, or sync) signal.

Compression A process by which a file is reduced in size by removing or encoding extraneous information. A lossy compression standard, such as JPEG, cannot regenerate the original file content to the same degree of accuracy. A lossless compression standard, such as GIF, can restore all the original information.

Control Area In DVD terms, a portion of the Lead-In area that contains a single ECC block, containing key disc data, repeated 16 times.

Cross-Platform Development Authoring and formatting to ensure playback on more than one computer platform. For example, Macromedia Director can be used as a cross-platform development tool to allow playback under either Windows or Macintosh systems.

Cross-Interleaved Reed-Solomon Code Shortened to CIRC. An error correction technique specifically developed for use with Red Book audio data. This form of error control results in only one uncorrectable bit per every 10^9 bits. Additional correction codes used in Yellow Book to cover CD-ROM data reduce uncorrectable bit counts even further, down to one per 10^{13}.

Cue Sheet A sequence of tracks to be recorded in order when mastering a multitrack audio or mixed-mode compact disc.

d-characters A basic ISO 9660 character set that consists of the following characters: Capital A through Z; numerals 0 through 9; and underscore (_).

DAT Shortened form of Digital Audio Tape. A storage format for audio information on 4mm tape cartridges. Sony originated the techniques for sampling audio data and converting to a digital framework while designing their original DAT drives.

Glossary

These sampling and encoding techniques were than adapted to compact disc.

Data Area In CD-ROM terms, the beginning of the user data area on a CD-ROM. Under ISO 9660, this area starts at address 00:02:16. In DVD terms, the physical region residing between the Lead-In and Lead-Out areas where the actual data content of the disc appears.

Data Capture Those techniques for converting data from non-computer formats (video tape, photographs, line drawings, pages of text) into a digital form that can be processed by a computer.

Data Compression Reducing the storage space needed for data by compressing repeating information, such as a string of blank spaces in a text file or a block of pixels the same color in a graphic image.

Data Conversion Transferring information from one type of storage format into another. For example, a GIF graphic file can be converted into a BMP file, or a Word for Windows file can be converted into ASCII text.

Decoder A device used to convert a composite video signal into red, green, blue, and sync components.

Digitize To convert from an analog source to digital form. For example, you can digitize an analog waveform to create a digital representation consisting of a string of binary values.

Digitizer A device used to convert video signals into a digital format suitable for display on a computer screen. Digitizers typically take several seconds to convert an image, and therefore require that the video picture be perfectly still.

Direct Memory Access Shortened to DMA. A technique for providing rapid transfer of data between a storage device and computer memory without requiring processor intervention.

Directory A special-purpose file containing details about other files stored on the same medium. Directories are commonly used for all types of storage media, including diskettes, hard disk drives, CD-ROM discs, and magneto-optical cartridges.

Disc Array A CD-ROM playback device that provides simultaneous access to several discs. Multiple laser read heads provide rapid access to individual discs, making disc arrays useful for network applications.

Disc-at-Once A method of single-session recording in which the entire CD is written, from start to finish, without stopping the laser. The table of contents and Lead-in area are written initially, so this data must be compiled by the CD recorder software before the operation begins. The Disc-at-Once capability is necessary in order to prepare CD masters to submit to a replicator.

Disc Description Protocol Shortened to DDP. A protocol used to describe a compact disc at the sector level. DDP is used to ensure reliable and consistent mastering and is often the preferred format at replication facilities.

Glossary

Disc Key The necessary value that must be provided to descramble the title key contained on a DVD-Video disc.

Dither A technique for simulating more colors or shades of gray than are actually available in the current output device or monitor.

DLT Shortened form for Digital Linear Tape. A data storage system that uses tape-based serpentine recording to cartridges that offer 40GB capacities. DLT is one of the preferred methods for submitting files for DVD replication.

Dolby Digital A technique developed by Dolby Laboratories for encoding audio files using perceptual algorithms. Most DVD-Video discs utilize Dolby Digital for the stored audio content.

Dolby Pro Logic The circuitry or software technique used to extract and present the individual channels of information from a matrix-encoded audio stream.

Dolby Surround A technique devised by Dolby Laboratories for encoding surround sound audio channels so that they can be presented to a stereo system.

DV Shortened form for Digital Video. Typically applies to the standard developed jointly by Sony and JVC for their version of the digital videocassette.

DVD Shortened form for Digital Video Disc or Digital Versatile Disc. Applies to the audio-visual optical storage medium based on 120-millimeter discs.

DVD-Audio A storage medium for digital audio that offers improved bit depths and sampling rates in comparison to the CD. A number of additional enhancements have also been added to the DVD-A specification, which was unreleased as of press time.

DVD-R A form of writable DVD that employs a dye sublimation technique to record data. Discs in this format can only be written once.

DVD-RAM A form of writable DVD that uses phase-change technology, similar to the CD-RW format, to allow multiple write operation and erasures on the disc surface. DVD-RAM media is contained in a cartridge.

DVD-ROM The original DVD format that encompasses both DVD-Video and DVD-ROM (Read-Only Memory). These discs can only be read—they cannot be recorded. The standard includes a number of different data types and a file system: UDF.

DVD-Video A storage medium designed for playback of audio and video content on set-top devices known as DVD players. Content stored on DVD-Video discs can include MPEG video, Dolby Digital audio, MPEG audio, as well as other formats.

Dye Sublimation The method of recording data by focusing pulses from a laser beam onto an organic dye material, which records marks that can be read as pits.

Dynamic Range An audio term that distinguishes the difference between the softest parts and the highest volume parts of an audio signal.

Glossary

Eight-to-Fourteen Modulation (EFM) A technique for encoding data transitions on CD-ROM to avoid indecipherable combinations. Eight-bit bytes are converted to 14-bit bytes.

Electroforming The technique employed to produce a metal master disc from a glass master. An electroplating process coats the glass master, retaining the embedded pits. The resulting disc, sometimes called the father, can then be used to produce a mother and a series of stampers for actual disc pressing.

Electronic Publishing Converting a print version of a document or presentation to a digital representation. An electronically published document can usually be used in diverse ways, such as distributed it on CD-ROM or displaying it on the World-Wide Web.

Elementary Stream A coded bitstream, composed of groups of packets, that is used to transfer audio or video content.

Encoder A device that converts and RGB video signal into a composite video signal.

Enhanced CD A term commonly applied to a type of CD that combines audio content suitable for playback on a CD player, as well as computer readable content, often interactive multimedia presentations, that can be accessed when the disc is inserted into a computer's CD-ROM drive. A variety of implementations have been introduced, including CD Extra and CD Plus.

Enhanced IDE Shortened to EIDE. A term that is commonly applied to a number of extended specifications based on the original IDE (Integrated Device Electronics) specification. IDE was designed to allow various devices to communicate on a single bus using inexpensive electronics. EIDE extended the standard to improve data transfer rates and open the IDE bus up to a variety of devices, including CD-ROM and CD-RW drives, DVD-ROM drives, and high-capacity hard disk drives. The ATAPI command set is used to control certain devices that comply with this standard, such as optical disc drives.

Encryption A process by which data is secured by encoding it in a form that it cannot be read without being decrypted. For example, commercial software is sometimes distributed on CD-ROM in encrypted form so it can be accessed only if the user purchases it and obtains the key to decrypt it.

Error Correction Code Shortened to ECC. A means of representing the information in a string of data so that the data can be reconstructed if errors occur during transfer.

Error Detection Code Shortened to EDC. A technique for using 32 bits in each CD-ROM sector to ensure the integrity of the transferred data. The EDC can be used to detect errors that occur during transfer; the ECC can be used to make corrections to the flawed data.

Exabyte A standard format for storing digital information on 8mm tape cartridges. Exabyte tape cartridges are widely used for transferring the data to be used in CD-ROM mastering and replication.

Extended ASCII Refers to the ASCII characters that have values higher than 127. These values are less standardized than the first 128 characters, which are referred to as simply ASCII.

Fast SCSI An extension of the original SCSI-1 bus standard, Fast SCSI (also called SCSI-2) supports both 16-bit or 32-bit data transfers at rates up to 10MB per second.

Field The even- or odd-numbered scan lines that constitute one-half of a television picture.

File System An organization of the logical elements of a collection of data, such as the files and directories, so that they can be located on the physical media, segmented by sectors.

FireWire A high-speed data transfer method that is commonly used for video and audio content (such as transferring the output of digital camcorders). The IEEE 1394 specification formalizes the FireWire standard introduced by Apple. FireWire connections to disc recorders are becoming increasingly popular.

Firmware Application code residing within some form of read-only memory, such as a ROM, EPROM, or Flash memory.

Flicker The flashing effect seen at zones of abrupt change in color or brightness in a CRT display. The term also refers to the strobe effect that occurs when the refresh rate of a video display is low enough for the eye to perceive. Monitors plagued by flicker tend to cause fatigue and eyestrain.

Form 1 A subformat of CD-ROM Mode 2. Form 1 consists of a structure containing 2,048 bytes of user data within the sector. The data is preceded by a 12-byte synchronization zone, a 4-byte header, an 8-byte subheader. The data is followed by a 4-byte EDC value and 276-byte ECC value. Photo-CDs and Electronic Books use Form 1.

Form 2 A subformat of CD-ROM Mode 2. Form 2 consists of a structure containing 2,324 bytes of user data within the sector. The data is preceded by a 12-byte synchronization zone, a 4-byte header, an 8-byte subheader. The data is followed by a 4-byyte EDC value. Form 2 is used in applications where data correction does not need to be as rigorous, such as for video sequences or audio data.

Fragmentation The scattering of the individual parts of files throughout the surface of a hard disk. Accessing fragmented files takes longer, which is why disk optimization—to eliminate fragmentation—is generally recommended before beginning compact disc recording.

Frame The complete television picture, consisting of two interlaced fields (see interlaced video).

Frame grabber A device that digitizes video at real-time rates.

Genlock A technique for mixing two or more video signals and ensuring that they remain in step. Combining video signals without genlock results in distortion.

Glossary

Gigabyte Shortened to GB. A measurement of computer data consisting of 1,024 megabytes, or 1,073,741,824 bytes.

Glass Master The initial recording medium in disc replication. The glass master is treated with a photo-sensitive coating and the data is recorded using a laser beam. Treating the exposed glass master creates the pattern of pits.

Gold Disc The recordable media used in CD recorders. A gold disc is completely blank of data. Tracks are indicated upon a preconfigured spiral, usually called the pre-groove. A dye layer and a reflective gold layer interact when struck by the laser beam during recording; the gold layer can be seen through the optical-quality lacquer giving the gold disc its distinctive hue.

Green Book The standard established by Philips to expand the CD-ROM to CD-I applications.

Header Field A set of 4 bytes that appears at the beginning of each CD-ROM sector. The header field indicates the address of the sector and the mode in which the sector has been recorded. Addresses are expressed as logical block numbers.

HFS Hierarchical File System. A system of file organization used by Apple for diskette and hard disk storage. CD-ROMs can also be formatted to the HFS structure to make the compact disc appear the same as a diskette or hard drive to the user.

High Sierra Format The basis of the ISO-9660 standard, drafted by the CD-ROM Ad Hoc Advisory Committee in May of 1986. Newer CD-ROMs invariably use the ISO-9660 standard, but you may still find older compact discs formatted to the High Sierra Format.

Huffman Coding A form of compression that assigns variable-length codes to precise value sets. A variation of Huffman coding, used with MPEG, relies on fixed code tables. This method of compression is lossless.

Hybrid A disc on which one or more sessions have been recorded, but that has not been closed. The term hybrid is also sometimes used to refer to a disc that contains both an ISO-9660 file system and a Macintosh HFS structure.

Image An assembly of data and files that corresponds with the exact manner in which they will appear on CD-ROM. An image can be used to simulate CD-ROM playback. It can also be written to tape or CD-R to serve as the model for replication.

Index A marker inserted in a compact disc track. Generally used in audio tracks to indicate specific movements, up to 99 indexes can be placed in a single track. Index zero is reserved for encoding pauses.

Injection Molding A manufacturing technique used during replication of compact discs. Molten plastic is injected into a mold and cooled to produce the disc. A stamper embeds the data pattern onto the disc surface as a part of this process.

Interactive Media A presentation designed to be controlled by a user interacting with some form of interface. For example, the interface may contain buttons to branch to different parts of the presentation, play audio files, or display stored graphic images.

Interlaced Video The process by which two separate video fields form a television picture. In the NTSC format, the field consisting of the odd-numbered lines is drawn first, followed 1/60 second later by the field with the even-numbered lines. Although the two fields don't actually appear at the same time, the human brain interprets them as a single frame lasting 1/30 seconds.

Interleaving The alternate placement of audio and video data with computer data to permit faster access and closer synchronization of sound to onscreen displays. Interleaving is defined under Green Book.

International Standard Recording Code Shortened to ISRC. An optional code associated with an individual track. The ISRC value, located in the Q subcode channel, identifies a track by country of origin (2 ASCII characters), the year of creation (2 digits), owner (3 ASCII characters), and a serial number (5 digits).

IRQ Shortened form of Interrupt Request. A dedicated line that signals a request for processor attention by a particular device, such as a COM port or a controller board. Duplicated IRQ lines can cause system conflicts and crashes. Under some conditions, IRQ lines can be shared by more than one device, but generally it is better to assign a unique IRQ for each hardware device.

ISO 9660 The universal file format established for CD-ROMs as published by the International Standards Organization. ISO 9660 structure provides relative platform independence for the contents of a CD-ROM and allows them to be read by DOS, Windows, Macintosh, UNIX, and other computers.

ISO 9660 Image Sometimes called disc image. A representation of CD-ROM contents as a single large file, reflecting the logical format as well as the full set of programs and data.

Jewel Case A plastic enclosure with a hinged lid that protects a compact disc during storage. The jewel case usually includes a cardboard insert describing the title, disc contents, information about the artist or contents, and sales information. Also sometimes called a jewel box.

JPEG The Joint Photographic Experts Group. Represents a standard means of compression for photographic images.

JPEG standard A compression scheme formulated by the Joint Photographic Experts Group.

Jukebox A CD-ROM or DVD-ROM player that handles more than one disc. The jukebox contains a mechanism, similar to the music jukeboxes that handled 45rpm records, that locates and mounts a particular compact disc for reading. Jukeboxes typically hold between 6 to 100 compact discs.

Lacquer Coating A protectant that is used to seal the surface of a compact disc after the data pattern has been imprinted.

Land The level surface region of a compact disc, sometimes called a flat. Transi-

Glossary

tions between lands and pits are used to indicate the data that is stored on a CD-ROM.

Laserdisc An optical disc that stores analog data in FM format for playback of movies and interactive multimedia content. These discs range in size from 8 inches to 12 inches. Laserdiscs are still employed in video training applications and for high-quality movie viewing, although DVD-Video now offers superior video characteristics.

Latency The inherent delay experienced by the laser read head when locating specified data.

Layered Error Correction Code (LECC) A technique for correcting errors that cannot be handled by the CIRC. The LECC reprocesses detected errors and attempts to perform the correction using the EDC and ECC values.

Lead-In Area A designated region on a recordable disc that appears at the beginning of a session and remains blank until the session is closed. The Lead-In Area occupies as many as 4,500 sectors (approximately 1 minute or 9 megabytes of storage) and is written with the session's table of contents and whether the disc is multisession. If the disc has not been closed, the Lead-In Area also indicates the next address where writing can take place.

Lead-Out Area A designated region that indicates the end of the data area within a particular session. The Lead-Out Area does not contain any data values of any type. The Lead-Out Area occupies 6,750 sectors for the first session on a disc, and 2,250 sectors for any subsequent sessions.

Letterbox A technique for displaying films in their original format by placing black matte regions on the top and bottom of the image area. This allows a widescreen image to be placed on a standard TV with its 4:3 aspect ratio. DVD-Video players can typically apply this feature automatically.

Level A The highest quality level of ADPCM compression. Specified in the CD-ROM-XA and CD-I standards, Level A recording has a frequency range of 17KHz. Up to 148 minutes of stereo audio material can be contained on a single disc, or 296 minutes of monaural sound.

Level B The secondary level of ADPCM compression. A single disc can hold up to 296 minutes of stereo audio material or 592 minutes of monaural sound. Sound quality is generally equivalent to FM radio quality.

Level C The most compressed form of ADPCM (and the lowest quality). A single disc can hold about 540 minutes of stereo audio material or 1,140 minutes of monaural sound. Sound quality is sometimes likened to that of AM radio.

Logical Block An addressable unit on a compact disc. Each logical block, as described in ISO 9660, is assigned an identifier, referred to as the Logical Block Number (LBN). LBNs are assigned in sequence starting from block number 0. Logical Block Numbers are used to locate data stored on a disc.

Logical Format The translation of the physical sector layout of the disc into the logical arrangement of files and directories. A compact disc's logical file format as structured by ISO 9660 organizes data files in a manner that they can be accessed from a variety of computer platforms.

Magneto-Optical (MO) A form of rewritable optical storage that combines magnetic principles with laser read and write techniques. MO discs have a magnetic coating that records a phase-shift when struck by a laser beam. They can be returned to the equivalent of a blank unrecorded state by using a laser pulse of different intensity.

Mastering The physical act of etching data pits into the photoresistant layer of a glass master in preparation for creating a metal stamper.

Matrix Encoding A technique that allows a number of surround sound audio channels to be presented to a conventional stereo system. A mathematical model is used to extract the appropriate audio information and deliver it in the proper format for stereo.

Matte A portion of the screen which is blackened or otherwise covered to change the aspect ratio of an image being presented on a monitor or television screen. A matte is typically applied to the top and bottom of the screen when using the letterbox format in DVD-Video playback.

Megabyte (MB) One million bytes of computer information. One byte is 8 bits. One bit is a single 1 or a 0, the basis of computer binary arithmetic. Eight bits grouped together form a byte which counts (in binary) up to 2 to the power of 8 = 256.

MIDI Shortened form for Musical Instrument Digital Interface. A notational language used to store and playback musical sequences through one or more synthesizers. MIDI is often used to store the music for CD-ROM games; the quality of the playback is directly dependent on the type of synthesizer used. General MIDI is a standard that relates specific instrument sounds to specific patches to try to achieve some uniformity in computer-based MIDI applications.

Mixed-Mode Disc A compact disc that contains both Red Book audio information as well as Yellow Book computer data. All of track 1 is reserved for computer data; the following tracks contain the PCM audio.

Mosquitoes A reference to a form of distortion that occurs after video compression. Mosquitoes appear as patterns of fuzzy dots that often appear around sharp transitions in the image. This effect is more formally known as the Gibbs Effect.

Mount The act of making a connected storage devise visible and available to the host computer. For example, if you mount a CD-ROM, the computer can identify it and access the data contained in it.

MPEG A standard formulated by the Motion Picture Experts Group to perform high compression of video data for reproduction on a variety of media, including CD-ROM and DVD-ROM. The processor intensive algorithms used for coding and decoding the video data work best when

Glossary

coupled with specialized hardware designed to accelerate this process. MPEG-1 and MPEG-2 have proven popular outside of the United States for distributing films and videos on CD-ROM. MPEG-2 is now the standard interlaced video format used for DVD-Video.

MPEG Audio A compression technique that uses perceptual encoding for storage and delivery. MPEG-2 audio format includes individual multichannel audio content.

MSCDEX A system file that extends MS-DOS to support the recognition of CD-ROMs as standard DOS volumes.

Multiangle An option used in DVD-Video productions that allows the user to select one of several different viewing angles for the video content.

Multichannel An audio technique that employs separate channels of audio content that are directed to different speakers, often speakers configured to optimize a surround sound effect.

Multilanguage A capability of DVD-Video in which the production can contain indivdual sound tracks or subtitles that are presented in a number of different languages.

MultiRead A standard that was developed to allow standard CD-ROM drives and other playback devices, including DVD-ROM drives, to read various CD formats, including CD-RW discs. A similar standard, termed Super MultiRead, is being developed to provide similar compatibility for the different forms of DVDs.

Multisession A form of compact disc recording that allows data to be written over more than one session. The disc can be removed from the recorder between sessions and then replaced. The term also applies to the ability of a playback device to read a disc that has been recorded in this manner.

Noise An unwanted, meaningless component of a signal that is generated by the recording process or some other aspect of signal processing. Digital audio and video technologies are relatively free from noise in comparison with analog methods of signal processing.

NTSC The television standard currently in use in North America and Japan, an acronym for the National Television System Committee that created the standard. Detractors of the now-aging standard sometimes refer to it as "never the same color".

On-the-Fly The act of recording data to a compact disc by referencing pointers to the data files, rather than a fixed ISO-9660 image. The image of the disc contents in this type of application is usually referred to as a virtual image.

One-Off A single copy of a compact disc, generally recorded on a desktop CD recorder for testing or as the master for large-scale replication.

One-third Stroke The most valid method for accurately measuring CD-ROM performance. Data search and retrieval operations must take place over a minimum of one-third of the entire disc surface, rather than from tracks that are in close proximity.

Glossary

Optical Disc Any of the family of discs that relies on light[md]usually a laser beam[md]to read the data recorded on the disc.

Optical Head The term sometimes applied to the laser read mechanism of a CD-ROM drive or recorder.

OSTA Shortened form for Optical Storage Technology Association. A trade organization active in all forms of optical storage that helps craft standards and further the understanding of the underlying technologies.

OTP Shortened form for opposite track path. One variation of the data pattern used in a two-layer DVD disc in which the data begins near the center of the disc on the first layer and progresses to the second layer travelling from the outer edge to the inner. This technique is typically applied to very long programs designed for continuous playback.

Orange Book The standard that applies to recordable compact disc applications and magneto-optical recording. Orange Book was largely written by Sony and Philips.

Overhead Surface area on recordable CD media that is not used for storing audio or data. The Lead-In and Lead-Out areas required for multi-session recordings take up a number of megabytes of disc storage space and are one form of overhead.

Overlay The process by which computer graphics are combined with video, to add titles and animation to a scene, for example.

Pack A unit of MPEG packets contained in a DVD-Video playback stream. Packs consist of the contents of a DVD sector, containing 2048 bytes.

Packet In DVD-Video terms, a unit of storage that consists of a sequence of data bytes associated with an elementary stream. Packets are clustered into packs within the storage system.

PAL/SECAM These are foreign counterparts to the NTSC video standard. PAL (Phase Alternating Line) is primarily used in Western Europe while SECAM (Sequence de Couleurs avec Memoire) is used in France, Eastern Europe and the Soviet Union.

Path Table A collection of data that defines the directory hierarchy on an ISO 9660 disc. The path table allows rapid access to subdirectories without having to perform multiple seek operations. ISO 9660 specifies two distinct types of path tables, which contain the same information sorted in different ways. The L path table lists multibyte numeric values in L-byte order. The M path table lists multibyte numeric values in M-byte order.

Pause Encoding A required pause between tracks that must appear on a CD's first track and any time a track changes between audio and data. Pause encoding is indicated in the Q subcode channel as index zero.

PCI Shortened form of Peripheral Component Interconnect. A local bus standard developed by Intel for high-performance data transfers. Rates of up to 133MB per second can be achieved for short periods of time.

PCI Shortened form for presentation control information. A data stream used in DVD-Video that carries timing details and other data, such as the selection information, aspect ratio, and so on.

Perceptual Coding A technique for compressing data that relies on human perceptions to determine what information to remove. Data that is least likely to be noticed can be extracted, while data that is more readily perceived is maintained. This compression technique is lossy.

PES Shortened form of packetized elementary stream. A low-level stream composed of MPEG packets. A PES might include audio or video content as an elementary stream.

Phase-Change A recording technique used for CD-RW and DVD-RAM that uses a laser beam to alternately change the state of the recording material. The recording layer is heated at one temperature to bring it to a crystalline state (when erasing) and another temperature to return it to an amorphous state (when recording a pit).

PhotoCD A CD-ROM format originated by Kodak that specifies storage requirements for photographic images. PhotoCD includes elements of CD-ROM XA and Orange Book Hybrid Disc specifications.

Pit A tiny impression in the surface of a compact disc that shifts the phase of reflected light from a laser beam. Pits are surrounded by lands; data transitions are recorded when a change in the data surface from land to pit or from pit to land is detected. Pits form the patterns of data that compose the information carrying layer of a CD-ROM. The size of a pit is approximately 0.5 by 2.0 microns.

Pitch The radial distance separating tracks on a compact disc (typically 1.6 microns)

Pixel Short for picture element, equals one dot on a computer display. A multimedia standard size for display is 640 pixels across and 480 down.

Plug and Play A technique for automatic installation of hardware and software that supports dynamic self-configuration. I/O ports, IRQ lines, and DMA channels are automatically assigned to non-conflicting values. Windows 95/98/NT is one operating system designed around Plug and Play principles.

Polycarbonate The plastic-based material composing the substrate of a compact disc upon which the reflective metal surface is layered.

Post-Gap An area that appears at the end of a compact disc or between audio and data regions; by Yellow Book standards, the post-gap region extends a minimum of two seconds and includes the same formatting as the track that it follows.

Power Calibration Area A specified, reserved area close to the center of a recordable compact disc that lets the recorder adjust power levels for optimal laser writing. Some space must remain in this calibration area if the recorder is write any additional data to a disc.

Pre-Gap A section that appears at the start of a CD or between regions that switch from data and audio. Formatting of the pre-gap introduces the formatting to be used on the track to follow; the pre-gap area is identified through pause encoding.

Premastering The preparation of digital data for recording to an appropriate CD format. Premastering includes the partitioning the data into sectors, adding headers with sector addresses, adding synchronization information, and calculating and inserting error correction codes. Recordable CD systems usually combine the premastering (data preparation) step with the mastering (recording) step to produce a properly formatted one-off.

Premaster Tape A tape containing structured files that disc replicators use to produce the glass master. Although some replication services work from CD one-offs, many still employ equipment that works most effectively when driven by Exabyte tape or other tape-based media. The DDP format is often used for tapes submitted for mastering.

Program Memory Area An area reserved near the center of a compact disc to temporarily store starting addresses for individual tracks until the disc is finalized. This information becomes the basis for the disc table of contents.

Pulse Code Modulation Abbreviated as PCM. A technique for storing soundwaves as discrete digital values. CD-DA discs use PCM to store musical data. Some CD mastering programs can directly convert WAV or AIFF files to PCM form for recording on CD.

QuickTime An architecture designed by Apple Computer for presenting different types of digital media, including video, audio, and still images.

Raster Scan The pattern in which a video screen is scanned, usually from the upper-left corner to the bottom right.

Red Book The original compact disc specification designed for the storage and playback of audio information. Subsequent CD-ROM standards are based on the Red Book standard.

Redundancy The addition of data that makes error checking and correction possible when associated with a set of primary data. Redundancy techniques allow the reconstruction of information when a portion of the data is erroneously transferred.

Reed-Solomon Code An error correction code that uses algebraic principles to compensate for the types of errors that commonly occur with compact disc data. The Reed-Solomon Code is incorporated in the CIRC error correction region on compact discs.

Refresh Rate The number of times per second that a video screen is repainted. NTSC video is shown at 30 frames per second while PAL and SECAM are displayed at 25 frames per second.

Replication The physical process of creating multiple copies of compact discs from a

Glossary

stamper using injection molding techniques.

Retrieval The act of locating a specified piece of data, usually used in respect to accessing information in databases.

RGB NTSC This is NTSC color-composite video decoded into its red, green, blue, and sync components. Broadcast-quality systems tend to use this format because the image quality is higher.

S-Video A form of NTSC video in which the chrominance (color) and luminance (brightness) signals are separated. S-Video produces slightly higher-quality images than color-composite NTSC.

Sampling Rate The rate at which a continuous waveform is measured as it is converted into a series of digital samples. Compact disc audio is conventionally sampled at the rate of 44,100 times per second. Other forms of digital recording, such as ADPCM, support a number of different sampling rates, ranging from very high-quality sound to voice grade.

Scan Converter A device that converts between video formats such as NTSC and PAL. These devices are usually very expensive.

Scan Lines The individual lines in a video picture. They are composed of pixels in the computer world and analog signals in the video world. A set of scan lines makes up a field.

Sector A unit of storage on compact disc based on 2,352 bytes. Sectors, sometimes called blocks, on CD-ROM each have a unique address. Depending on the specification being followed when formatting the disc, the actual data byte count and the bytes devoted to error correction can vary.

Seek The physical operation associated with positioning the read/write head of a storage device in the proper location to read or write a particular piece of data. For CD-ROM applications, the seek operation generally requires also varying the rotational speed of the disc in relation to the radial position of the laser read head.

Seek Error The inability to identify and locate required data on a disc. Seek errors can be caused by surface irregularities, improper focusing of the laser read mechanism, or shock and vibration.

Servo Mechanism A motorized mechanical assembly that controls precise movements in response to voltage signals and a feedback circuit. CD-ROM drives and recorders use servo mechanisms to control the positioning of the laser head over the disc surface.

Session The data written to a compact disc during the span of a single recording. Sessions are identified by a lead-in area, containing session contents, and a lead-out area, indicating the close of the session.

SGML Shortened form of Standard Generalized Markup Language. An elaborate set of definitions specifying the formatting of documents intended for electronic distribution. SGML generally makes electronic publications accessible on a number of different computer platforms.

Glossary

Single-session Disc A compact disc upon which only one set of information has been recorded.

Small Computer System Interface Abbreviated to SCSI. An interface standard used for computer peripherals, such as scanners, hard disk drives, and CD recorders, that supports high-speed transfers over a commmon bus for up to seven devices per host adapter.

SMPTE Time Code A method of indicating precise time by means of recording signals used to synchronize various events. This technique was devised by the Society of Motion Picture Television Engineers and breaks down events to hours, minutes, seconds, and frames.

Stamper The metal plate created by an electroforming process from a mother, used in the injection molding of compact discs.

Subcode A series of codes, ranging from P through W, that indicate display and control information that can be stored in conjunction with audio information on compact disc. The compact disc table of contents employs P and Q subcodes. Other codes are used for storage of MIDI information or graphic images in relation to CD-I playback.

Substrate The primary material that give weight and form to a compact disc. The polycarbonate substrate of a recordable CD is treated with a layer of dye and gold reflective layer in preparation for recording.

Sync An electronic metronome used to keep video signals in step with one another. In the NTSC color-composite standard, sync is combined with the red, green, and blue signals. In RGB systems, the sync signal may exist separately or be combined with the green signal.

Synchronization Field Sometimes called the sync code, this 12-byte field enables the laser read mechanism to synchronize with disc rotation before reading data from a sector. The field itself consists of introductory and closing hexadecimal 00 values bracketing a series of hexadecimal FFs.

Table of Contents (Disc) A region near the center of a compact disc containing the total number of tracks, the starting location of each track, and the extent of the data area on the disc.

Thermal Recalibration A feature on many high-volume hard disk drives that dynamically adjusts and calibrates read and write operations based on temperature variations. Thermal recalibration cycles in the middle of recording a compact disc can interrupt and damage the recording.

Throughput The sum quantity of data that can be moved through a given data channel. Throughput measures the efficiency and capacity of data transfer within a system.

Track The physical path on which information is stored on both magnetic and optical storage media. On CD-ROMs, tracks are arranged in a spiral from the center of the disc. On hard disk drives, tracks appear in concentric circles and are independent of each other.

Glossary

Track-at-Once A single-session method of writing one or more tracks to a recordable disc in which the table of contents and Lead-In area are added after the Program Area is completed. The disc is fixed after the TOC and Lead-In are appended and no further tracks can be added. Discs created in this manner are not suitable for masters to use in mass replication. Compare to Disc-at-Once.

Transfer Rate A measure of the speed at which data can be moved from one point to another. Transfer rates are generally expressed in terms of kilobytes per second. A quad-speed recorder can write data at the rate of 600 kilobytes per second.

UDF Shortened form for Universal Disc Format. A random-access file system devised by OSTA for use on a variety of optical media, including CD-RW and DVD-ROM.

UDF Bridge A form of UDF that also includes backwards compatibility to earlier devices that rely on ISO 9660.

UltraWide SCSI A high-speed implementation of the SCSI standard that can achieve transfer rates up to 40MB per second using a 16-bit data path.

Universal Product Code Abbreviated to UPC. A 13-digit catalog number that can be written to a disc's table of contents for identification. Not all CD recorders support writing of a UPC (sometimes referred to as EAN).

VideoCD A variation of the CD-ROM standard that includes video data compressed using MPEG. This standard also includes the definition of Karaoke CD discs.

Virtual CD Player A simulation of a CD device driver that can be used for performance testing of a set of files intended for CD-ROM mastering. Files are generally accessed from an ISO image and the operating system responds during playback as if an actual CD player was connected.

Virtual Image A collection of files arranged in a particular order for either recording directly to compact disc or for creating an ISO 9660 image on hard disk in preparation for recording.

Volume A term that applies to a complete CD-ROM disc.

Volume Descriptors The disc contents as expressed in ISO 9660 format. The volume descriptors indicate the logical arrangment of a disc, details about the path table, creation times and dates, and information about the originator.

Waveform Audio A sound file that contains a representation of an analog signal in digital form. Depending on the sample rate at which the waveform was digitized, whether stereo is enabled, and the number of bits per sample, the sound file can range less than 1MB per minute of recording to more than 10MB. The larger the file size, the better the quality. This format, usually represented with a .WAV file extension, is popular in Microsoft Windows.

Wide SCSI An implementation of the SCSI standard using 16-bit data paths that can reach data transfer rates of 20MB per second.

Wizard A software utility designed to simplify the use of a particular task, such as preparing the files for premastering a CD-ROM.

Write Once, Read Many (WORM) A form of optical storage defined in Orange Book that records data primarily for archival uses. WORM drives designed for archival applications using 12-inch recordable optical discs are still manufactured by companies such as Plasmon.

Yellow Book The CD-ROM standard originally developed by Sony and Philips that expanded the Red Book uses of compact disc to include computer data as well as audio information.

Index

Numerics
100-BASE-TX 330
10-BASE-T 330
30-pin PowerBook
    SCSI connector 115
3D engines
    for games 385
3DO 483
3DO games 311
3M 197
415 Productions 369
4mm Digital Audio Tapes
    for file distribution 125
68-pin Fast/Wide SCSI
    connector 115
8/16 modulation 483
8mm EXAbyte 125
9-track tape 159

A
A.Pack Dolby AC-3 encoder 361
A/V model
    hard disk drive 126
absolute-time 483
AC-3 483
AC-3 audio 422
access time 33, 483
a-characters 483
ACID
    from Sonic Foundry 249
Acrobat 250
    bookmarks 258
    disc publishing techniques 253
Acrobat Catalog 241, 255
Acrobat Distiller 254, 255
Acrobat PDF Writer 242
Acrobat Reader 254, 373
active medium 25
active termination
    SCSI devices 114
ActiveX 372
    Xtra 372
ActixeX
    on Mac 373
Adaptec 152
    DirectCD 121
Adobe Illustrator 175
    for labels 397
ADPCM 56, 483
Advanced SCSI Programmer's Interface 483
Advanced Streaming Format 244
Affex 231
After Effects 340, 345, 380
AGC 484
alignment
    when printing 234
Allaire Corporation 290
alternate video clips 365
alternative fibers
    for packaging materials 453
Amazon.com 444

Index

Analog to Digital Converter 484
analog video 335
Anderson, Ray
    CEO of Interface 452
Andrews, Chris 7
ANSI-Labeled 484
ANSI-Unlabeled 484
Anthony, David 421
AOL
    authoring for 292
API toolkits
    for GEAR 189
Apple
    support for DVD 425
Apple Media Tool 289
applications
    for professional use 151
APS CD-RW 8x4x32 FireWire 137
APS DVD-RAM external SCSI Drive 137
architecture
    CD-ROM 79
    of CD-ROMs 38
archival properties
    of CD-ROM 34
archiving
    on DVD-R 206
    utilities 149
archiving to tape 32
archivists
    requirements for 218
artifact 484
artifacts
    in digital video 340
artwork
    applying 28
    guidelines for disc 407
    preparing for labels 396
    preparing label 411
ASACA 344
ASCII 484
aspect ratio 484
asset managers 342
Astarte 434

Astronauts 318
asynchronous strategy
    of duplication 217
asynchronous system
    duplicator 214
ATAPI 484
    I/O interface 116
ATAPI IDE 107, 109
    inexpensive connection method 117
A-time 81
A-to-D conversion 270
ATRAC 7
auction
    Dutch 445
auctions
    online 444
audio
    AC-3 422
    DVD 350
    interleaved 65
    on DVD 96
audio CD
    creating an original 269
    wizard for creating 278
Audio CD architecture 37
Audio Interchange File Format 245
audio origins
    of CD-ROM 62
audio specs
    of DVD-Video 347
authoring 148, 149, 484
    DVD-Audio 97
    DVD-Video titles 96
authoring systems
    for DVD 428
authoring tools 437
Authorize.net 443
Authorware 423
autoloader class
    examples of 226
autoloaders
    efficiency of 217

Index

auto-loading duplicator
    in an enterprise environment 211
automated backup 174
automated duplicators 221
    capabilities 225
Automatic Gain Control 201
autoplay 484
AutoPrinter 235
AutoProtect Demo Disc 203
AUTORUN.INF 204, 370
autostart function
    for Mac and Windows 369
autostart switch
    in Toast 370
AVI 484
Avid
    editing systems 358
Axis Communication
    design of ThinServer technology 328

B

background noise
    elimination of 2
backup
    automated 174
Backup NOW! 150
backup operations
    to CD-R 251
Baja 318
bandwidth 484
bandwidth limits
    of NTSC 347
Barlow, John Perry 311
BCA 484
Beatnik 317
bee-hive
    containing blank discs 220
Bell, Alexander Graham 2
Bennington College 422
BeOS 147
Bias' Peak 345
bi-refringence 485
bit 484
Bit Error Rate 485

Bit-for-Bit Verification 191
bits per pixel 485
black box
    reducing elements to a 21
Blackhawk Down 265
blank media 220
    definition of 213
    shrink-wrapped 221
bleeds
    on labels 407
block
    addressable unit 37
Block Error Rate 485
block-by-block copy 55
Blue Book 55
Blue Book Enhanced CD 283
Blue Sky Software 260
Book A 485
Book B 485
Book C 485
Book D 485
Book E 485
bookmarks
    in Acrobat 258
Books
    DVD 87
bootable disc 181
British motorcycles 438, 445
Broderbund 309
buffer 485
buffer size
    as a factor 129
buffers
    in CD recorders 129
Bumpass Hell file format 67
bundling
    of recorder applications 152
burning a disc
    definition of 149
bus mastering 485
business uses
    disc recording 321
Buzzsaw 190

Index

Buzzsaw 4.0 151

C

C-3D 331
cabling
    SCSI 114
cache 485
cache buffer 49
caching
    of data 126
caddy 485
Canto Cumulus 342
capacitive storage 4
Card Discs 26
Catalog
    Acrobat 241, 255
CAV 486
Cave Dog 308
CD 106
    layers 29
CD Architect 274
CD duplication
    environmental considerations 232
CD duplicators
    origins of 211
    pricing of 128
    small publishing ventures 457
CD E100 328
CD formats
    to end users 147
CD image 155
CD jukebox 322
CD Net 329
CD Net servers
    protocols supported 330
CD recorder
    SCSI connections to 112
    selecting a 127
    support for incremental writes 55
    test utilities 182
CD recorder applications
    bundled 152
CD recorders
    least expensive hardware 437

    onboard buffers 129
    power calibration cycle 132
    pricing of 128
    read speed 130
CD recording
    on the Macintosh 108
CD Studio 140, 158
CD technology
    understanding 20
CD+G 60
CDCyclone 223
CD-DA 60, 486
    uses for 60
CD-I 67, 486
CD-i 6
CD-I Bridge 70, 486
CD-I hardware
    release of 69
CD-I Ready 69
CD-Manager 5 326
CD-PROM 84, 204, 486
CD-R 486
    debates over lifespans 195
    hybrid 203
    media costs 196
    used by components 152
    used for backup 251
CD-R FS packet writing 122
CD-R Gold Ultima Recordable 205
CD-R80 discs
    extended capacity 198
CD-recorder application
    selecting a 193
CD-ROM 486
    architecture 38
    audio origins 62
    data access slowness 33
    data storage 35
    early uses of 65
    embedded control buttons 365
    family tree 54
    file placement 36
    file tables 36

Index

for press kits 245
Green Book 67
historical 447
HTML content 264
interactive music for 305
logical and physical components 37
one-off for file submissions 394
origins of 5
physical sectors 57
playability in DVD-ROM drives 88
recycling of 453
Red Book 60
solar power resources 439
specialized references 438
storage capacity 27
support for applications 39
text storage 65
using Flash for 375
Video CD 73
virtual image of 49
White Book 70
Yellow Book standard 63
CD-ROM architecture 79
CD-ROM drive
    reading multi-session discs 75
CD-ROM drives
    high performance 33
    lasers used in 25
    XA compatibility 66
CD-ROM standards
    consideration of 51
    maintaining compatibility 54
CD-ROM XA 61, 63, 486
    compatibility 64
    finalization of 65
CD-ROMs
    creating 22
CD-RTOS 69
CD-RW 8, 19, 106, 108, 157, 200
    crystalline layer 200
    direct overwrite 131
    formatting the media 201
    layers in disc 199

media 199
CD-RW drive
    portable 135
CD-RW media
    six layers 118
CDs
    networking 323
    networking collections of 323
CD-WO 486
CD-Writer Plus 171
    wizard 164
CD-Writer Plus M820e
    from HP 135
    portable recorder 110
Cedar autoloading duplicator 228
cell 486
Centronics 50-pin
    connector 115
CeQuadrat 202
challenge key 486
Chan, Peter 382
channel 486
CineMatch
    personalized movie service 461
Cinepak 340
CIRC 62, 64
Circuit City 455
Circular Buffer Read-Ahead 487
clamping area 487
cleanliness
    when recording 232
Close Disc 487
Close Session 487
CLV 487
codec 339
coherent light 25
Cold Fusion 287
ColorScribe 9000 230
compact disc
    correction codes 3
    first shipping 4
    formatting of 119
    life-span 3

Index

origins of 2
size of 26
Compact Disc - Write Once
erasable characteristics 118
Compact Disc Digital Audio system
proposal for 4
Compact Disc Programmable Read-Only Memory 84
compact discs
signal-to-noise ratio 3
compatibility
of CD-ROM standards 54
components
disc recording 152
composite video 487
compression 487
of digital video 339
computer interface
selecting 107
Computer Output to Laser Disc (COLD) 324
Conan O'Brien 286
configurations
duplicator 214
Constant Angular Velocity 35
Constant Linear Velocity 36
Continuous Composite Write-Once Read-Many 208
continuous spiral,disc data pattern 35
control area 487
copiers
disc 211, 221
Corel Draw 253
corporate DVD playback 355
corporate uses of DVD 333
corporate video
delivery options 335
corporate video production
recommendations 366
correction codes
on a music CD 3
cost factors
optical recording 324

CP/M 67
Craig Associates International 230
Creative Digital Research 183
Cross-Interleaved Reed-Solomon Code 487
cross-platform development 487
cross-platform disc
file organization 82
CRW6416sxz drive
from Yamaha 124
crystalline layer
of CD-RW 200
CRZ140S/C
CD-RW drive from Sony 134
cue sheet 487
Cyanine 197
cyanine dye 47
cycle time
definition of 213

D

Das Boot 354
DAT 7, 154, 488
DAT recorder 270
data
caching 126
flow issues 49
pits and lands 25
data area 488
data capture 488
data compression 488
data conversion 488
data format
choosing a 20
data recording with light 1
data storage
on CD-ROM 35
on DVD 39
data storage techniques
DVD-ROM 89
data transfer rate,of CD-ROMs 33
data transfer rates
for DVD 48
data types
as supported by standards 52

Index

databases
    resource 439
DataLifePlus media 205
data-transfer techniques 49
Dave Brubeck Quartet 270
DB-25 connector 115
d-characters 487
DDP format 187
decoder 488
defragmenting
    hard disk drive 127
defragmenting a disk drive 49
delivery options
    for corporate video 335
description of digital video 337
design considerations
    for CDs 404
design docs
    value of 383
development of standards 53
DeVry Technical School 429
digital audio
    multichannel 92
Digital Audio Disc Committee 4
Digital Audio Disk Corporation (DADC) Recycling Program 454
Digital Audio Tape 159
Digital Cinema Mode 355
digital computer data
    importance of data integrity 63
digital images
    storage of 250
Digital Librarian 344
Digital Picture Mode
    DVD 355
digital sampling 62
Digital Versatile Disc 87
digital video
    artifacts 340
    description of 337
    methods of compression 339
    production process 340
    storage of 338

    transferring to computer 342
Digital Video cameras 456
Digital Works 448
    press releases through 446
digitizer 488
digitizing board 338
Direct Memory Access 488
Direct Music 308, 316
direct overwrite
    for CD-RW 131
DirectCD 150
    formatting 168
    from Adaptec 121
Director 285, 289, 333, 364, 366, 372, 374, 377
directory 488
directory indexing 51
DirectShow 426
disc
    layers in CD-RW 199
disc analyzers 214
    definition of 212
disc array 488
disc artwork guidelines 407
disc autoloaders
    definition of 212
disc copiers 211, 221
    characteristics of 222
Disc Description Protocol 488
disc duplication 214
    uses for 214
disc formatting
    considerations 119
disc geography control
    in Toast 379
disc key 489
disc labeling 213
    time required for 218
Disc Makers 310, 400
disc manufacturing 399
disc printers
    definition of 212
    two categories of 219

Index

disc printing
    inkjet 219
    wax transfer 219
disc publishing
    capabilities 212
disc recorder software
    types of 149
disc recorders
    support for 184
disc recording
    components 152
    hardware considerations 125
    platform issues 115
    practical applications 239
disc recording applications
    examples of 171
    support for recorders 159
disc recording components 150
disc recording process
    through software interface 146
disc recording software 145
disc recording tools
    basic 149
disc recycling 453
disc sleeves 237
Disc Wizard 164
Disc-at-Once 133, 151, 155, 161, 201, 275, 488
DiscMakers 283
disc-recording towers 154
discs
    affects of humidity 232
    packaging 399, 409
    pre-screened 227
    printing issues 233
    silkscreened 234
    Web-enabled label design 237
DiscZerver 323
Distiller
    Acrobat 254
distribute files
    to a replicator 124
dither 489

DIVX 459
    failure of 455
Dixon-Ticonderoga Redi Sharp Plus 210
DLT 124, 148, 154, 159, 189, 434, 489
    for file submissions 394
Dolby Digital 40, 489
Dolby Pro Logic 489
Dolby ProLogic 347
Dolby Surround 489
Dolby, Thomas 314, 317
Dreamweaver 371
DroidWorks 380
Droidworks
    simultaneous development 385
DTS 97
duotones
    for labels 400
duplication
    asynchronous strategy 217
    creating an image master 216
    definition of 212
    DVD 237
    on-demand disc publishing 214
    techniques 238
    unattended production 225
    workflow issues 216
    workgroup publishing 215
duplicator
    auto-loading 211
    configurations 214
duplicator robotics 211
duplicators
    automated 221, 225
    CD 211
    tower 221
    tower enclosures 211
Dutch auction 445
DV 489
DV cameras
    connection of 341
DVCam 339
DVCPro 339

Index

DVD 106, 489
    backwards compatibility 40
    biggest advantage of 458
    books 87
    channel data transfer rate 90
    corporate uses of 333
    data storage 39
    data transfer rates 48
    delivery methods 337
    disc rental lifespan 459
    distribution 464
    film storage 44
    guru 366
    in test markets 458
    innovative projects for 428
    media 205
    navigation 348
    presentations 348
    security features of 348
    selecting a format 367
    single-layer 41
    standard 87
    writable forms 46
DVD authoring packages 147
DVD equipment
    maturing 22
DVD Fusion 362
DVD player
    set-top box 347
DVD players 88
    compatibility of 434
DVD premastering applications 148
DVD standards
    evolution of 87
DVD Studio 140
DVD technology
    understanding 20
DVD video
    advantages of 356
DVD videos
    producing in-house 357
DVD+RW 46, 106, 119
DVD-10 43
DVD-18 39, 43
    layers 44
DVD-5 41
DVD-9 42
    layers 43
DVD-Audio 489
    authoring 97
    description of 96
DVDAuthorQUICK 362
DVDelight 359
DVDirector 359
DVDirector Pro 361
DVDit! 246, 362, 438
DVD-R 46, 98, 105, 106, 160, 489
    estimated media lifespan 102
    first-generation media 48
    for archiving 206
    for authoring 98
    for general 98
    for premastering 47
    high cost of recorders 101
    technology for independent filmmakers 438
    track pitch 99
    units 133
    uses 102
    write once characteristics 118
    write strategy 47
DVD-R media 238
DVD-RAM 19, 46, 106, 489
    5.2GB disc cartridges 330
    advantages of delivery format 363
    burning a disc 363
    exchanging digital video files 247
    for source files 433
    media 207
    uses 104
    write laser 103
DVD-ROM 119, 489
    backward compatibility of drives 45
    books 88
    data sector 89
    data storage techniques 89

Index

data types supported 90
discs produced with DVD-R recorders 102
filename translations 90
market trends 426
partition 91
playback for wide audience 395
playback issues 23
summary of books 88
uses for 424
Web-enabled 245
DVD-ROM drives
    compatibility with CD-ROM 88
DVD-ROM towers 240
DVD-ROMs
    creating 22
DVD-RW 46
DVDs
    networking collections of 323
DVD-Video 489
    authoring for 96
    compressed content 40
    description of 92
    disc structure 334
    encoding 430
    file formats 93
    independent releases 438
    navigating content 94
    playback 91
    sector on DVD-Audio disc 97
    stability of 424
dye
    colors 196
dye formulations
    for recordable media 195
    licensing of 197
dye sublimation 489
Dynamic HTML 264
dynamic range 490

E

early adopters
    of CD recording 146
Easy CD Creator 56, 177, 178

Easy CD Creator 4 150
Easy-CD Creator 156, 159
eBay 444
ECC 64
ECMA 168 163
ecommerce
    Web storefronts 448
ECP 109
EDC 64
Edison, Thomas 2
editing systems
    Avid 358
efficiency
    of autoloaders 217
EFM 61
EIDE 107
Eight-to-Fourteen Modulation 29, 61, 490
electroforming 490
electromagnetic methods
    of storage 31
electronic particle beams
    for data storage 331
electronic publishing 490
    tools 456
elementary stream 490
embedded controllers 221
encoder 490
encryption 490
Enhanced CD 490
enhanced CD
    creating an 279
Enhanced CDs 91
    real estate 288
Enhanced IDE 490
entrepreneurial ventures
    nature of 458
EPP 109
Epson
    printers 230
error correction
    Red Book 62
Error Correction Code 64, 490
Error Detection Code 64, 490

Index

evolution of recordable-CD software 146
Exabyte 491
Exabyte tape 159
Extended ASCII 491
extended capacity discs 198
Extended IDE
    added features 116
Extensis Portfolio 342

F

family tree
    CD-ROM 54
Fast SCSI 491
FastSCSI 112
Federal Law Enforcement Training Center 424
field 491
file formats
    native 56
File Manager 155
file organization on cross-platform discs 82
file placement
    on a CD-ROM 36
file placement on disc 56
file system 491
    for CD-ROMs 156
filename translations
    on DVD-ROM 90
filenaming conventions
    for Mac and Windows 370
files
    communicating structure to replicator 398
    preparing for a replicator 394
    selection of 155
Final Cut Pro 342, 345
Fire-Wire 246
FireWire 108, 491
    interface 110
Fireworks 375, 377
firmware 491
    upgrading through Flash ROM 130
Flabby Rode 312
Flash 374, 375, 379

Flash ROM 155
    for firmware upgrading 130
flicker 491
flourescent inks
    on labels 405
Fluorescent Multi-layer (FM)
    recording technique 331
FM-ROM 331
footage
    original 336
    stock 337
Form 1 491
Form 2 491
formatting
    of CD-RW media 201
forms creation 256
four-color printing
    for optical discs 397
fragmentation 491
frame 491
    data division 29
frame grabber 491
FrameMaker 255, 257
Frankfurt Group 79
Frankfurt, Germany 78
Fraunhofer MP3
    audio compression 184
From Alice to Ocean 239

G

G4 microprocessor 147
gallium-arsenide lasers 25
games
    first MIDI sound track for 306
gamma
    optimized for a monitor 365
Gear DVD Pro 153
GEAR PRO DVD 57, 185
GEAR Pro DVD 151
Genesis 247
genlock 492
geographic placement
    of files 56
Gigabyte 492

515

Index

glass master 27, 492
glass masters 189
gold disc 492
gold layer
    in recordable CDs 29
GoTo.com 446, 448
    search term bidding 442
Graboyes, Blaine 421
graphics accelerators 385
Green Book 67, 492
    data types 68
    implementation 69
Guggenheim Museum 421
guidelines for handling media 232
Gutenberg's movable-type press 437
Gypsy swing bands 441

H

hard disk drive
    for recording 126
hard disk drives
    thermal calibration 126
hardware considerations
    to disc recording 125
hardware installation
    guidelines 141
HD50 connector 115
HD-ROM drives 331
header field 492
hemp
    industrial 246, 450
HEURIS MPEG Professional 340
HEURIS Power Professional MPEG export engine 358
Hewlett-Packard 177
HFS 492
Hi8 339
Hi-Fi
    ADPCM 68
High Sierra File Format 5, 67, 156, 492
high-definition televisions 92
history
    of recordable CD 7
Hitachi 4

holographic lenses 40
host adapter
    SCSI 111
host adapters
    SCSI bus widths 112
    termination of 113
host computer 106
    for disc recorder application 145
    requirement for 105
    selecting a 123
HotBurn 151, 270, 275
hot-melt method
    for bonding disc substrates 41
hotspot creation
    in RoboHelp 262
HP CD-Writer+ 8200 167
HP Disaster Recovery 171
HP Simple Trax 174
HP SureStore CD-Writer Plus M820e 135
HP35470A DAT drive 189
HTML
    rollover creation 375
HTML editor
    for disc publishing 263
HTML Help 260
Huffman coding 492
humidity
    dangers of disc sticking 232
Humongous 308
hybrid 492
hybrid CD-R 203
HyCD Publisher 184
HyCD, Inc. 183
hyperlinks 256

I

I/O applications
    for SCSI 112
I/O interface
    importance of 20
IBM PC
    release of 67
IE-1284 standard 109

Index

image 492
    virtual 55
image master
    duplication 216
ImageAligner 235
ImageAutomator 227
Incat Systems 178
independent developers
    opportunities for 438
    permission marketing for 441
index 492
indexing
    in RoboHelp 262
    of text 255
industrial hemp 450
injection molding 492
ink
    water-based 233
inkjet disc printers 219
inkjet print engines 213
inks
    proofing 400
Inmedia Slides and Sound 259
installation
    hardware guidelines 141
Intel Imaging Technology 86
interactive forms
    in Acrobat 256
interactive media 493
interactive multimedia 5
interface considerations
    software 155
interface options
    for CD and DVD recorders 109
interlaced video 493
interleaved audio 65
interleaving 493
Internal Revenue Service
    tax forms 253
International Standard Recording Code 493
International Standards Organization 5
Internet
    channels for reaching an audience 441
    for DVD rentals 464
    reaching a specialized audience 448
Internet Explorer 372
interruption marketing 440
IRQ 493
ISO 13346 90
ISO 18925-1999 218
ISO 9600 322
ISO 9660 5, 54, 56, 79, 153, 156, 161, 163, 181, 183, 187, 326, 370, 395, 493
    extensions for Orange Book 80
    image 127
    implementation 79
    Level 1 157
    Level 2 158
    Level 3 158
    meeting conventions 116
    resolution of problems 83
    uses 79
ISO 9660 image 493

J

Java Virtual Machine 261
JavaScript 254, 256, 317
Jazz Impressions of Japan 270
Jedi Knight 386
    game engine 384
jewel boxes 220, 237
jewel case 412, 493
jewel case insert
    layout of 181
jewel cases
    inserts 397
job queuing 225
Johnny Brennan 279
Joliet 181, 183
    file limitations 179
Joliet standard 370
JPEG 260, 493
JPEG standard 493
jukebox 493
JVC 4

K

KanguruCD Duplicator 223

Index

kenaf 246, 450
Kerr effect 208
King of the Hill 312
Kodak 198, 199, 207
Kodak multi-session 162

L

label
    paper 220
    pre-screened 219
    printing a 218
    proof of 402
labels
    metallic inks 404
    paper 236
    trapping 401
LaCie 344
LaCie DVD-RAM drive 363
LaCie Ltd. 177
lacquer
    protective coating 27
lacquer coating 494
Lake Tahoe, California 67
land 494
Land Pre-Pits 205
laser
    gallium-arsenide 25
    invention of 24
    power of 198
    red 205
    use of a 2
laser power calibration 131
laser principles 25
Laserdisc 39
laserdisc 494
LaserFile 410
LaserWriter
    proofs 404
last-session-first 163
latency 494
Layer 0
    logical layer 37
Layered Error Correction Code (LECC) 494
LCD projector 355

Lead-In Area 74
Lead-In area 494
Lead-Out Area 75
Lead-Out area 494
Level 1
    ISO 9660 157
Level 2
    ISO 9660 158
Level 3
    ISO 9660 158
Level A 494
Level B 494
Level C 494
LF-D101 DVD-RAM drive 362
licensing conditions
    on package 416
life-span
    of compact discs 3
lifespan
    of DVD rentals 459
lifespans
    of recorded discs 195
light
    for digital video 346
    for recording data 1
Lingo 376
Linux 147
lithographic printing
    for artwork 28
LiveStage 365
logical block 494
logical format 495
logical formatting 37
look-and-feel
    of applications 156
Low Voltage Differential I/O interface 112
LPP 205
Lucas Learning 381
LucasArts 308, 309, 316, 369, 380
    test facility 386

M

Mac development
    for CD-ROMs 380

Index

Macintosh
    CD recording 108
    software tools for 147
Mackie 1202 mixer 271
Macromedia 152
Macromedia Dreamweaver 263
Macromedia site
    on CD-ROM 378
MacWarehouse
    founder of 460
magnetic-optical
    media lifespans 208
magneto-optical (MO) 495
magneto-optical cartridges 103, 207
Magneto-Optical drive 31
magneto-optical drives 8
magneto-optical media
    changes to file structure 80
MakeDisc 140
manufacturing CD-ROMs
    cost reductions 32
manufacturing discs 399
Marantz CD-R610mkII 123
markets
    niche 448
Marquee
    Neflix program 461
mastering 495
matrix encoding 495
matte 495
Maxell 207
Maxoptix 331
media
    blank 220
    CD-R costs 196
    CD-RW 213
    designed for high-speed applications 198
    dual-sided recordable 205
    durability 459
    DVD 205
    DVD-R 48, 238
    DVD-RAM 207
    handling guidelines 232
    high-speed recording 204
    pre-labeled 230, 234
    printable surfaces 209
    quality of 232
    recordable 195, 213
    training in use of 233
Media 100 362
Media Cleaner Pro 345, 358
media lifespan
    DVD-R 102
media lifespans
    magneto-optical 208
media tests
    independent 196
Media100
    editing systems 358
MediaFORM 226
MediaPath MA32+ 326
Megabyte 495
memory management 146
Merchandizer
    Web store hosting 443
Meridian Data Systems 329
Metal Azo 197
metal azo 196
metallic inks
    use of in labels 404
Michaud, Collette 381
MicroBoards, Inc. 228
Microsoft Front Page 263
Microsoft Word 175
Microtech 225, 227, 234
MicroUDF 93
MIDI 312, 495
    expressiveness of 307
    General 306
    working in 305
MIDI Manufacturers Association 306
MIDI-XG 306
Minerva 433
Minerva Impression 362, 434

Index

MiniDisc 7
miniDV 339
Mitsubishi Chemical Company 197
Mitsui 85, 203
Mixed Mode 282
Mixed Mode standard 63
mixed-mode disc 495
Mode 1
    Yellow Book 66
Mode 2
    Yellow Book 66
moiré pattern 400
moiré patterns 408
mosquitoes 495
mount 495
MP3 269, 447, 456
    encoders 204
MPEG 495
    encoding process 358
MPEG audio 496
MPEG compression tools 127
MPEG decoder 91
MPEG-1 7, 56, 249, 432
MPEG-2 40, 56, 92, 338, 339, 345, 357, 358, 363, 364, 422
    one-trick pony 366
MPEG-4 339
MSCDEX 496
multiangle 496
multichannel 496
multilanguage 496
multimedia content
    on CD-ROM 33
multimedia interaction
    using CD-ROM XA 63
multimedia sounds 377
multi-platform access 51
multiple-session recording
    through Orange Book 75
MultiRead 46, 103, 119, 157, 201, 496
    compatibility issues 203
multi-session
    Kodak 162

multi-session compatibility 496
    issues 77
multi-session disc
    identifying the first session 78
multi-session overhead 78
multi-session reading 120
multi-session recording 162
multi-session write compatibility 22
multithreaded I/O 147
multi-volume recording 76
music-only CDs 310
MYST 239

N

NAS 322, 327
NASA 162
native file formats 56
navigation
    on DVD-V 94
Netbox 344
Netflix founder
    Marc Randolph 457
NetReady server 327
network access
    of tower duplicators 225
network interface cards
    disabling when recording 142
network storage solutions 325
networking
    of optical discs 323
New Media Center
    in Bennington, VT 422
newsletter
    for promotional purposes 442
Nichia Chemical Industries 330
noise 496
    in music recordings 2
Norsam 331
Novell NCP
    support for 326
Novell NetWare
    support for 229
NSM
    jukeboxes 344

Index

NTSC 92, 335, 347, 496
    bandwidth limits 347

O

on-demand disc publishing 214
on-demand publishing 214
one-off 496
one-third stroke 496
online assistance
    for recording applications 164
ONP 163
on-screen display
    DVD 352
on-the-fly 496
OPC 132
Open New Program bit 163
opt in
    giving permission 441
optical data storage devices
    family of 31
optical disc 497
optical disc equipment
    businesses uses 321
optical discs
    conserving resources 449
    for version control 323
    packaging for 409
optical head 497
Optical Media International 7
optical pickup
    for DVD 45
optical recording
    cost factors 324
optical recording gear
    selecting 105
optical recording techniques 19
optical recording timeline 6
Optical Storage Technology Association 84
Optical Super Density 208
Optical Super Density format (OSD) 331
Optimum Power Calibration 47
Optura camera 341
Orange Book 163, 497
    Chapter 11 204
    inescapable facts 117
    laser power calibration requirement 131
    multi-session reading and writing 120
Orange Book Part III 199
original footage 336
origins of the compact disc 2
OSTA 90, 497
OTP 497
overhead 497
    to multi-session recording 78
overlay 497

P

P and Q subcodes 274
pack 497
package design
    choosing a 410
package types
    illustrated 412
packaging
    as part of duplication workflow 216
    automatic 218
    discs 399
    protective characteristics 410
    prototype of 411
    recycling considerations 449
packaging materials
    alternative fibers 453
packaging options
    for discs 409
packet 497
packet writing 120, 153
    fixed and variable 122
PacketCD
    from CeQuadrat 121
Packetized Elementary Streams 426
PageMaker 255
PAL 92, 335
PAL/SECAM 497
Panasonic DVD-A320 355
Panasonic LF-D101 DVD-RAM 344
Panasonic portable Palm Theatre 429
paper consumption
    reducing 450

Index

paper label
    for disc printing 220
paper labels 236
paper sleeve 416
paperless office
    myth of the 450
parallel light
    laser 2
parallel port
    interface 109
path table 497
pause encoding 497
PC Card 110
PC card
    interconnection to recorder 108
PCI 498
PCM 60, 271, 276
PDF
    launching files 376
PDF Writer 254
perceptual coding 498
PerfectWriter 231, 235
performance
    obtaining best 22
    of drives 33
performance issues 24
performance simulation 160
permission marketing 440, 448
personal computers
    minimum system requirements for recording 123
PES 498
phase-change 498
phase-change techniques
    for storing data 8
phase-change technology 103
Philips 4, 6, 199
Photo CD 54, 70, 77
    specification 71
PhotoCD 162, 498
photons
    patterns of movement 25
Phthalocyanine 197

phthalocyanine 196
physical formatting 37
physical sectors
    on CD-ROM 57
Picture CD 85
    use of 85
Pinnacle Micro 177
Pioneer 206
pit 498
pitch 498
pits and lands
    data storage 25
pixel 498
Plasmon 325
Plasmon AutoTower 327
Plastic Recycling, Inc. 454
plastic substrate 27
platform
    issues when recording 115
platforms
    choice of 147
playback
    achieving satisfactory 161
plug and play 498
PMS colors
    on labels 407
PMS inks 400
polycarbonate 498
polycarbonate substrates
    of DVD 47
polychrome printing 219
portability
    of SCSI devices 115
post-gap 498
PostScript 254, 255
Power Calibration Area 498
power calibration cycle
    for CD recorders 132
PowerPoint presentation
    putting on CD 23
PowerPoint presentations 366
PowerPrinter 231

Index

practical applications
  of disc recording 239
preemptive multitasking 146
pre-gap 499
pre-labeled media 230, 234
premaster tape 499
premastering 145, 148, 499
premastering applications
  professional 149
premastering to tape 189
Premiere 342, 345
pre-screened discs 227
presentations
  lecturer support 364
  playback devices 355
  using DVD 348
press kits
  on disc 245
press releases
  through Digital Works 446
pressing a disc
  definition of 149
Primary Volume Descriptor 81
Primera 227, 228
printable media
  definition of 213
printable surfaces
  media 209
printing
  four-color for optical discs 397
  issues 233
  spot 403
PrintShop Multimedia Organizer
  from Broderbund 250
Prism 231
production of DVD-Video
  in-house 357
production process
  for digital video 340
production speed
  of a CD duplicator 213
professional premastering applications 149
Program Calibration Area 74

Program Memory Area 74, 499
progress gauges
  for recording 156
Project BBQ 305, 317
protective characteristics
  of packaging 410
Protegé 235
prototype
  of packaging 411
Pulse Code Modulation 60, 499
Putt Putt game 308

Q
QIC 159
QuAC file 295
Quac files 289
Quake 385
Quantum 329
queuing
  of recording jobs 225
QuickTime 149, 317, 338, 362, 364, 365, 379, 425, 499
  definition of 339
QuickTime licensing 266
QuickTime's MoviePlayer 333

R
Randloph, Marc 457
random access
  to data 32
raster scan 499
rating
  DVD 350
RCA connectors 270
read speed
  of CD recorders 130
Reader
  Acrobat 373
RealSystem G2 245
recommendations
  for producing corporate video 366
recordable CD
  Frankfurt discussions 78
  history of 7
  layers 29

Index

recordable CD modules 147
recordable DVD
    options 106
recordable media 195
recordable-CD software
    evolution of 146
recorder software
    getting benefits from 154
recorder support
    in applications 159
recording issues
    for optical discs 117
recording on-the-fly
    virtual image 160
recording,optical 19
recycling
    discs 453
    of optical discs 246
    of packaging 449
Red Book 60, 499
    compatibility 307
    data transfer rates when recording 143
    error correction 62
    implementation 60
    Mixed Mode 60
    original disc standard 51
red laser 98
redundancy 499
Reed-Solomon Code 499
references
    on CD-ROM 438
reflective disc surface 26
refresh rate 499
replication 500
    definition of 212
    file preparation 394
    overview 394
replication services
    market for 393
research
    on digital storage 4
resistors
    termination 113

resource databases 439
retrieval 500
rewritable media
    lack of suitability for mastering 119
RGB colors
    for labels 401
RGB NTSC 500
Ricoh 197
Ricoh Corporation 177
Right Angle, Inc. 279
Rimage 219, 225, 229, 231, 234, 235
Rimage Corporation 85
Robèrt, Karl-Henrik 451
RoboHelp
    hotspot creation 262
    indexing 262
    printed documentation from 261
RoboHelp HTML 2000 260
robotics
    in duplicators 211
Rock Ridge Extensions 79
Rock Ridge extensions 158, 185
Roland Sound Canvas 306
rollovers
    creating 375
RS-232 serial connections 111
RS-232-C
    on DVD player 424
Run Length Limited
    clocking method 29
Running Optical Power Calibration 132, 196
S
sales of 88
sampling
    sound 62
sampling rate 500
Sanford Sharpie Permanent Marker 210
Sanger, George 305
Saving Private Ryan 462
scan converter 500
scan lines 500
SciNet, Inc. 326

Index

screen saver
    turning off 108
SCSI 107, 108, 109
    advantages of 110
    cabling 114
    daisy chaining 112
    host adapter 111
    IDs 114
    portability of devices 115
    support in PCs 111
    termination 113
    transfer rates 111
    variations 112
    version supported 131
SCSI host adapter
    dedicated 160
SCSI host adapters
    using two 50
SCSI-I
    connector 111
search terms
    bidding for 442
    bidding on 446
SECAM 335
sector 500
sector-by-sector checking
    of disc content 379
seek 500
seek error 500
selecting a CD recorder 127
selecting a CD-recorder application 193
selecting a computer interface 107
selecting files 155
selecting recording equipment 105
self-clocking mechanism 29
Sequential CD-RW
    formatting option 202
servo mechanism 500
    for positioning laser 32
session 500
set-top playback equipment 67
SGML 500
shovelware 5

signal-to-noise ratio
    of compact discs 3
silkscreen process
    for artwork 28
silkscreened discs 234
silk-screening 42
    registration for 401
simulation
    data access 160
    of disc recording 159
    performance 160
single-session compact disc
    table of contents 119
Single-session Disc 501
single-session recording 161
    through Orange Book 75
Siren 152
sleeves
    disc 237
slide show
    on CD-ROM 259
Small Computer System Interface 501
Smart and Friendly 177
Smart and Friendly, Inc. 152
Smart PhotoShop 362
SmartCD 150
SMPTE time code 501
software application
    for premastering 145
software features
    for disc recorders 153
software interface considerations 155
solar power
    use in Interface 452
Sonic Solutions 246, 432
Sony 4, 152, 199
Sony Spressa Professional CRX140S/C 134
Sorensen 365
Sound Canvas
    Roland 306
sound editing software
    PC-based 105
Sound Forge 249, 271, 345

Index

Sound Forge 4.5 270
Sound Forge XP 4.5 276
SoundEdit 345
sounds
    for multimedia 377
speed 128
    recording 128
spindles
    of media 220
spiral
    data pattern 27
spiral groove
    in blank media 29
spiral pregroove 47
spot printing 403
sputtering 41
stamper 501
standalone recorder 105
standards
    development of 53
Star Craft 313
Star Wars 381
stock footage 337
storage of digital images 250
storage solutions
    network 325
StorPoint server 328
subcode 501
substrate 501
subtitle
    DVD 351
Sun Microsystems 261
Super MultiRead 119, 203
SURF.COM 311
SurroundSound 359
S-VHS 339
S-Video 92, 500
Svoboda, Lance 400
sync 501
synchronization field 501
SyQuest cartridges 125
system conditions
    for The Natural Step 451

System Use Sharing Protocol 79

T

table of contents 501
tape backup, for archiving 32
tape blocking factors 187
tape systems
    support for 189
target audience
    for packaging 409
T-cal cycle 126
TDK 197, 207
Team Fat 309
Teenage Mutant Ninja Turtles 283
TenXpert 323
TeraCart 344
termination
    of SCSI devices 113
terminology
    duplication 212
    of recording software 148
Terran Interactive 358
test utilities
    CD recorder 182
testing
    of discs 374
text storage
    on CD-ROM 65
The Blair Witch Project 456
The Education of a CD-ROM Publisher 7
The Fat Man 305
The Jerky Boys 285
The Learning Company 455
The Matrix 353
The Natural Step 451
    system conditions 451
The Seventh Guest 306
The Stock Market 260
The Ventures 318
thermal calibration
    on hard disk drives 126
thermal recalibration 501
Thin Servers 322

Index

throughput 501
    definition of 212
timeline
    of optical recording 6
Titanic 462
TNS 451
Toast 56, 295
    autostart switch 370
    disc geography control 379
Toast DVD 360, 433
tools
    for electronic publishing 456
tower configurations
    software for 223
tower duplicators 221
    examples of 223
    suitability for companies 223
tower enclosures
    duplicators 211
Trace Digital 220, 230, 231
track 501
Track-at-Once 132, 275, 502
tracks
    maximum of 99 61
    on a CD-ROM 36
training
    course on the Web 442
transfer rate 502
trapping
    for CD labels 401
    for labels 408
tripod
    use of 346
TSR applications 142
turnaround time
    for disc production 215
twin-laser pickup 45

U

UDF 84, 90, 121, 157, 185, 188, 201, 502
    supported platforms 91
UDF Bridge 323, 395, 502
UDF standard
    Appendix 6.9 93

used for packet writing 120
Ultima 80 Media 199
Ultra DMA
    extension of EIDE 117
Ultra SCSI 112
Ultra160/m SCSI 112
Ultra2 SCSI 112
UltraWide SCSI 112, 502
unattended production
    of discs 225
understanding CD technology 20
understanding DVD technology 20
Uninterruptible Power Supply 125
Universal Disc Format 84
Universal Product Code 502
UNIX 147, 157, 183, 190
USB 107, 108, 109
uses for DVD-RAM 104
UV light exposure
    effect on dyes 196
UV method
    for bonding disc substrates 41

V

Variable Bit Rate encoding 365
VCR
    heir-apparent to 459
Vegas Pro 249
    projects 243
Verbatim 197, 198, 202
Veritas 204
video
    analog 335, 338
    corporate 335
video applications
    in the corporate environment 333
Video CD 69, 73
video clips
    using alternate 365
video compression 339
video object block 93
Video Object Files 422
video storage techniques
    double-sided 51

527

Index

video transition effects
    use of 346
VIDEO_TS.IFO 93
videocassette
    as a playback medium 337
VideoCD 249, 502
VideoCD 2.0 40
video-on-demand 458
vinyl records 2
vinyl restoration 105
    process of 270
vinyl sleeve 415
virtual CD player 502
virtual disc driver
    creating 160
virtual image 55, 127, 502
    recording 160
Virtual Reality
    capability of QuickTime 366
virus protection program
    interference with recording 108
VOB files 434
volume 502
volume descriptors 502
Voyager 455

W

water-based ink 233
waveform audio 502
wax transfer
    disc printing 219
wax-transfer printers 234
    media surfaces for 234
WebHelp 260, 261
White Book 70
    data types 70
    origination of 71
    uses of 70
Wide SCSI 503
widescreen televisions 352
Williams, John 312
Windows 2000 147, 149, 153
Windows 3.1
    support for 371
Windows 95
    suitability for recording 146
Windows 98 147
Windows Media Audio 245
Windows Media Player 426
Windows NT 225
Wing Commander 312
WinOnCD 151, 164
wizard 503
wizards 164
    as software guides 146
workflow issues
    for disc duplication 216
workgroup publishing
    to disc 215
Working Group 4 97
Working Group 6 47
WORM drive 31, 331
WORM drives 325
writable DVD 46
Write Once, Read Many 503
write strategy
    for DVD-R 47
Write-Once, Read Many drives 325

X

XA-compatibility 66
Xing Technologies 91

Y

Yahoo 441
Yamaha 306
Yellow Book 58, 503
    four-layer architecture 64
    implementation 64
    modes 66
    origins of 63
    origins of the format 5
Young Minds, Inc
    UNIX support 140

Z

Zefiro Acoustics ZA-2 271
zShops 448
    on Amazon.com 444
Zuma Digital 421

DISK WARRANTY

This software is protected by both United States copyright law and international copyright treaty provision. You must treat this software just like a book, except that you may copy it into a computer in order to be used and you may make archival copies of the software for the sole purpose of backing up our software and protecting your investment from loss.

By saying "just like a book," McGraw-Hill means, for example, that this software may be used by any number of people and may be freely moved from one computer location to another, so long as there is no possibility of its being used at one location or on one computer while it also is being used at another. Just as a book cannot be read by two different people in two different places at the same time, neither can the software be used by two different people in two different places at the same time (unless, of course, McGraw-Hill's copyright is being violated).

LIMITED WARRANTY

McGraw-Hill takes great care to provide you with top-quality software, thoroughly checked to prevent virus infections. McGraw-Hill warrants the physical diskette(s) contained herein to be free of defects in materials and workmanship for a period of sixty days from the purchase date. If McGraw-Hill receives written notification within the warranty period of defects in materials or workmanship, and such notification is determined by McGraw-Hill to be correct, McGraw-Hill will replace the defective diskette(s). Send requests to:

> McGraw-Hill
> Customer Services
> P.O. Box 545
> Blacklick, OH 43004-0545

The entire and exclusive liability and remedy for breach of this Limited Warranty shall be limited to replacement of defective diskette(s) and shall not include or extend to any claim for or right to cover any other damages, including but not limited to, loss of profit, data, or use of the software, or special, incidental, or consequential damages or other similar claims, even if McGraw-Hill has been specifically advised of the possibility of such damages. In no event will McGraw-Hill's liability for any damages to you or any other person ever exceed the lower of suggested list price or actual price paid for the license to use the software, regardless of any form of the claim.

McGRAW-HILL SPECIFICALLY DISCLAIMS ALL OTHER WARRANTIES, EXPRESS OR IMPLIED, INCLUDING, BUT NOT LIMITED TO, ANY IMPLIED WARRANTY OF MERCHANTABILITY OR FITNESS FOR A PARTICULAR PURPOSE.

Specifically, McGraw-Hill makes no representation or warranty that the software is fit for any particular purpose and any implied warranty of merchantability is limited to the sixty-day duration of the Limited Warranty covering the physical diskette(s) only (and not the software) and is otherwise expressly and specifically disclaimed.

This limited warranty gives you specific legal rights; you may have others which may vary from state to state. Some states do not allow the exclusion of incidental or consequential damages, or the limitation on how long an implied warranty lasts, so some of the above may not apply to you.